Second Edition

ENGINEERING DESIGN GRAPHICS

Sketching, Modeling & Visualization

James M. Leake • Jacob L. Borgerson

Embry-Riddle Aeronautical University

COLLEGE OF ENGINEERING

Engineering Fundamentals

Wiley Custom Learning Solutions

Cover Image © Embry-Riddle Aeronautical University.

Cover Design by Michelle A. Harvey.

Printing identification and country of origin will either be included on this page and/or the end of the book. In addition, if the ISBN on this page and the back cover do not match, the ISBN on the back cover should be considered the correct ISBN.

This custom textbook may include materials submitted by the Author for publication by John Wiley & Sons, Inc. This material has not been edited by Wiley, and the Author is solely responsible for its content.

To order books or for customer service, please call 1(800)-CALL-WILEY (225-5945).

Printed in the United States of America.

ISBN 978-1-119-25275-7
Printed and bound by LSC Communications

10 9 8 7 6

CONTENTS

CONTENTS

ENGINEERING DESIGN GRAPHICS

SKETCHING, MODELING, AND VISUALIZATION

Second Edition

James M. Leake

Department of Industrial & Enterprise Systems Engineering
University of Illinois at Urbana-Champaign

with special contributions by

Jacob L. Borgerson

Paradigm Consultants, Inc.
Houston, Texas

WILEY

John Wiley & Sons, Inc.

EXECUTIVE PUBLISHER	Don Fowley
ASSOCIATE PUBLISHER	Dan Sayre
EDITORIAL ASSISTANT	Charlotte Cerf
SENIOR PRODUCTION EDITOR	Sujin Hong
TEXT AND COVER DESIGNER	Madelyn Lesure
COVER PHOTO	

Catapult (Adam Fabianski, Eric Mason, Alyssa Cast, Haotian Wu)
Windmill (Tori Ammon, Siddhant Anand, Augusto Canario, Dan Malsom, Matthew McClone)
Hose nozzle (Collin Statler, Chris Huab, Harrison Meyer)

This book was set in InDesign by Thomson Digital Inc. and printed and bound by RRD Von Hoffmann. The cover was printed by RRD Von Hoffmann.

This book is printed on acid free paper. ∞

Founded in 1807, John Wiley & Sons, Inc. has been a valued source of knowledge and understanding for more than 200 years, helping people around the world meet their needs and fulfill their aspirations. Our company is built on a foundation of principles that include responsibility to the communities we serve and where we live and work. In 2008, we launched a Corporate Citizenship Initiative, a global effort to address the environmental, social, economic, and ethical challenges we face in our business. Among the issues we are addressing are carbon impact, paper specifications and procurement, ethical conduct within our business and among our vendors, and community and charitable support. For more information, please visit our website: www.wiley.com/go/citizenship.

ISBN 978-1-118-07888-4

Printed in the United States of America

10 9 8 7 6 5 4 3 2 1

James M. Leake
dedicates this work to
Stephanie, and to the
relationship that we share

Jacob L. Borgerson
dedicates this work to
Erin

☐ PREFACE

The traditional first-year engineering graphics course has undergone significant change in the past quarter century. Although the emergence of computer-aided design (CAD) and the expansion of the graphics curriculum to include design are perhaps the most significant developments, more recent trends include a movement away from 2D CAD and toward 3D parametric solid modeling, an increased emphasis on freehand sketching at the expense of instrument drawing, a greater focus on the development of spatial visualization skills, and an expansion of the curriculum to include the latest developments in design technology. All of this has occurred despite a strong countervailing trend to de-emphasize graphics in order to accommodate other material in the four-year undergraduate engineering curriculum.

The aim of this book, then, is to provide a clear, concise treatment of the essential topics included in a modern engineering design graphics course. Projection theory provides the instructional framework, and freehand sketching the means for learning the important graphical concepts at the core of this work. The book includes several hundred sketching problems, all serving to develop the student's ability to use sketching for ideation and communication, as well as a means to develop critical visualization skills. New to this second edition are the additions of 38 worksheets containing more than 80 sketching problems. By encouraging students to work directly within the book, these worksheets make it easier to gain additional sketching experience. Also new to this second edition are two detailed example problems in Chapter 5 that focus on the development of visualization skills.

Engineering design serves to bracket the graphical content of the book with an introductory chapter on the engineering design process and a later chapter on product dissection. Material contained in the first chapter is based on introductory material found in leading engineering design textbooks. The chapter on product dissection concludes with thumbnail images of student projects. An extensive list of the products and devices that have been successfully reverse engineered is also included. Typically, the team obtains a commercial product, which is then dissected and reverse engineered. In other appropriate technology and history of technology projects where the product is unavailable, the student teams work from drawings, photographs, written descriptions, and so on.

A chapter on computer-aided product design software, with an emphasis on parametric solid modeling, is also included. The chapter is designed to complement, rather than replace, instructional materials for a specific CAD package. The chapter provides an overview of different kinds of CAD software, as well as general modeling concepts shared by all parametric modelers. Also in this chapter and new to the second edition is a discussion of nonuniform rational B-spline (NURBS) modeling. Parametric modeling is limited in its ability to create the freeform, sculpted geometry so popular in modern product design. In order to create these organic shapes, NURBS modeling is required. The development of NURBS is discussed, starting with physical splines. The important relationship between Bézier curves and B-splines is described. The chapter concludes with a discussion of NURBS surfaces, continuity, and curvature.

A chapter on perspective projections and sketching has been included because it reflects the way that engineering graphics has traditionally been taught at the University of Illinois at Urbana-Champaign (UIUC), starting from the general (e.g., perspective projection) and moving to the specific (e.g., multiview orthographic projection).

This second edition includes three new chapters: Reverse Engineering Tools, Digital

Simulation Tools, and Concept Design Tools. These chapters address some of the latest developments in design technology. The chapter on reverse engineering tools includes sections on 3D scanning and rapid prototyping. Both scanner hardware and reverse engineering software tools are discussed. While the scanner hardware is used to digitize a physical object, reverse engineering software is used to convert the scanner output (a point cloud) into either a polygon mesh or a CAD file. Rapid prototyping completes this cycle, converting a digital file into a physical prototype.

After an initial discussion of the benefits of conducting analysis early in the design process, both finite element stress analysis and kinematic analysis are discussed in the chapter on digital simulation tools. The chapter on concept design tools begins with sections on innovation, industrial design, and concept design. Concept design tools are then discussed, including digital sketching, direct modeling, and freeform modeling.

Key features of the book include the following:

- A succinct, scaled-down approach, with important concepts distilled to their essence

- Hundreds of sketching problems, including dozens of worksheet problems, to help students learn the language of technical graphics and develop their sketching, visualization, and modeling skills

- Assembly problems requiring a wide range of modeling tools, not just extrusions and revolutions

- Lots of visualization materials: sections on multiview visualization and the section view construction process are included, as are missing view problems, problems that require students to mentally rotate and then sketch a different pictorial view of the object, problems that require students to find a partial auxiliary, missing, and pictorial view when two views are given, as well as section view problems

- A strong student focus, with many examples showing what students can produce in an engineering design graphics course

- A chapter on engineering design that reflects the thinking of leading engineering design educators

- A chapter on product dissection, something unique to engineering design graphics textbooks

- A unified planar projection theory framework that provides a common basis for understanding the relationships between different kinds of sketches (e.g., perspective, oblique, isometric, multiview) and also serves as an introduction to the study of computer graphics

- Several detailed multistep example sketching problems that provide students with problem-solving procedural templates

- Significant coverage of such important trends and technologies as 3D scanning, rapid prototyping, digital sketching, direct modeling, FEA, kinematic analysis, and NURBS modeling

Much of the book's content, in particular, chapters 2 through 6, 8, and 13, is strongly influenced by a system of teaching engineering graphics that has developed over the years in the Department of General Engineering at UIUC. In particular, I would like to acknowledge the work of my immediate predecessor, Michael H. Pleck. Hallmarks of this approach include a focus on planar projection theory, starting from general case perspective projections and advancing to more specific projection types, as well as an emphasis on spatial visualization problems. A special thanks goes out to the many UIUC students who have made significant contributions to the content of this work. The book's co-author, Jacob Borgerson, is responsible for the many fine problems and worksheets that are included in the book's end-of-chapter exercises, as well as for his careful reading of and thoughtful comments on the text.

Our thanks go out to the book's many reviewers, including: Brian Brady, Ferris State University; Randy Emert, Clemson University; Andrea Giorgioni, New Jersey Institute of Technology; Davyda Hammond, Germanna Community College; Ghodrat Karami, North Dakota State University; Michael Keefe,

University of Delaware; Robert D. Knecht, Colorado School of Mines; Soo-Yen Lee, Central Michigan University; Anthony Maxwell, Buck's County Community College; Patrick McCuistion, Ohio University; Ramarathnam Narasimhan, University of Miami, College of Engineering; Jeff Raquet, University of North Carolina—Charlotte; and Ken Youssefi, University of California, Berkeley/San Jose State University.

The inspiration for certain chapters deserves special mention. Chapter 1 on Engineering Design is based on the introductory chapters from some of the best books on engineering design, including those of G. Pahl and W. Beitz, George Dieter, Rudolph Eggert, and Clive Dym. Chapter 9, Reverse Engineering Tools, owes much to the collection of essays, *Reverse Engineering: An Industrial Perspective*, edited by Vinesh Raja and Kiran J. Fernandes. Chapter 12 on Product Dissection is largely based on the work of Sheri Sheppard, Kevin Otto and Kristin Wood, and Ronald Barr.

James M. Leake
Urbana, Illinois
March 2012

□ CONTENTS

CHAPTER

2

FREEHAND SKETCHING

INTRODUCTION

Prior to the introduction of the personal computer in the early 1980s and the accompanying introduction of computer-aided design (CAD) software packages such as AutoCAD® shortly thereafter, most engineering drawings were executed manually using equipment like drafting machines, T-squares, triangles, and compasses. Today, however, nearly all engineering drawings are executed using a CAD system. This sea change in the way technical drawings are produced has had a major impact on the engineering graphics curriculum. Instrument drawing (with T-squares, triangles, and so on), for example, has largely been replaced with *freehand sketching*.

Engineering is a creative endeavor with roots that can be traced back to great Italian Renaissance artists like Leonardo, Michelangelo, Raffaello, and Donatello. Although it is certainly true that engineering is technology driven, it is important not to lose sight of the discipline's rich graphical and creative traditions. Great engineering is evidenced as much through the ability to communicate ideas via freehand sketching as it is by manipulating differential equations, by making a computer "sing," or for that matter, by presenting carefully reasoned, well-crafted prose.

While some of us are blessed with the natural ability to draw, most are not. All of us, though, can improve in our ability to communicate, document, and visualize using freehand sketches. All that is required is practice.

Freehand sketching has a number of important uses. Engineers often need to make sketches out in the field. These sketches are subsequently converted to CAD back at the office, in order to document modifications that are to be made to an existing structure. Sketching is also a just-in-time communication tool used by engineers, designers, and craftspeople, as well as with clients and supervisors. Freehand sketching is used creatively for brainstorming ideas, inventing, and exploring alternatives. Figure 2-1 on page 24, for example, shows sketches done by engineering graphics students as part of their concept design projects. Either sketching or instrument drawing is essential in order to practice the language of engineering graphics. In her work, Sorby[1] has amply demonstrated that freehand sketching is an excellent way to improve spatial visualization skills.

Engineers use technical sketches to create rough, preliminary drawings that represent the main features of a product or structure. Freehand sketches should not be sloppy. Above all, careful attention should be paid to the proportions of the sketch. Although not drawn to a specific scale, freehand sketches should appear to be proportionally accurate to the eye.

SKETCHING TOOLS AND MATERIALS

All that is really needed for sketching is paper, pencils, and an eraser. Most any kind of paper will do for sketching; the reverse side of already-used paper works well. Rectangular grid paper is

---

[1] S. Sorby, Developing 3-D Spatial Visualization Skills, *Engineering Design Graphics Journal,* Vol. 63, No. 2, Spring 1999.

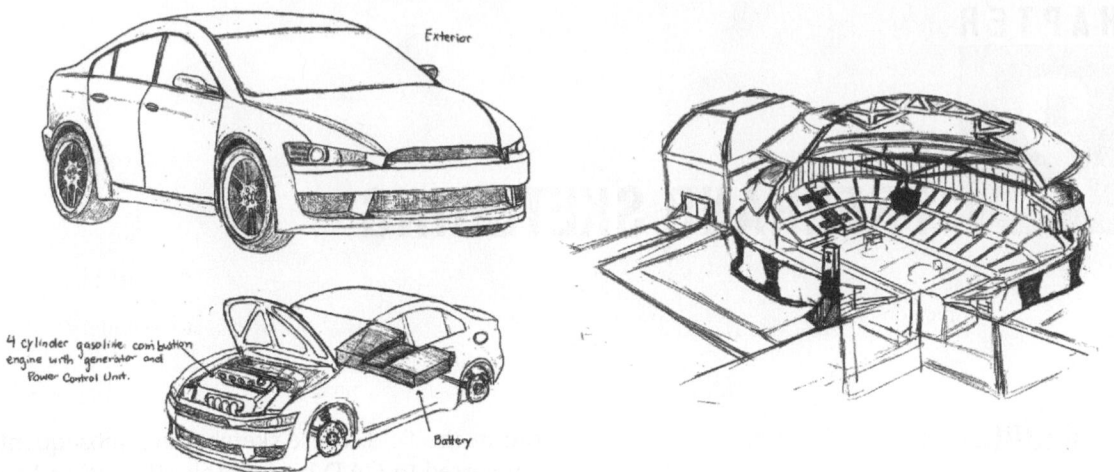

Figure 2-1 Design project brainstorming sketches (Courtesy of Jonathan Schmid, Donjin Lee)

frequently used for sketching. Engineering firms often use custom-ordered A (8½″ × 11″) or A4 metric size tablets of rectangular grid paper. The grid is actually on the back side of the paper, but it is dark enough to be visible on the front side. The paper also includes a custom title block. This paper is useful for combining text, calculations, and sketches.

Isometric grid paper, shown in Figure 2-2, is helpful when one is first learning to create isometric sketches. This type of grid paper is not typically encountered in engineering design firms.

Most engineers use fine-line mechanical pencils, like those shown in Figure 2-3, both for sketching

and for calculations. Each mechanical pencil is designed to hold a specific lead size. Common lead sizes include 0.3 mm, 0.5 mm, 0.7 mm, and 0.9 mm. The lead size is the diameter of the lead. The best size for general usage is 0.5 mm, whereas 0.7 mm is good for bolder stokes. An important advantage of mechanical over wooden pencils is that mechanical pencils do not require sharpening.

Pencil lead is available in different hardness grades, as shown in Table 2-1. The harder the lead, the lighter, sharper, and crisper the resulting line. Soft lead, on the other hand, results in darker lines that tend to smear easily. Medium-grade leads (3H, 2H, H, F, HB, and B) are good for general-purpose drafting, and HB is the most commonly used lead grade.

Finally, a good eraser that does not tend to smudge is an important sketching tool. See Figure 2-4.

Figure 2-2 Isometric grid paper

Figure 2-3 Fine-line mechanical pencils

Table 2-1 **Pencil lead grades**		
Range	Grades	Purpose
Hard	9H, 8H, 7H, 6H, 5H, 4H	Accuracy, precision
Medium	3H, 2H, H, F, **HB,** B	General purpose
Soft	2B, 3B, 4B, 5B, 6B, 7B	Artistic rendering
	harder → softer	

Figure 2-4 Eraser (© Datacraft Co Ltd./Getty Images, Inc.)

■ SKETCHING TECHNIQUES

Line Techniques

A freehand sketch should begin with proportionally laid out *construction lines*. Construction lines are thin and drawn lightly; they serve to guide the lines to follow. All other lines should be dark, crisp, and of uniform thickness; they can be sketched directly over the construction lines. See Figure 2-5, where construction lines are first used to lay out the proportions of the sketch, and then the features are indicated using bold lines. If drawn correctly, construction lines need not be erased. It is a common mistake to draw construction lines that are too dark, making them hard to distinguish from other lines.

In addition to the relative lightness or darkness of a line, two line widths, thick and thin, are used in engineering sketches and drawings. Continuous lines indicating visible object edges are thick, while hidden, center, and construction lines are thin. See Figure 2-6 for a brief summary

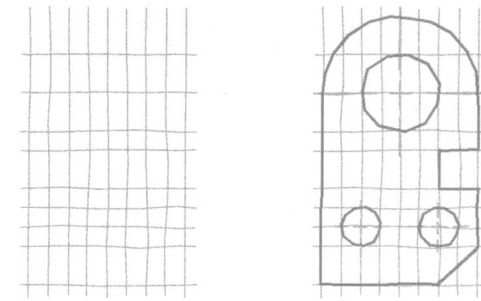

Figure 2-5 Construction lines with bold lines drawn over

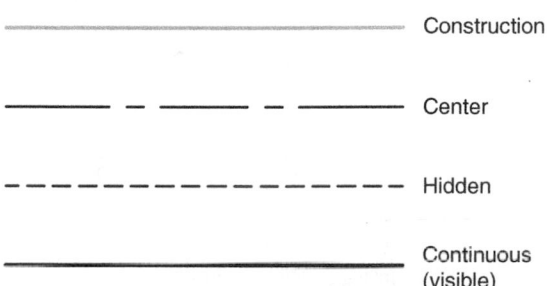

Figure 2-6 Common types of sketching lines

of the characteristics of the most commonly used lines.

A more thorough discussion of the line styles employed in CAD and manual engineering drawings is provided at the end of the chapter.

Sketching Straight Lines

To sketch a straight line, start by marking both end points. Next, place the pencil point at one of these end points. Keeping your eye fixed on the point to which the line is being drawn, sketch a light line with a single stroke. Finally, darken the line. See Figure 2-7 on page 26.

To sketch an especially long line, use several short overlapping strokes, and then darken the line, as shown in Figure 2-8 on page 26. Slight wiggles are okay as long as the resulting line is straight. Occasional gaps are also acceptable. See Figure 2-9 on page 26. If the resulting line tails off and is not straight, the pencil is probably being gripped too tightly. See Figure 2-10 on page 26 for an example.

For right-handers, horizontal line strokes are typically made from left to right, as shown in

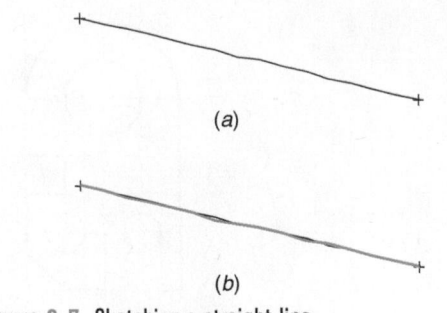

(a)

(b)

Figure 2-7 Sketching a straight line

(a)

(b)

Figure 2-8 Sketching a long straight line

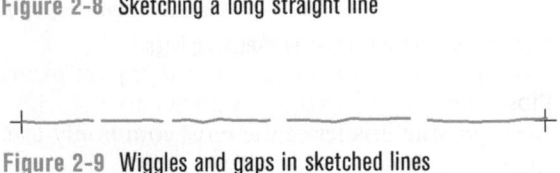

Figure 2-9 Wiggles and gaps in sketched lines

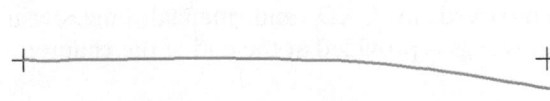

Figure 2-10 Sketched line tailing off due to overly firm grip

Keep eye on
end point

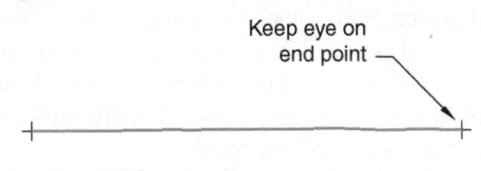

Figure 2-11 Sketching horizontal lines

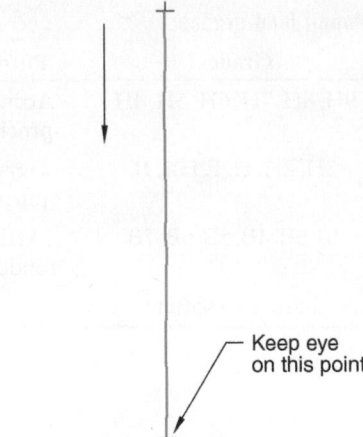

Keep eye
on this point

Figure 2-12 Sketching vertical lines

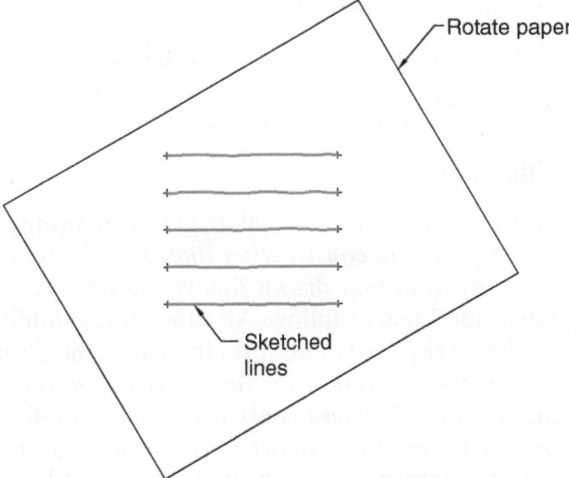

Rotate paper

Sketched
lines

Figure 2-13 Rotating paper to sketch inclined lines

Figure 2-11. Vertical lines are usually drawn from top to bottom, as seen in Figure 2-12. Inclined lines at certain orientations can be difficult to draw, in which case the paper can be rotated to a more comfortable position and then drawn, as shown in Figure 2-13. In fact, for lines drawn at any orientation, it is a good idea to rotate the paper to a comfortable position and then sketch the line.

Sketching Circles

A number of different techniques can be used to sketch a circle. In the trammel method, a piece of scrap paper is used to locate points on the circumference of the circle. On a straight edge of the paper, mark two points at a distance equal to the radius of the circle. With one point at the

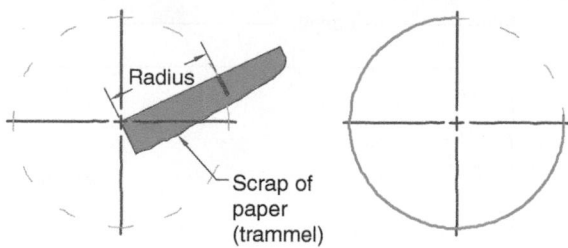

Figure 2-14 Trammel method used to sketch circles

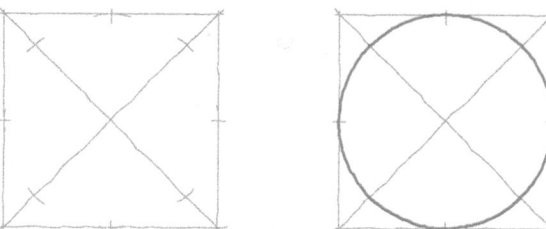

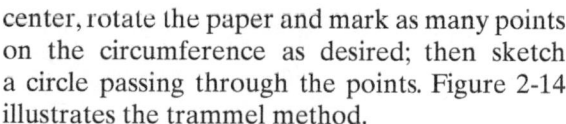

Figure 2-15 Square method used to sketch circles

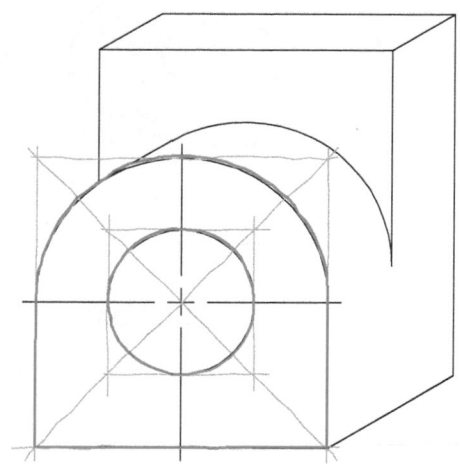

Figure 2-16 Circles and arcs sketched on an oblique pictorial

center, rotate the paper and mark as many points on the circumference as desired; then sketch a circle passing through the points. Figure 2-14 illustrates the trammel method.

In the square method, the enclosing square of the circle is first sketched. Next the midpoints of the sides of the square are marked. The midpoints are used as the quadrants of the circle. A circle passing through the quadrant points and tangent to the sides of the square is then sketched. See Figure 2-15 for an example of a circle constructed using the square method. Note that by constructing the diagonals of the square and marking the radius along the diagonals, you can add four more points on the circumference of the circle.

Either of these methods can be used to sketch a circle or arc feature in a *pictorial sketch* (see Chapter 3), as long as the face on which the circle or arc appears is parallel to the view plane. Figure 2-16 shows an example of this for an *oblique* pictorial sketch.

Sketching Ellipses

The rectangle method can be used to sketch an ellipse. First construct the enclosing rectangle, and then locate the midpoints along the sides of the rectangle. Sketch the ellipse passing the midpoints

and tangent to the sides of bounding rectangle. See Figure 2-17 for an example of this method.

In a pictorial sketch a circle will appear as an ellipse, unless the circle is parallel to the view plane. To sketch such an ellipse, first construct the enclosing parallelogram. The ellipse will pass through the midpoints of the sides of the parallelogram and will be tangent to the sides of the parallelogram. Figure 2-18 on page 28 shows an example of the construction of an ellipse on an *isometric* pictorial sketch.

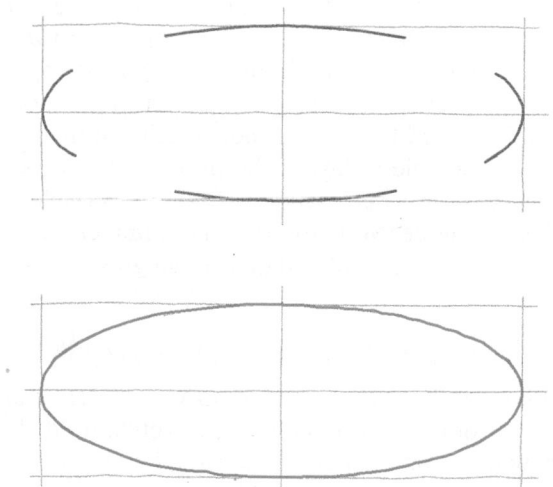

Figure 2-17 Rectangular method used to sketch ellipses

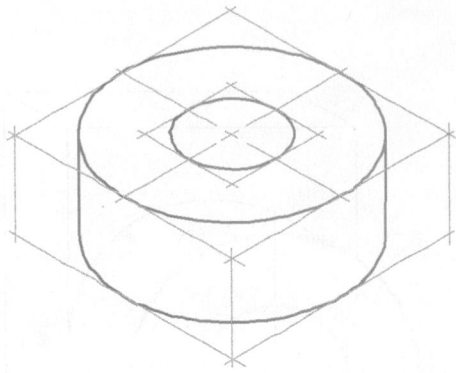

Figure 2-18 Parallelogram method used to sketch ellipses

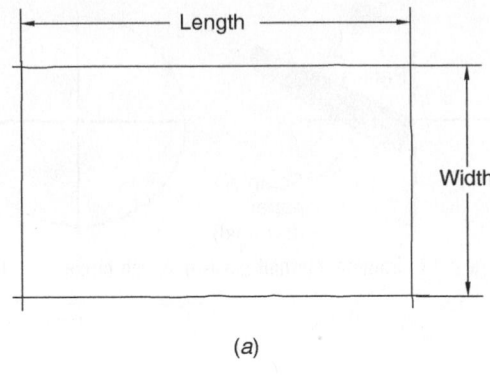

(a)

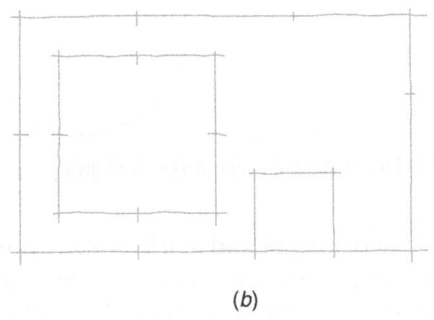

(b)

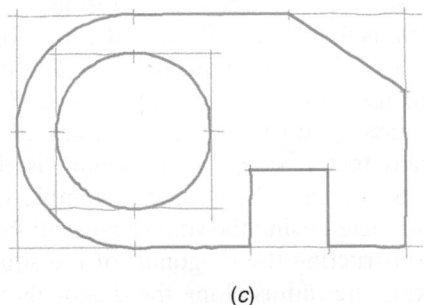

(c)

Figure 2-19 Proportioning with principal dimensions

▌ PROPORTIONING

Although freehand sketches are not drawn to scale, it is important to maintain the relative proportions between the principal dimensions of the object. To accomplish this, first estimate the proportional relationship between the object's principal dimensions and lightly block them in, as shown in Figure 2-19a. The estimated dimensions of each feature should then be proportioned with respect to established dimensions, as seen in Figure 2-19b. Work at developing the ability to divide a line in half by eye; the halves can then be further divided into fourths. Finally, use bold lines to fully define the object, as in Figure 2-19c.

While it is assumed that scales are not used when sketching, a *trammel* may be used to improve proportional accuracy. Decide on a unit of length, and then transfer it to the trammel, or straight-edged scrap paper. Additional unit-length graduations can be added to the trammel, which can then be used as a scale to layout the proportions of each object feature. See Figure 2-20 for an example of this technique. Note also that the trammel can be folded to obtain half and quarter lengths.

Estimating Dimensions of Actual Objects

It is sometimes necessary to make a sketch of an actual object. To make such a sketch, hold the pencil at arm's length, with the pencil between the eye and the object. Using the pencil as a sight, establish a proportional relationship between an object edge and the length of the pencil. Do this by aligning the end of the pencil with one end of the object edge. Move your thumb along the pencil until it coincides with the other end of the edge. Now use this proportion to estimate the lengths of other object features. See Figure 2-21 for an example of this technique. A similar technique can be used to estimate angles, as shown in Figure 2-22.

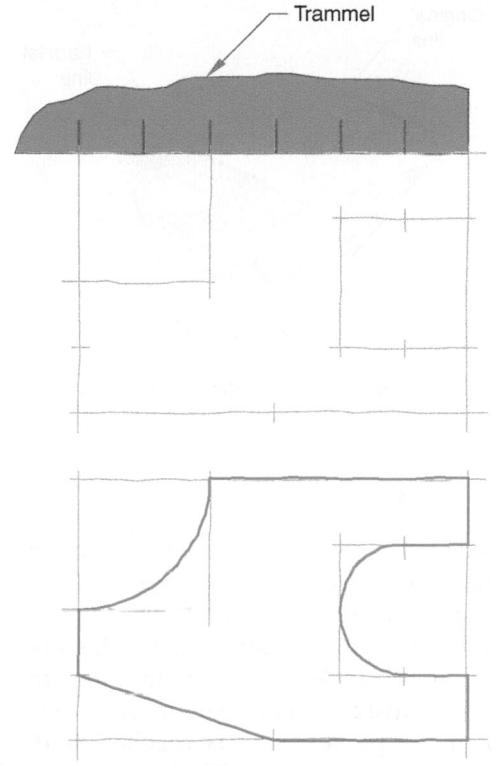

Figure 2-20 Using a trammel to improve proportions

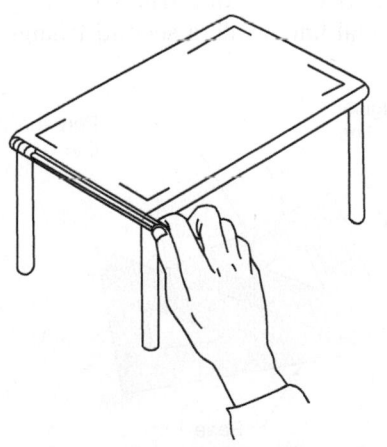

Figure 2-21 Using a pencil to estimate proportional lengths

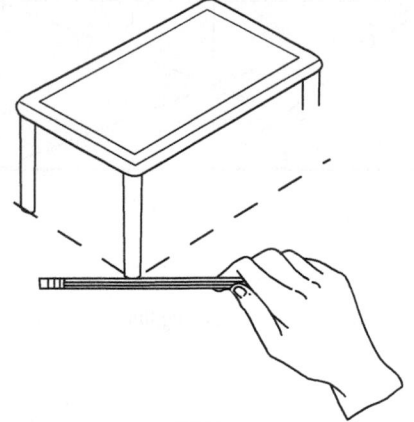

Figure 2-22 Using a pencil to estimate angles

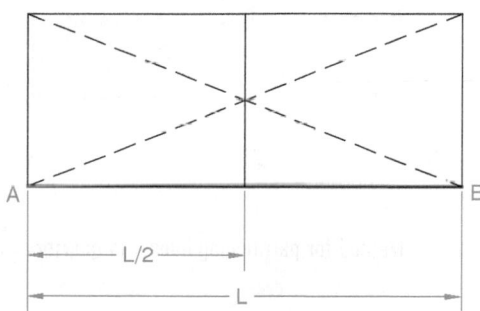

Figure 2-23 Method for partitioning lines

Partitioning Lines

The following procedure may be used to divide a line into fractional parts. To subdivide the line AB shown in Figure 2-23, construct a rectangle on the line AB. Next draw both of the diagonals of the rectangle. Pass a line perpendicular to AB that passes through the intersection of the diagonals. This line divides AB in half.

To partition AB into thirds, sketch another diagonal from one corner of the original rectangle to the midline on the opposite side. Now sketch a line perpendicular to AB that passes through the point where the two diagonals (full diagonal and the half diagonal) intersect. This determines a one-third length of AB, as seen in Figure 2-24 on page 30. Repeating this same process will further divide the line into fourths, fifths, and so on. Figure 2-25 on page 30 shows this.

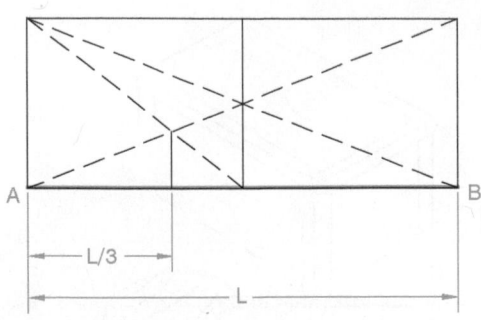

Figure 2-24 Method for partitioning lines into thirds

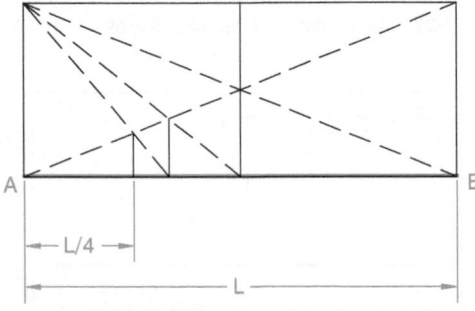

Figure 2-25 Method for partitioning lines into quarters

■ INSTRUMENT USAGE–TRIANGLES

Although drafting machines and T-squares have fallen into disuse, the construction of parallel and perpendicular lines using triangles remains a useful skill. Every engineer should possess both a 45° triangle and a 30°–60° triangle. These triangles are shown in Figure 2-26.

Figure 2-26 30° – 60° and 45° triangles

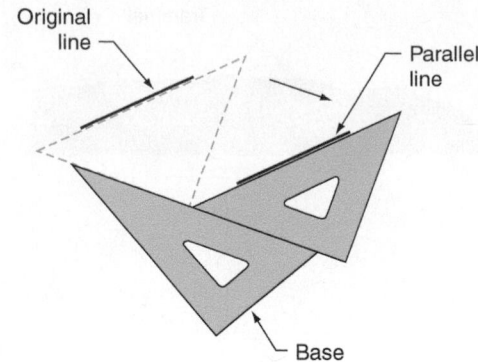

Figure 2-27 Using triangles to draw parallel lines

Parallel Lines

In order to draw a line parallel to another line using two triangles, align the hypotenuse of one triangle with the original line. Next place the hypotenuse of the second triangle along the leg of the first triangle. Now slide the first triangle along the fixed second triangle until the location of the parallel line is reached, and construct the line. Figure 2-27 demonstrates this technique.

Perpendicular Lines

Triangles are also useful for drawing one line perpendicular to another. There are a couple of ways to accomplish this. In the first method, shown in Figure 2-28, one leg of a triangle is aligned with the original line, while a second triangle is used

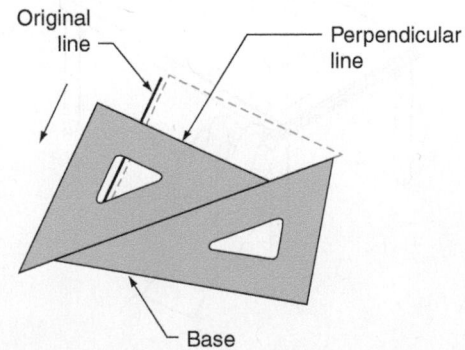

Figure 2-28 Using triangles to draw perpendicular lines: method 1

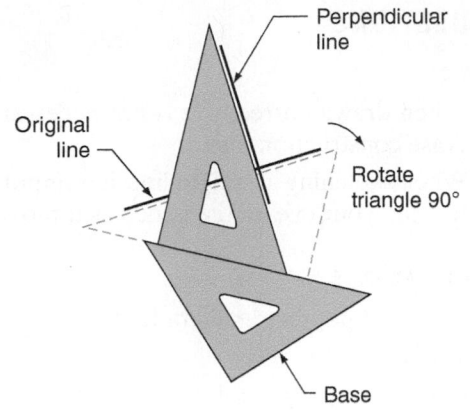

Figure 2-29 Using triangles to draw perpendicular lines: method 2

as a base. Sliding the first triangle along the base triangle, you can use the other leg of the first triangle to draw the perpendicular line. In the second method, shown in Figure 2-29, the first triangle is rotated 90°, rather than slid, prior to drawing the perpendicular line.

LINE STYLES

Line styles, sometimes referred to as the alphabet of lines, describe the size, spacing, construction, and application of the various lines used in CAD and manual engineering drawings. These conventions are established in ASME Y14.2M-1992, Line Conventions and Lettering.

Two line widths are recommended on engineering drawings: thick and thin. The ratio of these line widths should be approximately 2:1. The thin line width is recommended to be a minimum of 0.3 mm, with a recommended minimum thick line width of 0.6 mm.

Figure 2-30 shows the various types of lines and their widths. *Visible lines* are used to show

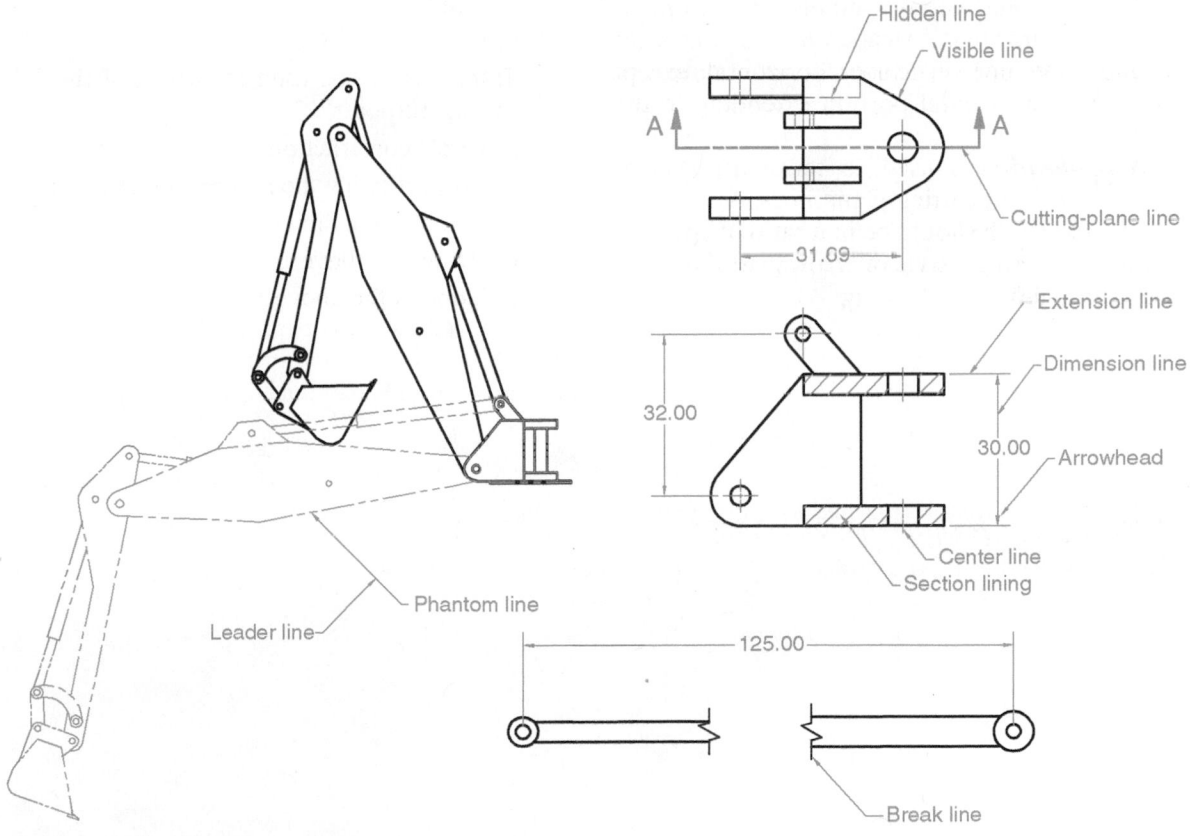

Figure 2-30 Line styles

the visible edges and contours of objects. **Hidden lines** are used to show the hidden edges and contours of objects. **Center lines** are used to represent axes or center planes of symmetrical parts and features, as well as bolt circles and paths of motion. **Phantom lines** are most commonly used to show alternative positions of moving parts. **Cutting-plane lines** are used to indicate the location of cutting planes for section views. The ends of the lines are at 90 degrees and are terminated by arrowheads to indicate the viewing direction. **Section lining** is used to indicate the cut surfaces of an object in a section view. **Break lines** are used when complete views are not needed.

Dimension lines are used to indicate the extent and direction of a dimension and are terminated with uniform arrowheads. **Extension lines**, used in combination with dimension lines, indicate the point or line on the drawing to which the dimension applies. **Leader lines** are used to direct notes, dimensions, symbols, and part numbers on a drawing. A leader line is a straight inclined line, not vertical or horizontal, except for a short horizontal portion extending to the note.

Arrowheads are used to terminate dimension, leader, and cutting-plane lines. Arrowhead length and width should be in a ratio of approximately 3:1. A single style of arrowhead should be used throughout the drawing.

▌QUESTIONS

TRUE OR FALSE

1. When drawn correctly, it is not necessary to erase construction lines.
2. When sketching a straight line, it is important to keep your eye on the pencil as it moves.

MULTIPLE CHOICE

3. The most popular lead grade is:
   a. 2H
   b. H
   c. F
   d. HB
   e. B
4. Which line type is drawn more thickly than the others?
   a. Construction
   b. Continuous
   c. Hidden
   d. Center
5. Trammels can be used for which of the following purposes?
   a. Circle construction
   b. Proportioning the features in a sketch
   c. Transferring dimensions
   d. All of the above
   e. None of the above

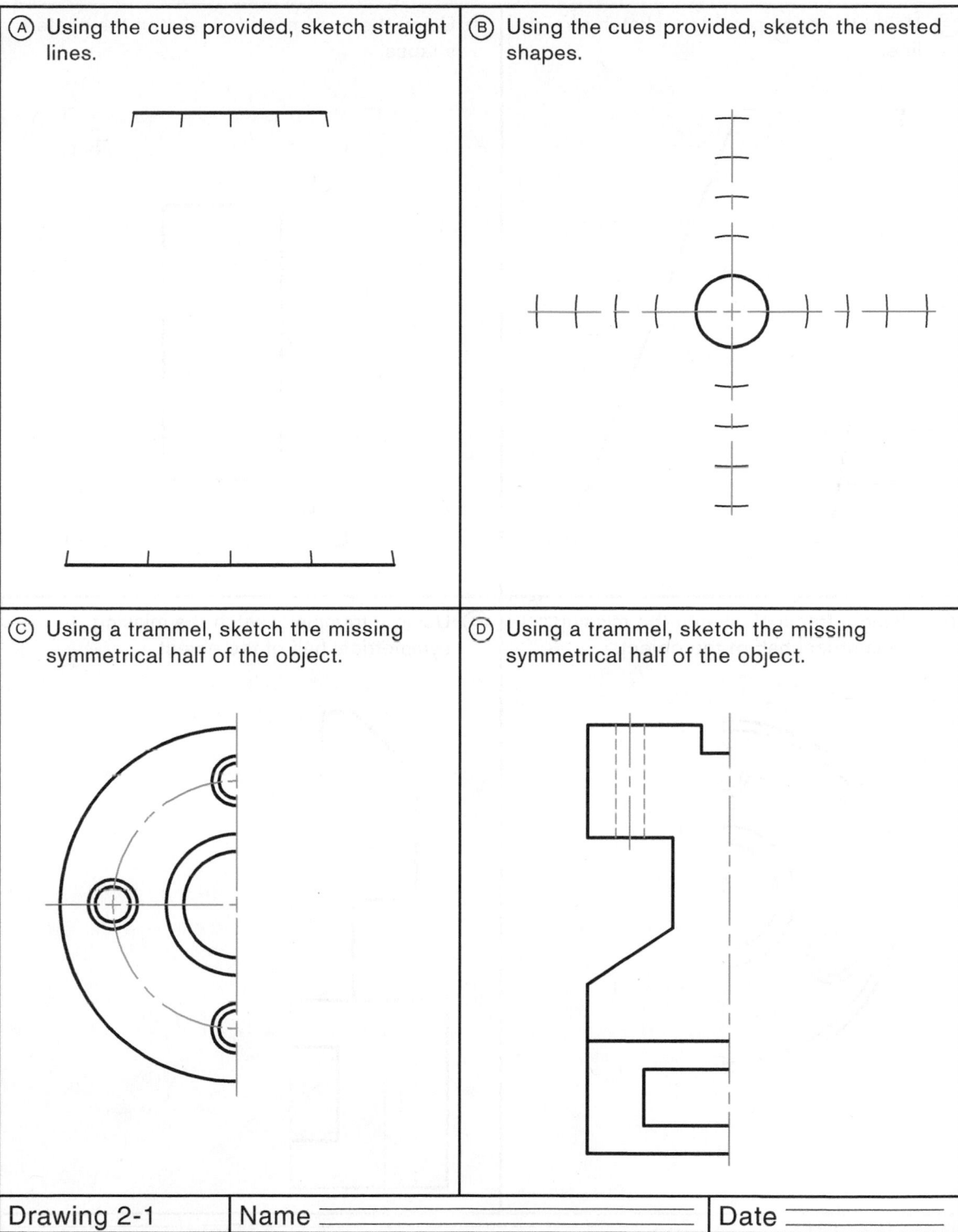

Ⓐ Using the cues provided, sketch straight lines.

Ⓑ Using the cues provided, sketch the nested shapes.

Ⓒ Using a trammel, sketch the missing symmetrical half of the object.

Ⓓ Using a trammel, sketch the missing symmetrical half of the object.

| Drawing 2-1 | Name | Date |

(A) Using the cues provided, sketch straight lines.

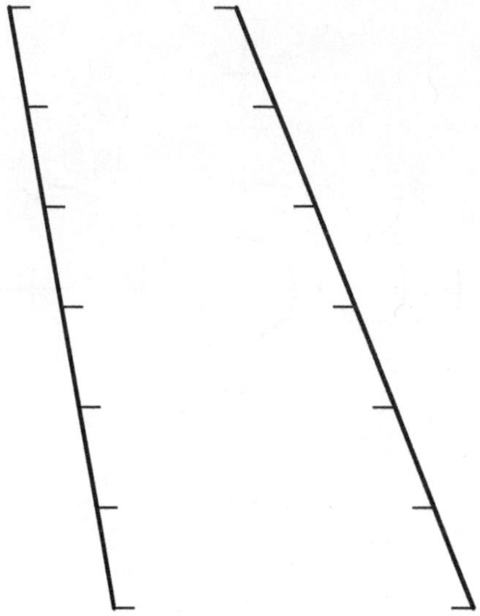

(B) Using the cues provided, sketch the nested shapes.

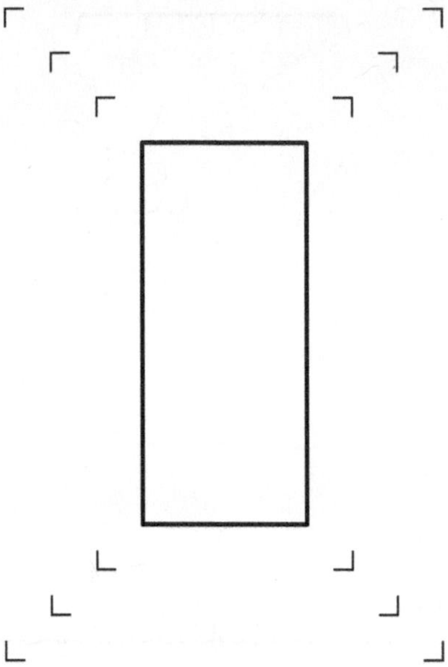

(C) Using a trammel, sketch the missing symmetrical half of the object.

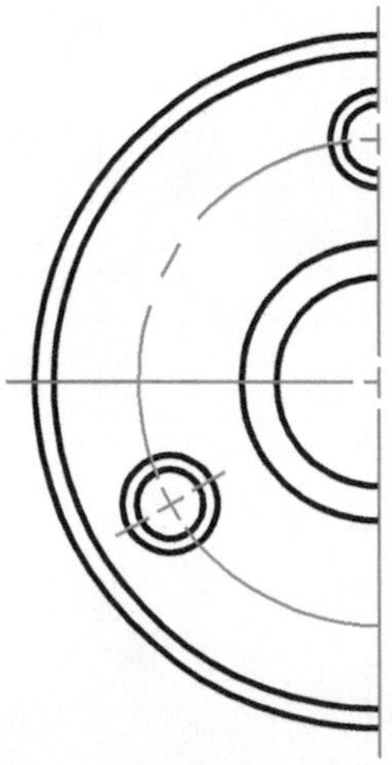

(D) Using a trammel, sketch the missing symmetrical half of the object.

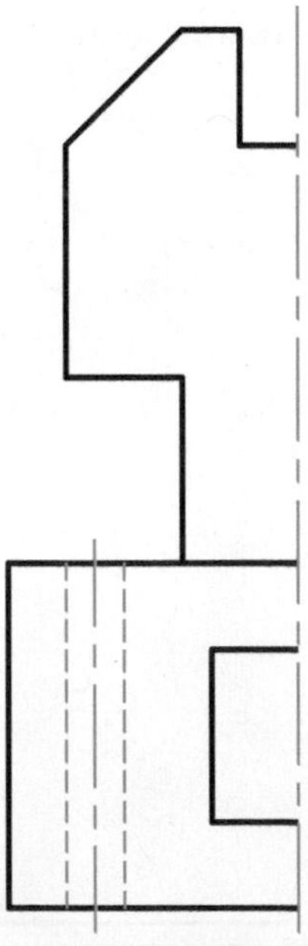

Drawing 2-2   Name ══════════════════   Date ═════════

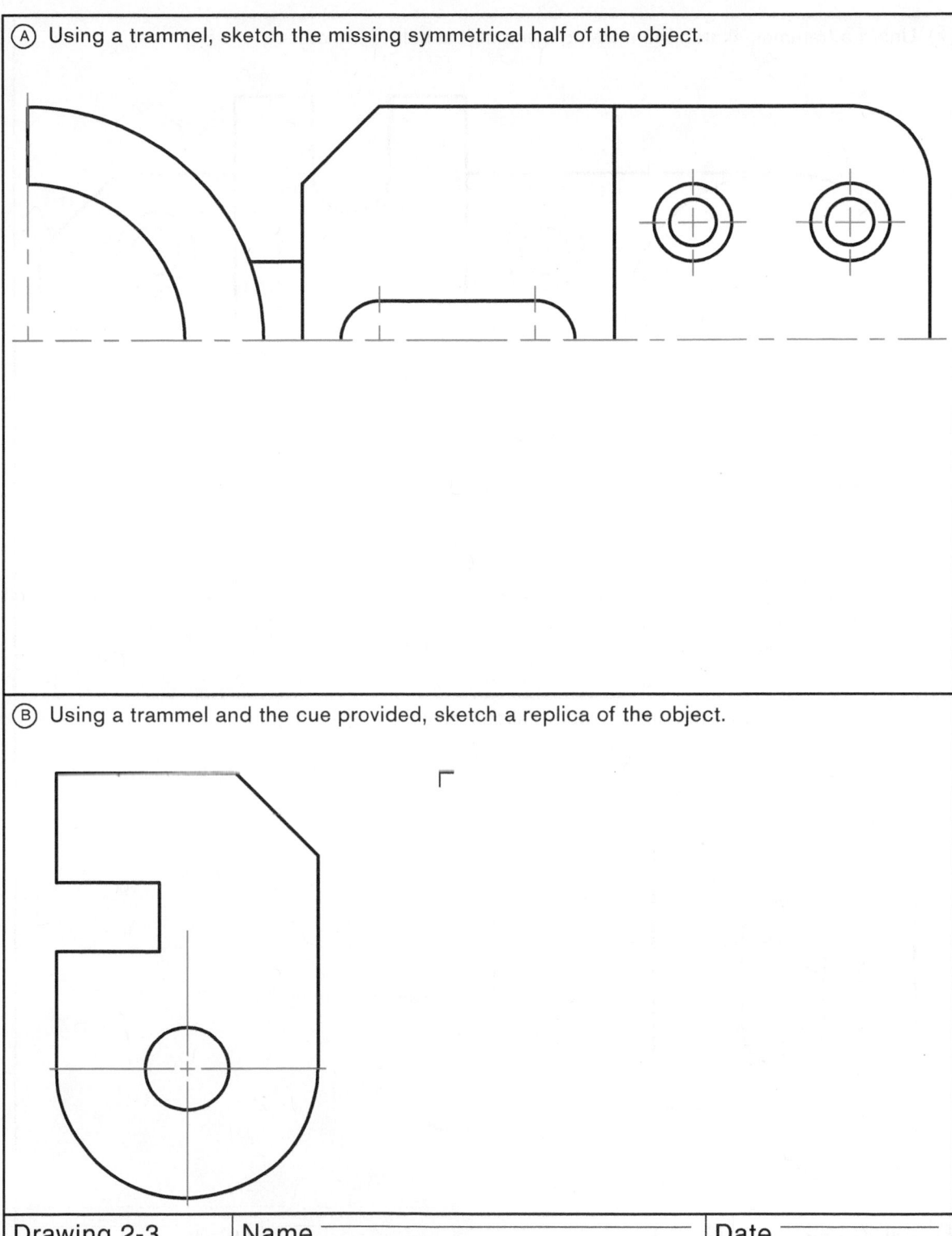

Ⓐ Using a trammel, sketch the missing symmetrical half of the object.

Ⓑ Using a trammel and the cue provided, sketch a replica of the object.

| Drawing 2-3 | Name | Date |

(A) Using a trammel, sketch the missing symmetrical half of the object.

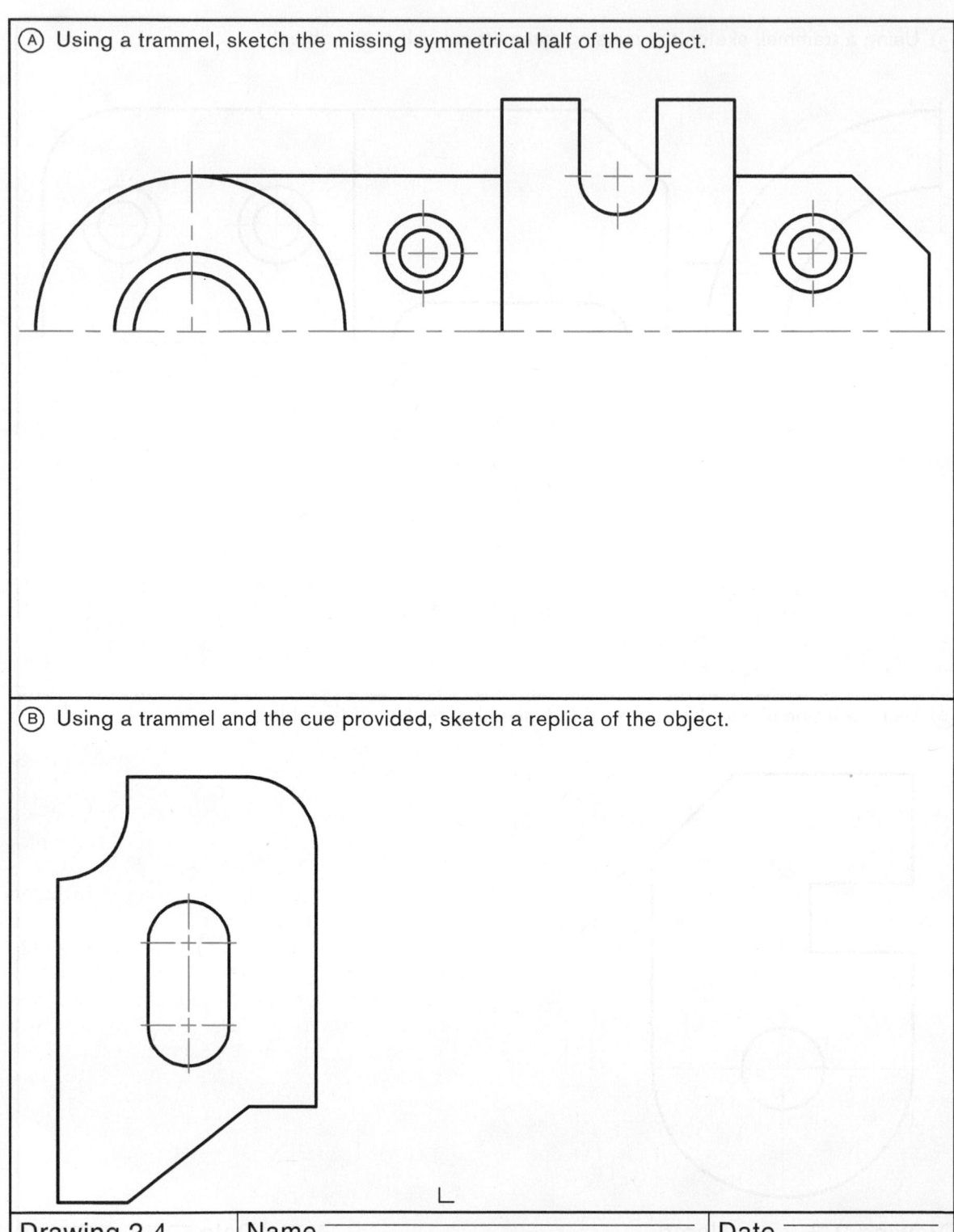

(B) Using a trammel and the cue provided, sketch a replica of the object.

L

| Drawing 2-4 | Name | Date |

3 PLANAR PROJECTIONS AND PICTORIAL VIEWS

■ PLANAR PROJECTIONS

Introduction

Projection is the process of reproducing a spatial object on a plane, curved surface, or line by projecting its points. Common examples of projection include photography, where a 3D scene is projected onto a 2D medium, and map projection, where the earth is projected onto a cylinder, a cone, or a plane in order to create a map. ***Planar projection*** figures prominently in both engineering and computer graphics. For our purposes, a projection is a mapping of a three-dimensional (3D) space onto a two-dimensional subspace (i.e., a plane). The word *projection* also refers to the two-dimensional (2D) image resulting from such a mapping.

Every planar projection includes the following elements:

- The 3D object (or scene) to be projected
- Sight lines (called projectors) passing through each point on the object
- A 2D projection plane[1]
- The projected 2D image that is formed on the projection plane

These elements are indicated in Figure 3-1. The projection is formed by plotting piercing points created by the intersection of the projectors with the projection plane. By mapping these points onto the projection plane, the 2D image is

---

[1] Although commonly represented as a bounded rectangle, the projection plane is infinite in extent.

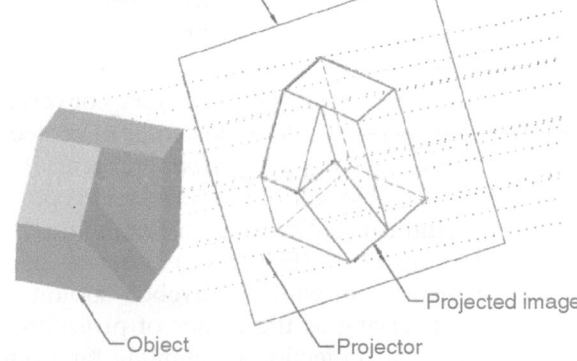

Figure 3-1 Elements of a planar projection

formed. In effect, three-dimensional information is collapsed onto a plane.

The Albrecht Dürer drawing shown in Figure 3-2 on page 38 further illustrates these projection elements. The lute lying on the table is the ***object***, the piece of string is a (movable) ***projector***, and the frame through which the artist looks is the ***projection plane***. Two movable threads are mounted on the frame, allowing the artist to identify a single piercing point. A piece of paper hinged to the frame serves as the basis for the ***projected image***. Once the projected point is transferred to the paper, the artist's assistant moves the string to another point on the lute, and the process is repeated.

Classification of Planar Projections: Projector Characteristics

Planar projections are initially classified according to the characteristics of their projectors. In a

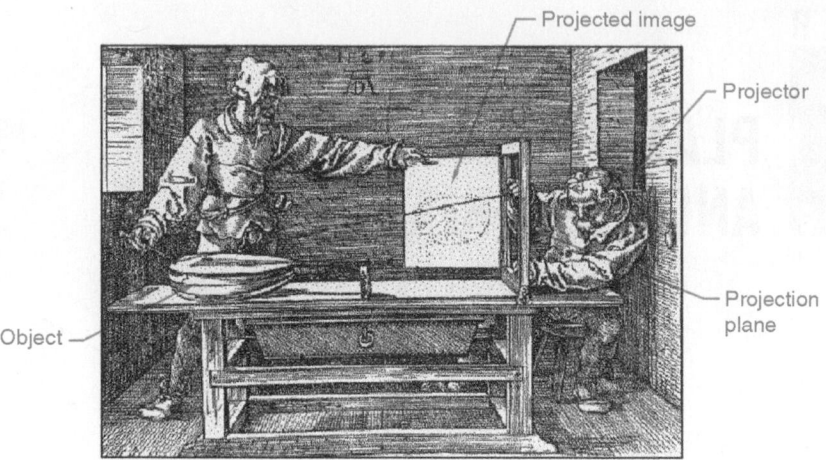

Figure 3-2 Albrecht Dürer, *Artist Drawing a Lute*, 1525

perspective projection, the projectors converge to a single viewpoint called the center of projection (CP). The *center of projection* represents the position of the observer of the scene and is positioned at a finite distance from the object. Dürer's *Artist Drawing a Lute* (Figure 3-2) illustrates perspective projection, with the eyebolt mounted on the wall serving as the center of projection. If the center of projection is infinitely far from the object, the projectors will be parallel to one another. In this case a *parallel projection* results. Figure 3-3 compares the projectors in both a perspective and a parallel projection, where in this case the projected object is simply a vertical line.

Preliminary Definitions

Before moving on to a more detailed discussion of the different kinds of planar projections, it is useful to introduce some additional terminology. Figure 3-4 shows a cut block, typical to the objects used throughout the course of this work. An object's *principal enclosing box (PEB)* just contains the object, and consequently its dimensions are the maximum width, depth, and height of the object. These are referred to as the *principal dimensions* of the object. Mutually perpendicular axes corresponding to the edges of the PEB are referred to as the *principal axes* of the

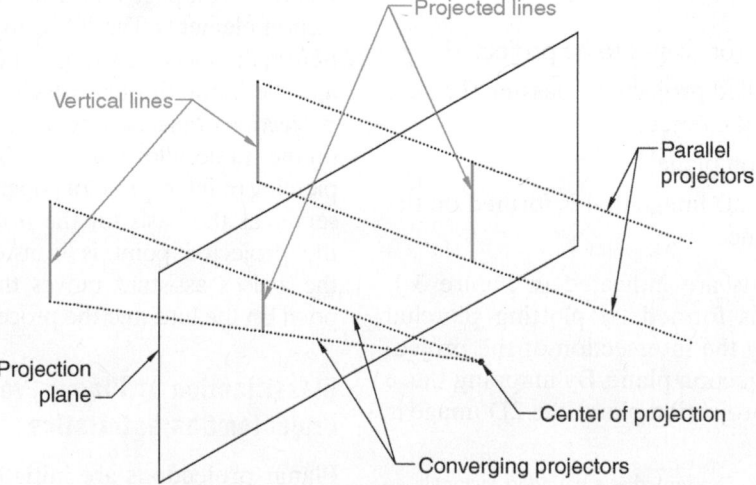

Figure 3-3 Comparison of projectors for perspective and parallel projection

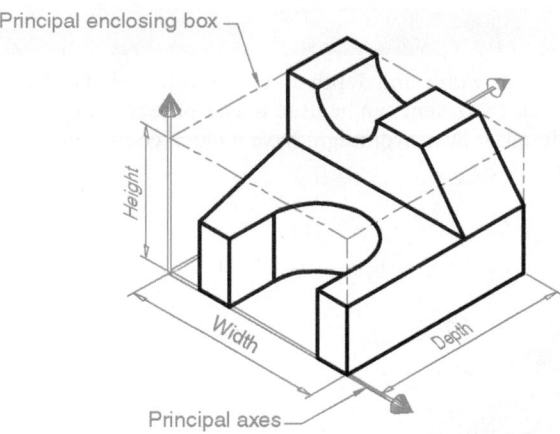

Figure 3-4 PEB, principal dimensions, and principal axes

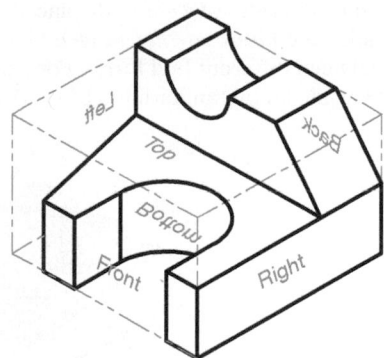

Figure 3-5 Principal planes

object. The PEB is also referred to as a ***bounding box***; these terms can be used interchangeably.

The faces of the PEB are referred to as the ***principal planes*** or ***faces*** of the object. As shown in Figure 3-5, these planes are categorized as being either frontal (front, back), horizontal (top, bottom), or profile (right, left), for a total of six.

Foreshortening is an important graphical concept related to projection theory. A dictionary definition of ***foreshorten*** is to shorten by proportionately contracting in the direction of depth so that an illusion of projection or extension in space

is obtained. To demonstrate foreshortening, close one eye and hold your hand in front of you, with your palm perpendicular to your line of sight. Now curl your fingers toward you until they are parallel to your line of sight. The image that you see of your fingers is foreshortened.

The term ***pictorial*** is used to indicate a kind of projection, view, drawing, or sketch that includes all three dimensions and thus creates the illusion of depth. The principal types of pictorial views are perspective, oblique, and axonometric. Figure 3-6 shows four different pictorial views

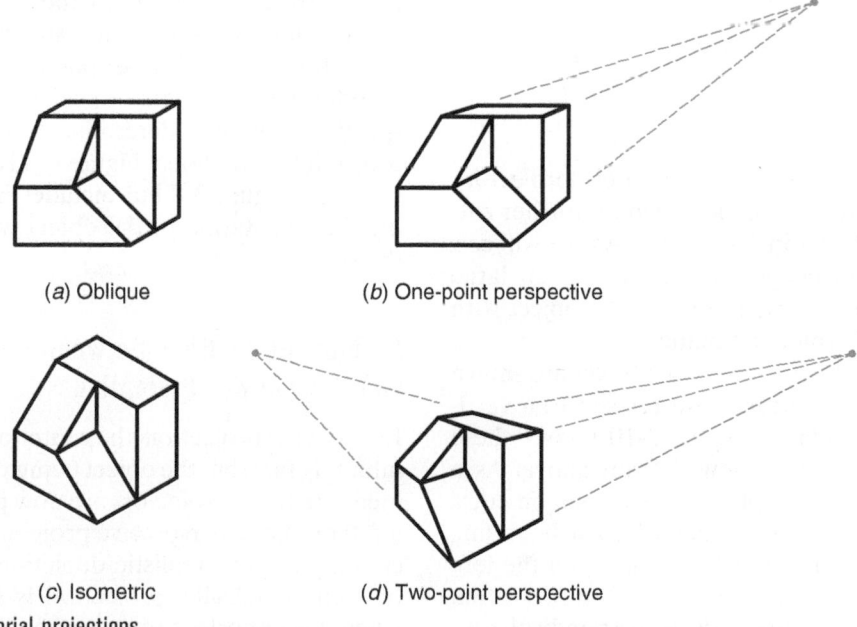

(a) Oblique

(b) One-point perspective

(c) Isometric

(d) Two-point perspective

Figure 3-6 Pictorial projections

In ship design, the ratio of a ship's displaced volume to its length, width, and depth up to the waterline (i.e., its PEB) is called the ***block coefficient*** (see Figure 3-7). The block coefficient can be used to compare the relative fineness between different hull forms. For example, a speedboat or destroyer might have a block coefficient of about 0.38, while an oil tanker has a block coefficient of about 0.80.

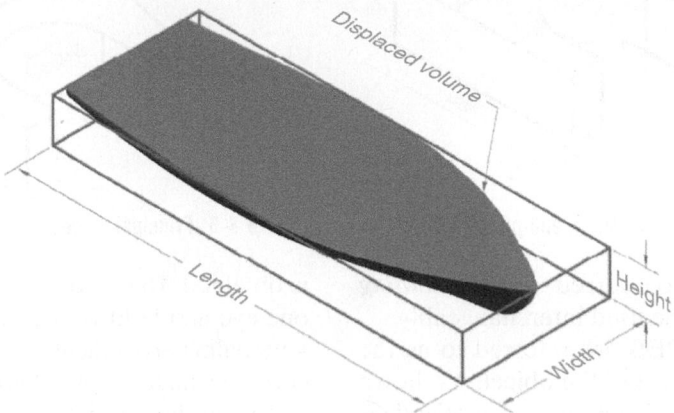

Figure 3-7 Block coefficient

of the same object. The two on the left (oblique and isometric) are parallel projections, while the two on the right (one-point and two-point) are perspective projections.

Classification of Planar Projections: Orientation of Object with Respect to Projection Plane

Beyond the characteristics of their projectors, planar geometric projections can be further categorized as shown in Figure 3-8. As we will see, these planar projection subclasses are in large part based on the orientation of the object with respect to the projection plane.

In Figure 3-9, three bounding boxes are shown in different orientations with respect to a vertical projection plane. Figure 3-10 shows these same elements when viewed from above. As a result, the projection plane is now seen on edge. Note that the principal axes of each bounding box are also shown. For box A note on the left, two principal axes are parallel to the projection plane, while the third axis is perpendicular to

the projection plane. In the case of box B in the center, one principal axis (i.e., vertical) is parallel to the projection plane, while the other two are inclined (i.e., neither parallel nor perpendicular) to the projection plane. In the last case, box C on the right, all three axes are inclined to the projection plane. These possible orientations, together with the type of projection, either perspective or parallel, determine most of the planar projection categories. Figure 3-11 on page 42 is based on Figure 3-8 but includes images depicting the orientation of the object with respect to the projection plane.

Further Distinctions Between Parallel and Perspective Projections

In a parallel projection, the center of projection is infinitely far from the object being projected. This means that the projectors are now parallel to one another. While perspective projection is useful in creating a more realistic depiction of an object or scene, a parallel projection is typically used when it is important to preserve the dimensional

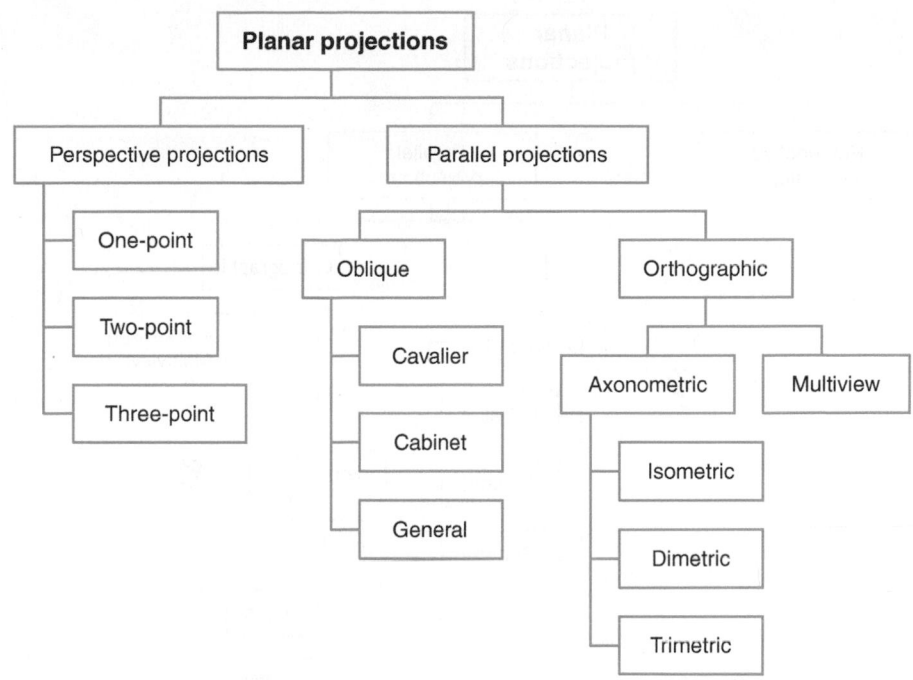

Figure 3-8 Planar geometric projection classes

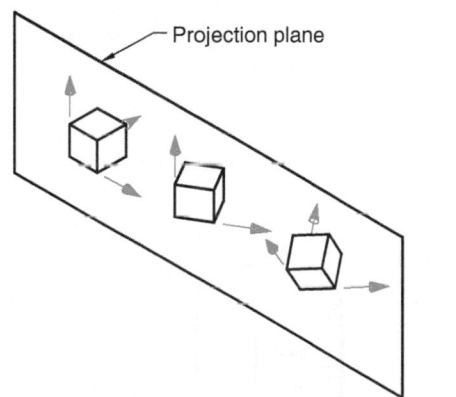

Figure 3-9 Orientation of object with respect to projection plane

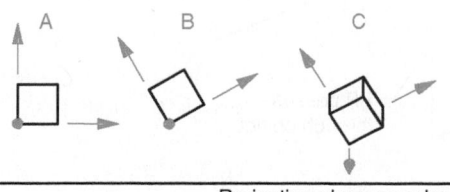

Figure 3-10 Orientation of object with respect to projection plane as seen from above

properties of the object. As an example, compare the projections employing parallel projectors in Figure 3-12 on page 42 with the converging projectors used in Figure 3-13 on page 43. In both cases, the projected object face is parallel to the projection plane. Note that parallel projection preserves both the size and the shape of the object face, whereas perspective projection preserves only the shape of the object face (and that the shape may even be inverted!), depending on the location of the projection plane.

In a parallel projection, parallel object edges remain parallel when projected. Unlike perspective projection, there are no *vanishing points*. Figure 3-14 on page 43 shows a parallel projection and a perspective projection of the same object. Note that whereas the receding edges of the parallel projection on the left remain parallel, the corresponding edges on the perspective projection on the right converge to a vanishing point. Although the perspective projection provides a more realistic depiction of the object, the parallel projection, in preserving parallelism, is easier to scale and draw.

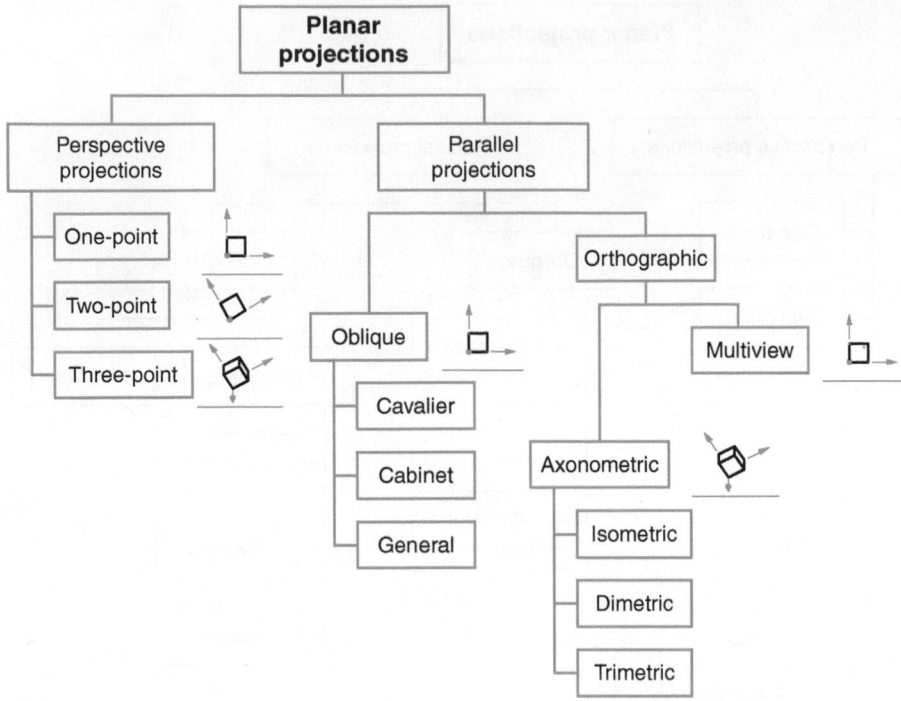

Figure 3-11 Projection classes with orientation shown

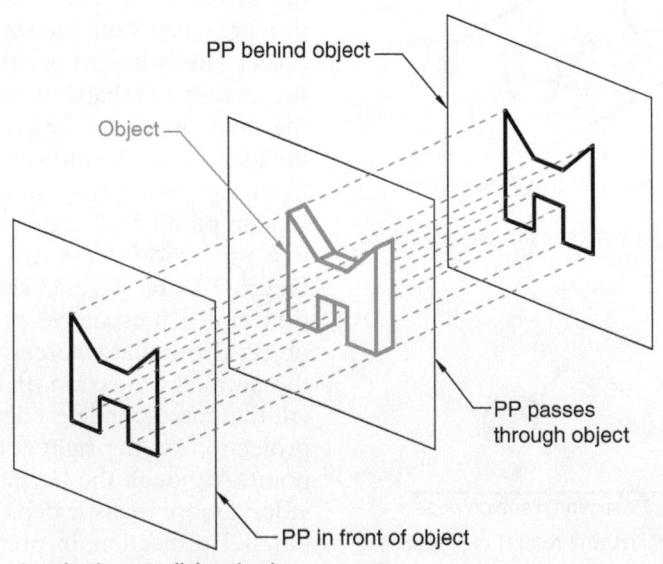

Figure 3-12 Projection plane location in a parallel projection

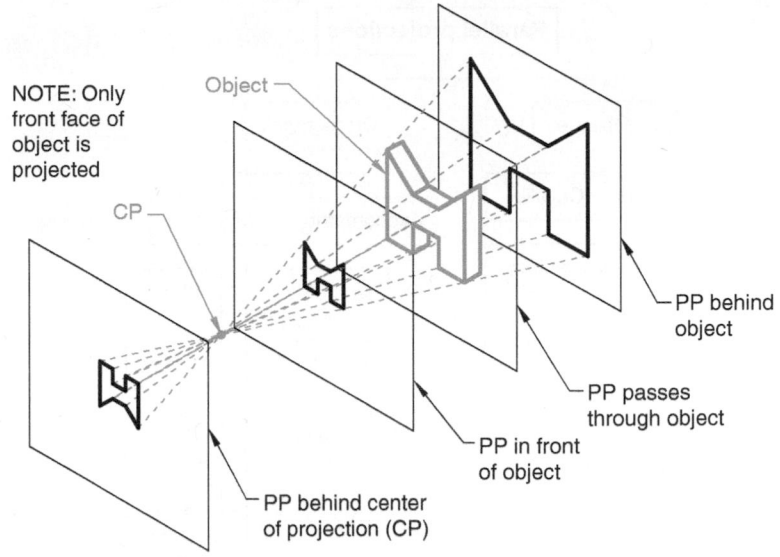

NOTE: Only front face of object is projected

Figure 3-13 Projection plane location in a perspective projection

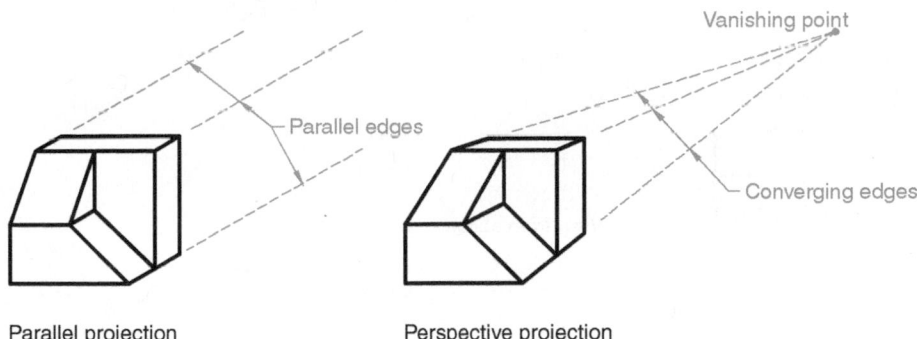

Parallel projection

Perspective projection

Figure 3-14 Projection plane location in a perspective projection

Though extremely important in computer graphics, perspective projection is not commonly employed in engineering. For this reason, perspective projection and perspective sketching are treated separately in Chapter 13.

Classes of Parallel Projections

Figure 3-15 on page 44 shows a breakdown of the different kinds of parallel projection techniques. Oblique and axonometric projections will be discussed in the remainder of this chapter. Multiview sketching is the subject of the following chapter.

■ OBLIQUE PROJECTIONS

Oblique drawings are traditionally employed when one object face is significantly more complicated than the other faces of the object.

Oblique Projection Geometry

Figure 3-16 on page 44 shows the geometric arrangement of an oblique projection. As seen in Figure 3-16a, parallel projectors intersect the projection plane at an oblique angle. In addition, one principal face of the object is parallel to the projection plane. In an oblique projection, the

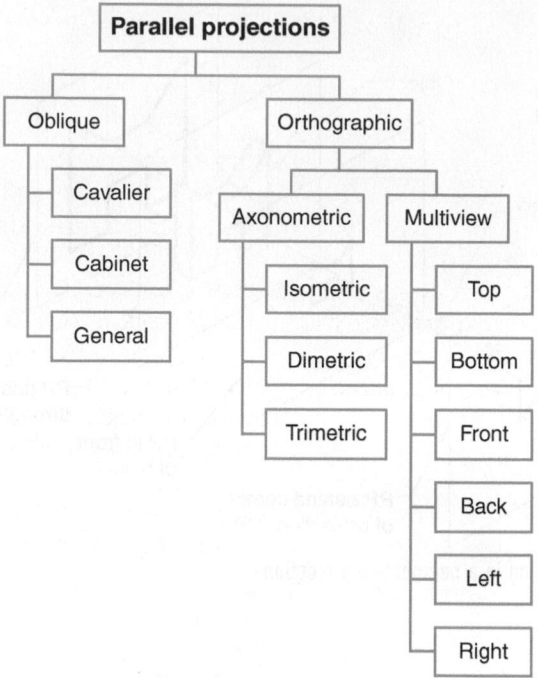

Figure 3-15 Projection plane location in a perspective projection

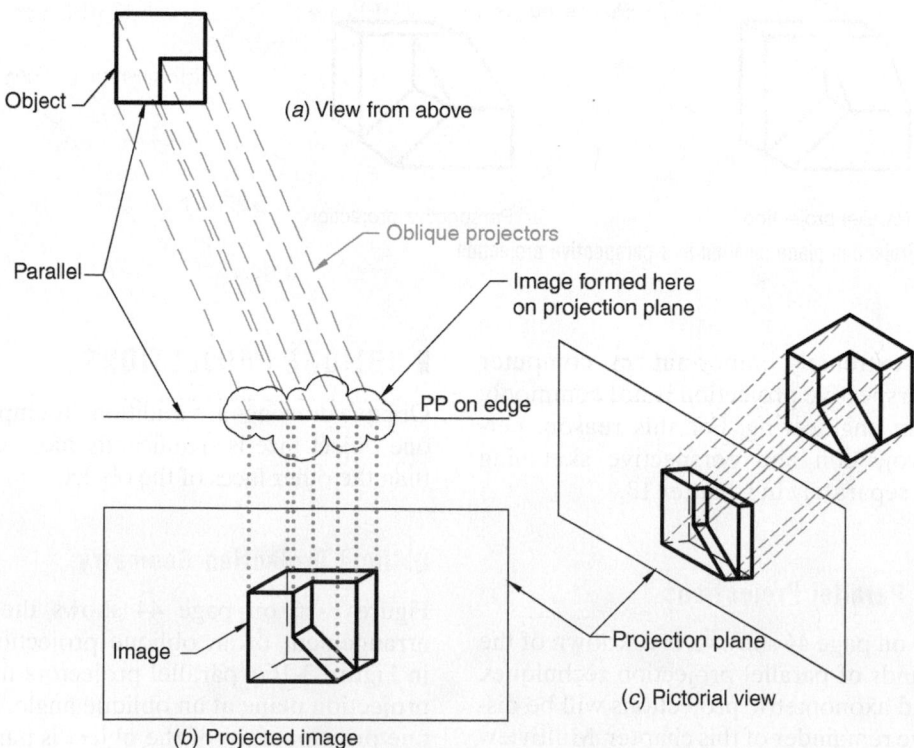

Object

Parallel

Oblique projectors

Image formed here
on projection plane

PP on edge

Image

(a) View from above

(b) Projected image

Projection plane

(c) Pictorial view

Figure 3-16 Projection plane location in a perspective projection

object face that is parallel to the projection plane is projected true size.

Oblique Projection Angle

Two angles can be used to describe the intersection of an oblique projector with a projection plane, as shown in Figure 3-17. An in-plane angle β measures the angle of rotation of the projector about the projection plane normal. The out-of-plane angle α is called the ***oblique projection angle***. The oblique projection angle determines the type of oblique projection: cavalier, cabinet, or general.

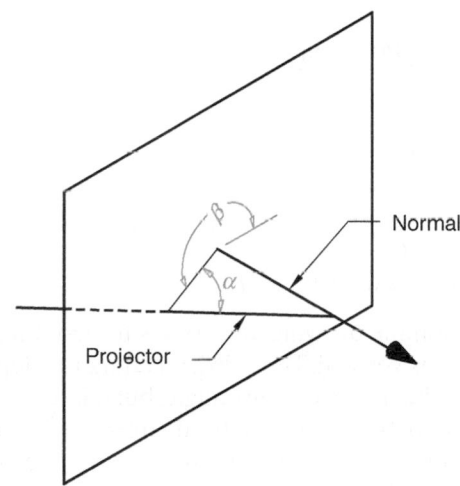

Figure 3-17 In-plane and out-of-plane angles defining an oblique projection

Classes of Oblique Projections

Figure 3-18 shows an oblique projection setup, where two identical cubes (shown in the top half of the figure) are projected onto a projection plane (in the bottom half of the figure). The front face of both cubes is projected true size. The oblique projection angle (α) used for the cube on the left is 45 degrees, and the resulting projection is called a ***cavalier oblique***. Note that the projection of this cube appears to be elongated in the receding (depth) axis direction. In fact, the measure of all of the edges of both the cube and its cavalier projection are identical. In a cavalier projection, the receding axis is not foreshortened; it is scaled the same as the other (horizontal, vertical) principal axes.

On the right half of Figure 3-18, the oblique projection angle (α) is approximately 63.43 degrees, resulting in a ***cabinet oblique*** projection. In comparing the projected lengths of this cube, you will find that the projected receding edge length is one-half that of the other (horizontal, vertical) edge lengths. The receding axis of a cabinet oblique is foreshortened to exactly one-half that of the other principal axes. Note also that the projection that results from a cabinet oblique appears to the eye to be more proportionally correct than the cavalier oblique. This is because, visually, we expect some foreshortening to occur along a receding axis.

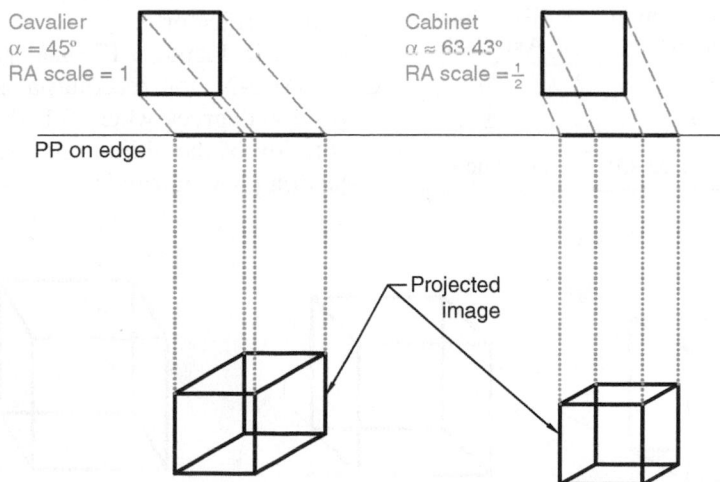

Figure 3-18 In-plane and out-of-plane angles defining an oblique projection

By collapsing the 3D oblique projection geometry into 2D, it is easy to understand why a 45 degree projection angle results in no foreshortening (i.e., 1:1 scale), while a ~63.43 degree angle scales a projected length to one-half of the original. Figure 3-19 shows a projection plane on edge and the object, a line of length L. The line is perpendicular to the projection plane, much like the cube edges (not parallel to the projection plane) in Figure 3-18 on page 45 are perpendicular to the projection plane. (*Note*: Tan 45° = 1, Cot 63.43° = $\frac{1}{2}$.)

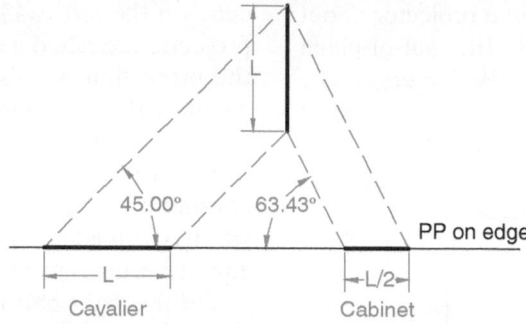

Figure 3-19 In-plane and out-of-plane angles defining an oblique projection

If an oblique projection angle between 45 and 63.43 degrees is used, the result is called a ***general oblique*** projection. On a general oblique, the receding axis is scaled between $\frac{1}{2}$ and 1. This would normally be done to improve the appearance of the projected image. Table 3-1 summarizes the different kinds of oblique projections.

Table 3-1 Classes of oblique projections

Type of Oblique	Oblique Projection Angle (α)	Receding Axis Scale
Cavalier	45°	1
Cabinet	63.43°	$\frac{1}{2}$
General	45° < angle < 63.43°	$\frac{1}{2}$ < scale < 1

Receding Axis Angle

In an oblique drawing, one axis is horizontal and another is vertical. The third, receding (or depth) axis can be inclined at any angle, but it is normally chosen to be 30, 45, or 60 degrees. This angle determines the relative emphasis of the receding planes on the projection, as seen in Figure 3-20. Note that the receding axis angle should not be confused with the oblique projection angle.

The receding axis angle is related to the in-plane projector angle β discussed earlier and shown in Figure 3-17 on page 45. As shown in Figure 3-21, the receding axis angle is equal to β – 180 degrees, where β is the in-plane angle of rotation of the oblique projector about the projection plane normal.

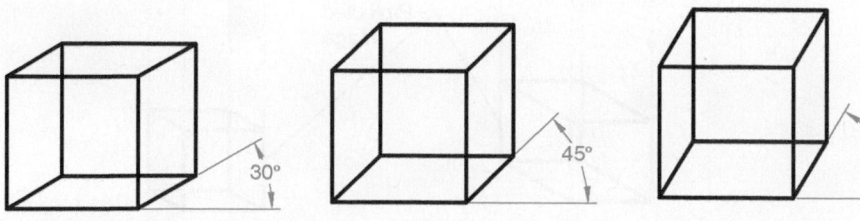

Figure 3-20 Receding axis angle

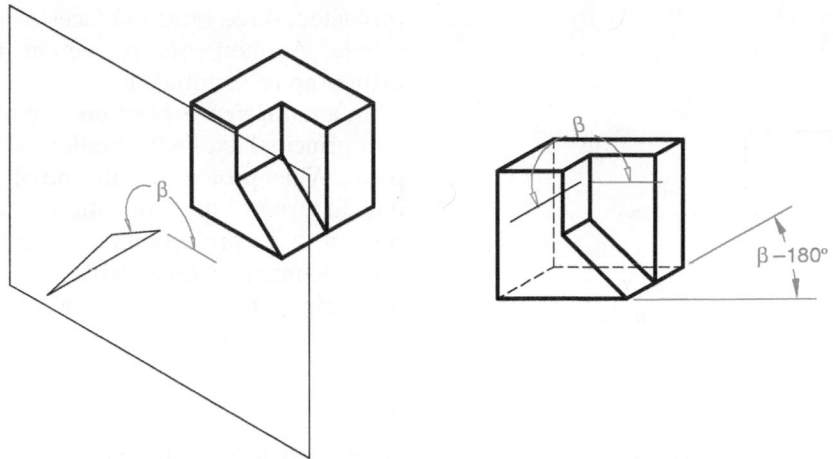

Figure 3-21 Relationship between the receding axis angle and in-plane projector angle β

■ ORTHOGRAPHIC PROJECTIONS

Orthographic projection is the most commonly used projection technique employed by engineers. CAD systems typically employ orthographic projection techniques, although the user is often provided with the option of changing to perspective projection.

Orthographic Projection Geometry

Orthographic projection is also a parallel projection technique, but it differs from oblique projection in that the parallel projectors are perpendicular (normal) to the projection plane. Figure 3-22 shows the parallel projectors and their relationship to the projection plane.

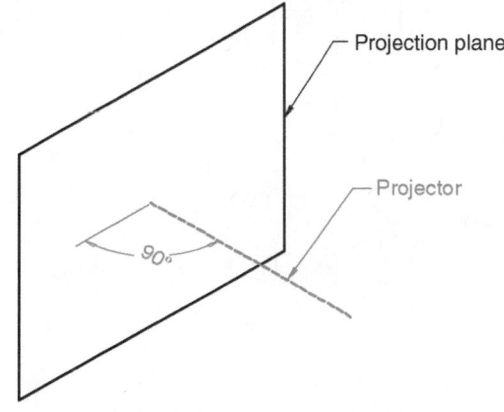

Figure 3-22 Orthographic projection geometry

Orthographic Projection Categories

Orthographic projections are subdivided according to the orientation of the object with respect to the projection plane. In an ***axonometric projection***, all three principal axes are inclined to the projection plane. Figure 3-23 shows this orientation, both as viewed from the front and as viewed from above. No axis is parallel (or perpendicular) to the projection plane. When

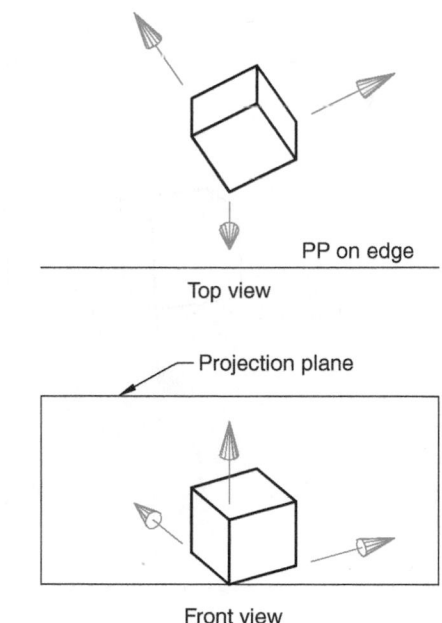

Figure 3-23 Object position in axonometric projection

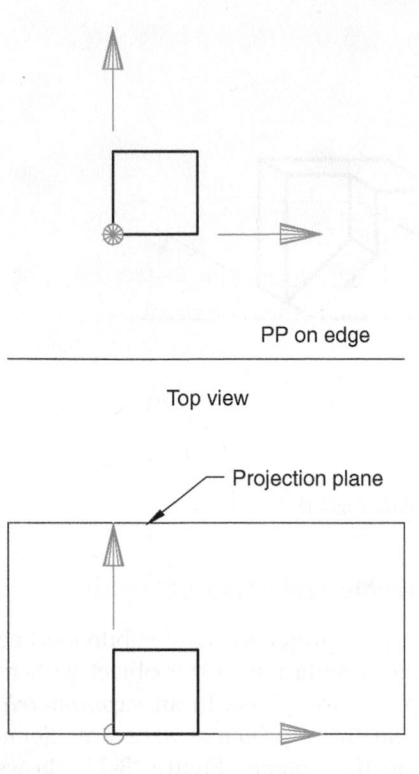

PP on edge

Top view

Projection plane

Front view

Figure 3-24 Object position in multiview projection

projected, three principal faces of the object are visible. Axonometric projection results in an orthographic pictorial view.

In a ***multiview projection***, one object face and two principal axes are parallel to the projection plane. When projected, only one object face is visible. In terms of the orientation of the object with respect to the projection plane, multiview projection is identical to oblique projection, as well as to one-point perspective projection. See Figure 3-24.

▌AXONOMETRIC PROJECTIONS

Axonometric projections are classified according to the angles made by the principal axes when projected onto the projection plane. The upper half of Figure 3-25 shows three different axonometric projections of a cube, along with an attached principal axis triad. In the lower portion of the figure, only the projected axes and the angles between them are shown.

In a ***trimetric projection***, depicted on the left in Figure 3-25, none of the angles between the projected principal axes are equal. The middle projection of Figure 3-25 is called a ***dimetric projection***,

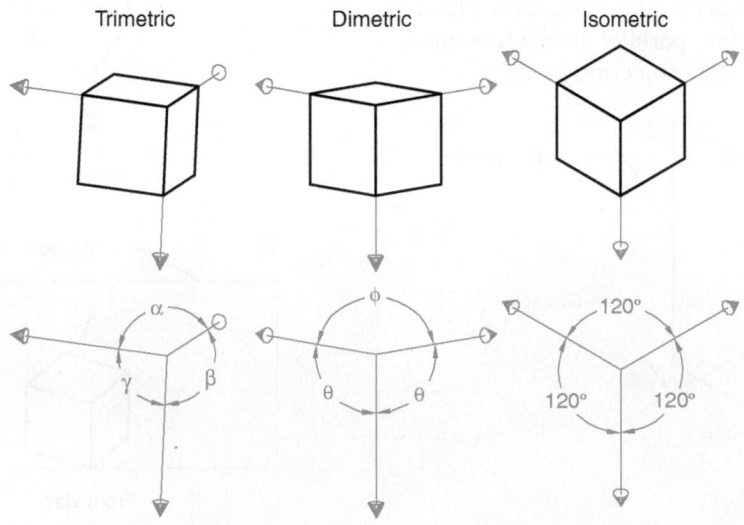

Figure 3-25 Axonometric projection classes

where two of the three angles are equal. In an ***isometric projection***, shown on the right in Figure 3-25, all three angles are equal.

Also note that, because all of the equal-length axes are inclined to the projection plane, all of their projections are foreshortened, as depicted in Figure 3-25. The amount of foreshortening is related to the projected angle. In the trimetric projection (on the left) all three axes are foreshortened by different amounts, and all three projected angles are different. Two of the three projected axes in the dimetric projection are foreshortened by the same amount, and the angles opposite these axes are also equal. In the isometric projection on the right, there is an equal amount of foreshortening along all three axes, and all of the angles are equal.

■ ISOMETRIC PROJECTIONS

An isometric projection is foreshortened, or scaled, equally along all three principal axis directions. This fact makes isometric projections particularly useful in engineering. As a pictorial, an isometric projection is relatively easy to visualize, and it is also good at preserving the dimensional properties of the object.

To understand how an object must be oriented in order to obtain an isometric projection, imagine the projection of a cube. Figure 3-26 on the left shows a trimetric view of the cube, on which a cube diagonal has been drawn. If the cube is rotated so that we look down the diagonal, as in Figure 3-26 on the right, we get an idea of how an isometric view is generated.

In most CAD systems an isometric view can be generated automatically. Figure 3-27 shows an example of a CAD viewing tool. Clicking on any diagonal arrow generates one of eight possible isometric views.

More specifically, an isometric view is generated as shown in Figure 3-28:

1. Start with one principal face of the object parallel to the projection plane.

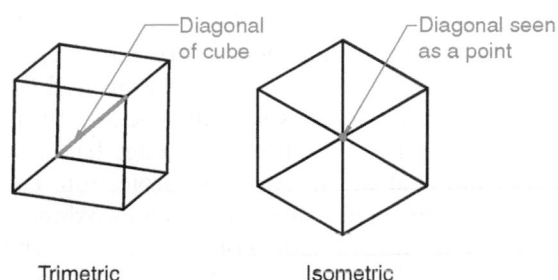

Figure 3-26 Looking down the diagonal of a cube

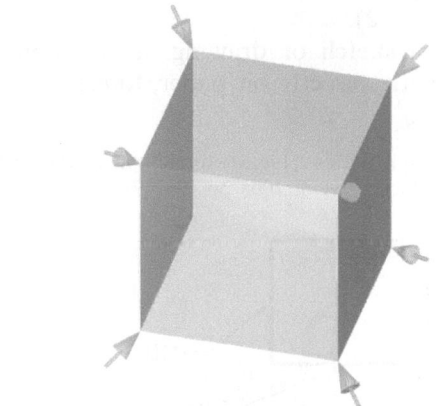

Figure 3-27 CAD system viewing tool

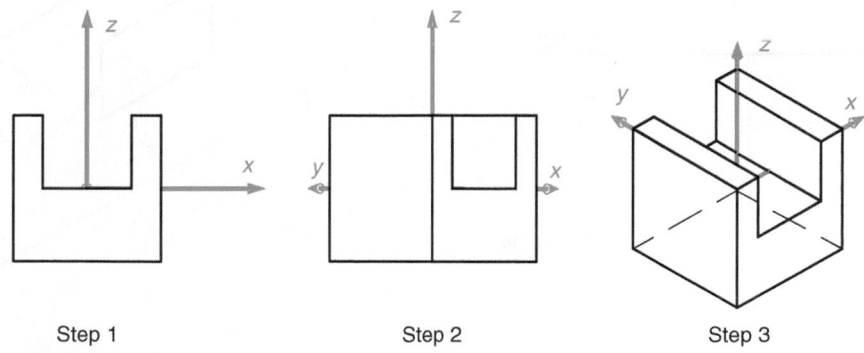

Figure 3-28 Object orientation for an isometric view

2. Rotate the object about a vertical axis $(45 \pm 90n)$ degrees, where n is an integer.

3. Rotate the object out of the horizontal plane by approximately ± 35.26 degrees.[2]

Isometric Drawings

As a result of the particular object orientation in an isometric view, all three principal axes are foreshortened by exactly the same amount. This means that if a *solid model* is created in a CAD system, and an isometric view of the object is then printed out at a 1:1 scale, the printed (i.e., *projected*) edge lengths will all be equal to one another, but less than the actual edge lengths. It turns out that in an isometric projection, each principal axis is foreshortened to approximately 82% of its true length. For this reason, when plotting an isometric view in a CAD system, it is possible to correct for this foreshortening by multiplying the desired scale by the reciprocal of 0.82 (i.e., ~1.22).

When a sketch or drawing of an isometric view is made directly on paper, foreshortening

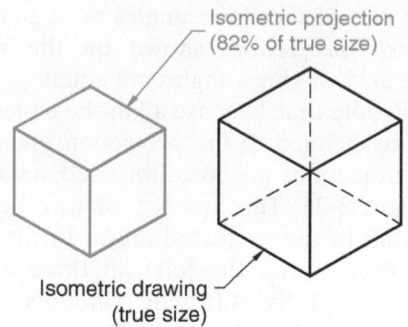

Figure 3-29 Comparison between an isometric projection and an isometric drawing

effects are typically ignored. An isometric projection without foreshortening is referred to as an ***isometric drawing***. Note that, as shown in Figure 3-29, an isometric drawing is larger than a true isometric projection.

Multiview Projections

An arrangement for multiview projection is characterized by the following elements (see Figure 3-30).

1. The projectors are parallel to one another.

2. The projectors are normal to the projection plane.

[2] For a derivation of this value, see Ibrahim Zeid, *Mastering CAD/CAM*, McGraw-Hill, 2005, pp. 496–497.

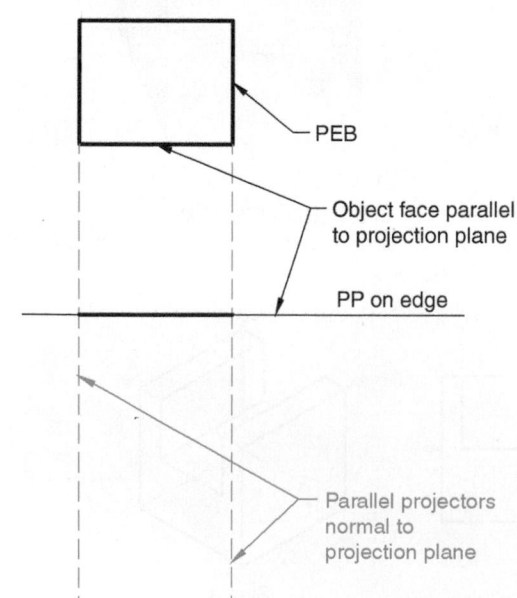

Figure 3-30 Multiview projection setup

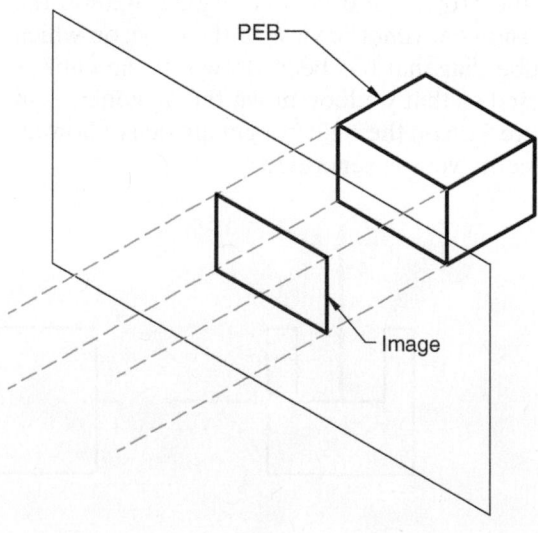

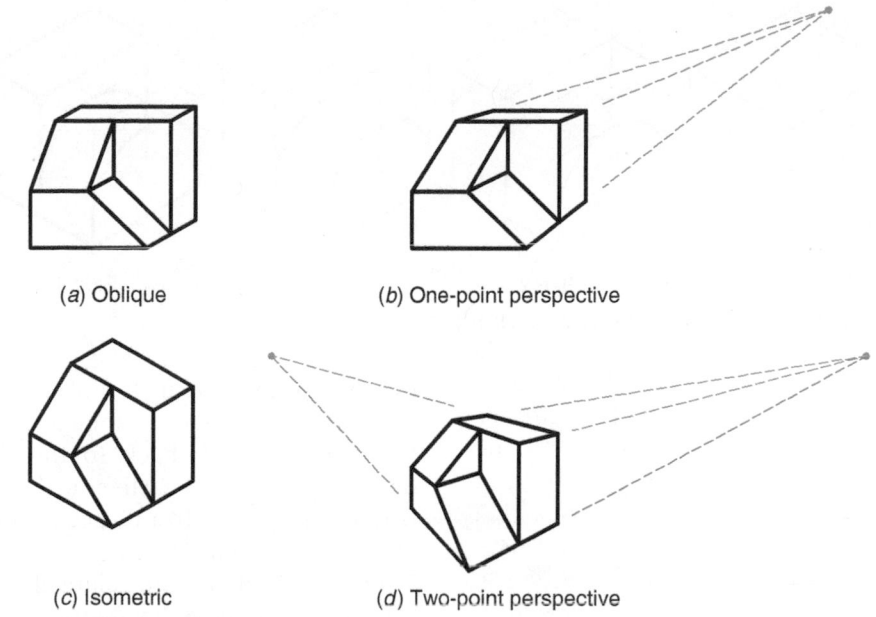

(a) Oblique

(b) One-point perspective

(c) Isometric

(d) Two-point perspective

Figure 3-6 (Repeated) Pictorial projections

3. The object is positioned with one principal face parallel to the projection plane.

As a direct result of these characteristics, multiview projections are good at preserving the object's dimensional information, but more than one view is necessary to fully describe the object.

In the following sections, different methods and techniques for constructing oblique and isometric pictorial sketches will be discussed. Multiview sketching is the subject of Chapter 4.

■ INTRODUCTION TO PICTORIAL SKETCHING

In a pictorial view, three principal faces of an object are visible. A *pictorial sketch* shows an object's height, width, and depth in a single view. Unlike the multiview orthographic sketches discussed in the following chapter, a pictorial sketch conveys the object's three-dimensional shape in a single view. In this chapter, parallel projection (oblique and isometric) pictorials are discussed. Chapter 13 includes a discussion of perspective

(both one-point and two-point) pictorials. Generally speaking, parallel projections preserve the object's dimensional properties, while perspective projections convey a strong sense of realism. Figure 3-6, which is reprinted above, shows how the same object would appear using these different pictorial projection techniques.

Regardless of the particular pictorial sketch being executed, the same general technique can be employed:

1. Using light construction lines, sketch a properly proportioned bounding box.

2. Continuing with construction lines, add feature details.

3. Starting with curved features, go bold.

4. Compete the sketch using bold lines.

This process is illustrated in Figure 3-31 on page 52 for an isometric pictorial.

Polyhedral shapes are frequently used for sketching. A *polyhedron* is a 3D solid bounded by a connected set of polygons, where every edge of a polygon belongs to just one other polygon. Polyhedral geometry consists of faces, edges, and vertices, as shown in Figure 3-32 on page 52.

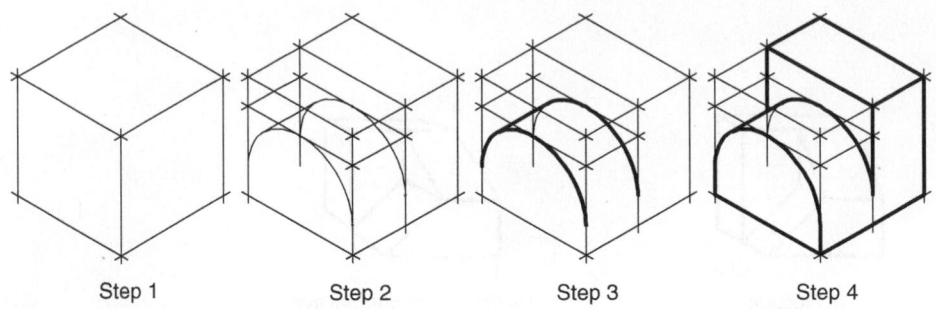

Step 1   Step 2   Step 3   Step 4

Figure 3-31 General procedure for pictorial (isometric) sketch

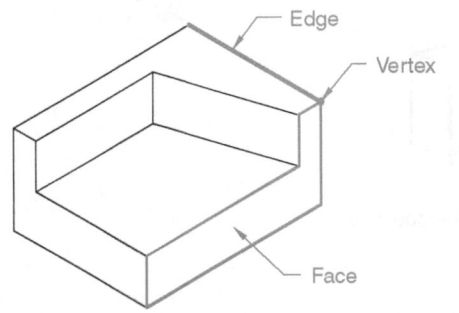

Figure 3-32 Cut block with vertex, edge, and face labeled

A *face* is a bounding surface on an object, whereas an *edge* serves as the intersection between two faces. Finally, a *vertex* is the end point of an edge.

▌ OBLIQUE SKETCHES

Introduction

An oblique sketch shows the true size and shape of one principal face of an object. Two receding principal faces are also depicted in order to complete the pictorial view. Traditionally the object is positioned such that its most complex face (i.e., irregular, curved edges, etc.) is shown true size. Relating back to the discussion of oblique projections earlier in this chapter, the true-size (typically front) face is oriented parallel to the projection plane.

Oblique sketches are relatively easy to draw. This is because complex features parallel to the projection plane project without distortion. For example, a circular edge will project as a circle, not an ellipse, as long as it is parallel to the projection plane.

Axis Orientation

In most oblique sketches, the object is oriented so that its front face is shown true size (parallel to the projection plane). Horizontal and vertical axes define the frontal plane. Width is measured along the horizontal axis, height along the vertical. The receding axis is then used to measure depth. The receding axis is typically drawn either up and to the left, or up and to the right. See Figure 3-33. The angle that the receding axis makes with the horizontal is usually drawn to be 30, 45, or 60 degrees.

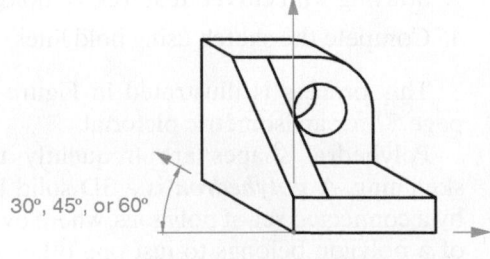

30°, 45°, or 60°

(*a*) Receding axis up and to left

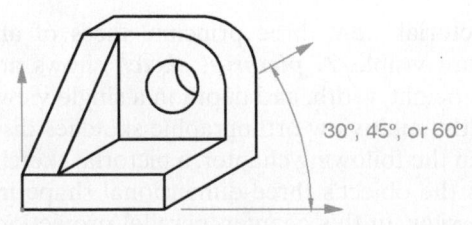

30°, 45°, or 60°

(*b*) Receding axis up and to right

Figure 3-33 Oblique sketches of the same object

Table 3-2	**Classes of oblique sketches**
Oblique Type	Receding Axis Scale
Cavalier	1
Cabinet	$\frac{1}{2}$
General	Between $\frac{1}{2}$ and 1

Receding Axis Scale

Recall, from the section Classes of Oblique Projections, that there are three different classes of oblique pictorials: cavalier, cabinet, and general. The type of oblique is determined by the amount of scaling that occurs along the receding axis.[3] See Table 3-2.

In a cavalier oblique, the same scale factor is used along all three axes: horizontal, vertical, and receding. The resulting pictorial appears to be too long in the depth direction because of the lack of foreshortening (see Figure 3-34a). To correct for this lack of foreshortening, a cabinet oblique

[3] Recall from the discussion of oblique projections earlier in the chapter that the amount of foreshortening along the receding axis is determined by the angle at which the parallel projectors pierce the projection plane. This angle is called the oblique projection angle (α). For a cavalier projection, α is 45 degrees; for a cabinet projection, α is ~64 degrees.

employs a receding axis scale of $\frac{1}{2}$. Although it is convenient and also improves upon the appearance of a cavalier oblique, a cabinet oblique can appear to have too much foreshortening. To compensate for this in order to obtain a more pleasing pictorial, a general oblique that scales the receding axis between that of the cavalier and that of the cabinet is sometimes used.

Object Orientation Guidelines

Two rules apply when selecting the front true-size face to be used in an oblique pictorial. The first rule has already been discussed—that the projection plane should be chosen so that it is parallel to the principal face containing the most complex (circular, curved) or irregular shape. This rule was employed, for example, in Figure 3-33, where the curved features are parallel to the projection plane. The second rule is to use the longest face as the front true size view. Figure 3-35 shows two oblique pictorials of an angle beam, oriented according to the two rules.

In the event that the two rules are in conflict, the first rule takes precedence. Figure 3-36a on page 54 is both easier to construct and less distorted than Figure 3-36b.

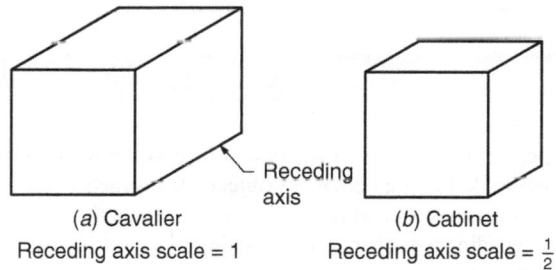

(a) Cavalier	(b) Cabinet	(c) General
Receding axis scale = 1	Receding axis scale = $\frac{1}{2}$	$\frac{1}{2}$ < Receding axis scale < 1

Figure 3-34 Cavalier, cabinet, and general obliques of a cube

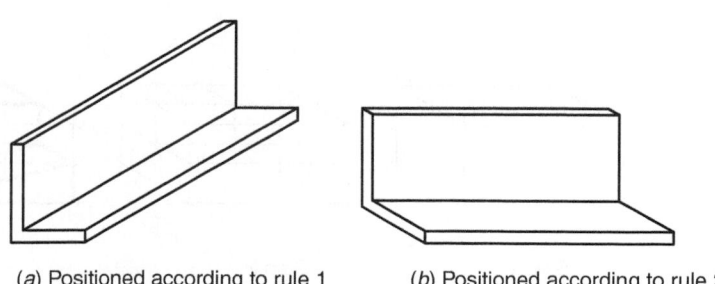

(a) Positioned according to rule 1	(b) Positioned according to rule 2

Figure 3-35 Object orientation for oblique pictorials; two rules

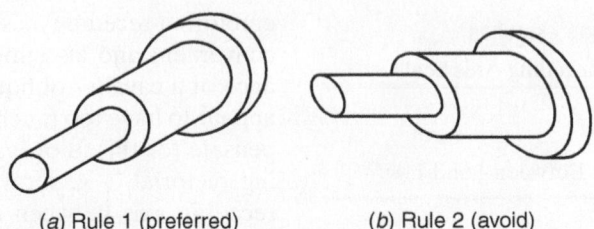

(a) Rule 1 (preferred)   (b) Rule 2 (avoid)

Figure 3-36 Object orientation for oblique pictorials; rule 1 preferred

Sketching procedure for a simple extruded shape (see Figure 3-37)

Drawing an oblique pictorial of an *extrusion* is particularly easy.

1. Sketch the extruded profile as a front view; select and lightly sketch the direction and angle of receding axis.
2. Sketch construction lines extending from other front-face vertices, parallel to the receding axis.
3. Determine the depth, and then complete the back face of the object.
4. Go bold.

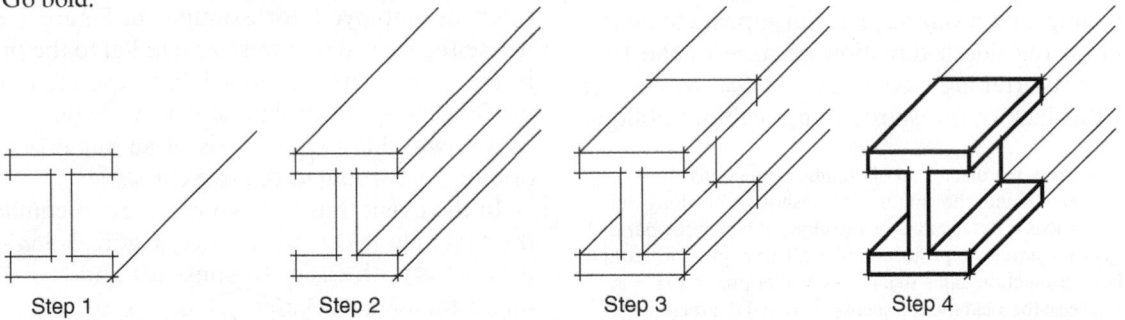

Step 1   Step 2   Step 3   Step 4

Figure 3-37 Oblique sketch of an extruded shape

Step-by-step cabinet oblique sketch example for a cut block (see Figure 3-38)

1. PEB construction:
   a. Identify the most *complex* (irregular shape, circular features, etc.) object face as the front view.
   b. Lightly sketch a properly proportioned rectangle that just encloses object's front face.
   c. Lightly sketch a receding axis (direction, angle) from one vertex of the front rectangle.
   d. Mark off the appropriate depth along the receding axis (for cabinet, scale $= \frac{1}{2}$).
   e. Complete the bounding box using construction lines.
2. Using construction lines, block in feature details.
3. Go bold.

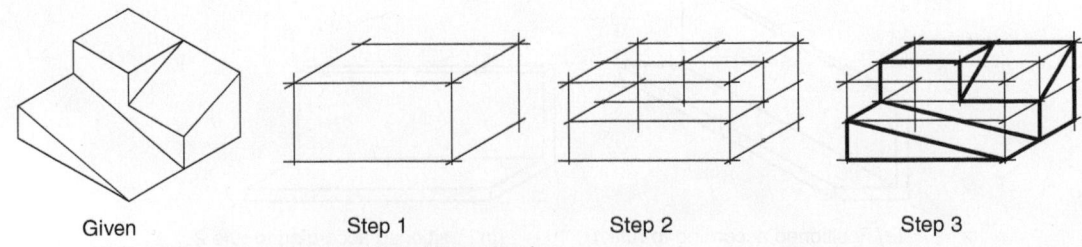

Given   Step 1   Step 2   Step 3

Figure 3-38 Multiple steps for cut block cabinet oblique sketch

1. PEB construction:
   a. Identify the most *complex* (irregular shape, circular features, etc.) object face as the front view.
   b. Lightly sketch a properly proportioned rectangle that just encloses object's front face.
   c. Lightly sketch a receding axis (direction, angle) from one vertex of the front rectangle.
   d. Mark off the appropriate depth along the receding axis (for cavalier, scale = 1).
   e. Still using construction lines, complete the bounding box.

2. Using construction lines, block in linear features.

3. Arc feature construction:
   a. Locate arc quadrants (front and back).
   b. Sketch the front face arc in bold.
   c. Use construction lines to sketch the back face arc.
   d. Sketch a line tangent to the two arcs in bold.
   e. Sketch the visible portion of the back arc in bold.

4. Hole feature construction (front face):
   a. Locate center and quadrant points.
   b. Sketch circle in bold.

5. Hole feature construction (back face):
   a. Locate center and quadrant points.
   b. Sketch circle using construction lines.
   c. Sketch visible portion of back edge in bold.

6. Go bold.

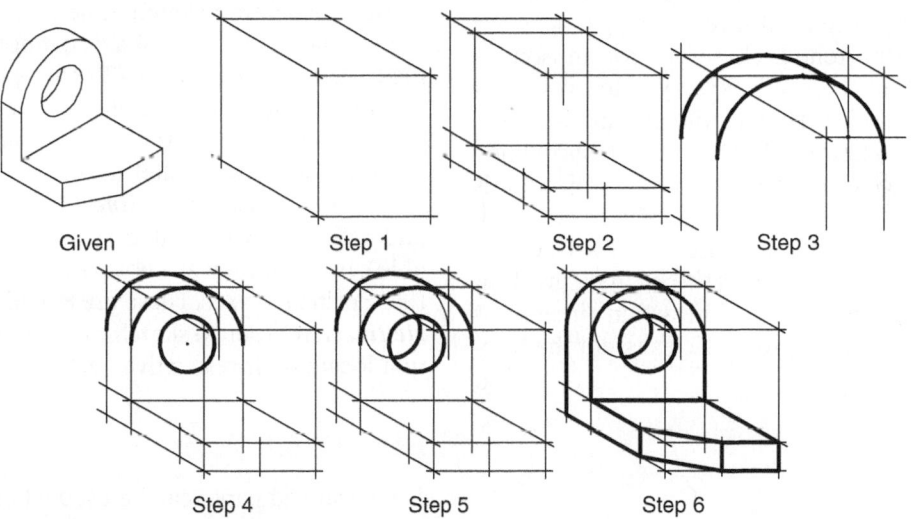

Figure 3-39 Multiple steps for cavalier oblique sketch of object with circular features

■ ISOMETRIC SKETCHES

Introduction

Similar to obliques, isometric pictorials are parallel projections from which dimensional information can be obtained. In addition, isometric views can easily be created in CAD systems, something that is not true of oblique projections.

Axis Orientation

In sketching an isometric, the principal axes are aligned as shown in Figure 3-40. One axis is vertical, and the other two are inclined at 30 degrees to the horizontal. Note that this 30 degree angle is a direct outcome of the object's orientation with respect to the projection plane used to generate an isometric projection, which was described in the earlier section on isometric projections.

Isometric Scaling

Recall from the discussion earlier in the chapter that the most important property of an isometric projection is that all three principal axes are equally foreshortened. Therefore, for an isometric pictorial, all object edges parallel to a principal axis are scalable and can be directly measured.

For an isometric sketch, the foreshortening that would occur in a true isometric projection is generally ignored. Imagine, for example, an actual rectangular prism that measures $3 \times 2 \times 1$. An isometric sketch of this prism is shown in Figure 3-41. When the sketch is created, the actual dimensions are laid out along the isometric axes, not the foreshortened lengths.

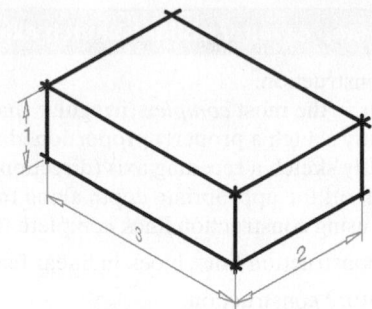

Figure 3-41 Isometric sketch of a rectangular prism

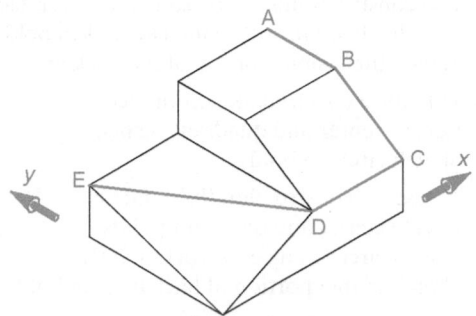

Figure 3-42 Isometric scaling

In an isometric sketch, object edges that are *not* parallel to a principal axis cannot be directly measured. In Figure 3-42, for example, the lengths of edges AB (parallel to the y axis) and CD (parallel to the x axis) can be directly measured. These lines parallel to a principal axis are referred to as **isometric lines**. Edge lengths BC and DE cannot be scaled (i.e., are not measurable), because they are not parallel to any of the three principal axes. These are called **nonisometric lines**. In order to sketch these edges, we must first locate their respective vertices.

Isometric Grid Paper

Isometric grid paper can be used when one is first learning to make isometric sketches. Figure 3-43 shows an example of isometric grid paper, along with superimposed isometric coordinate axes. Isometric grid paper consists of three sets of intersecting parallel lines: vertical, up and to the right at 30 degrees to horizontal, and up and to left at 30 degrees to horizontal. Figure 3-44 shows an isometric sketch of a cut block using isometric grid paper.

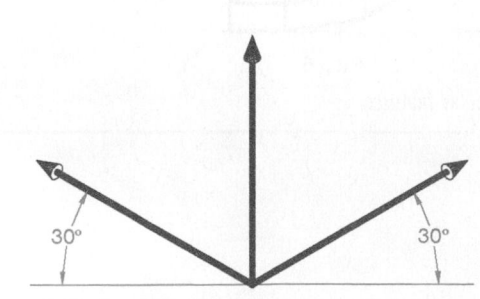

Figure 3-40 Isometric sketch axes

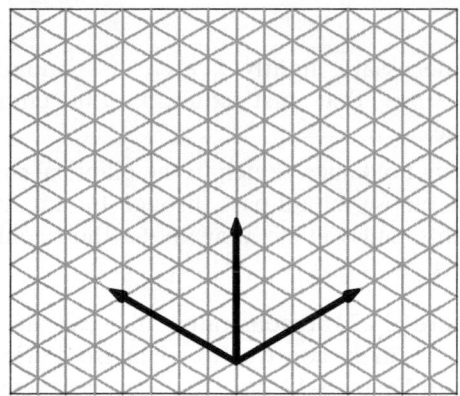

Figure 3-43 Isometric grid paper with superimposed isometric coordinate axes

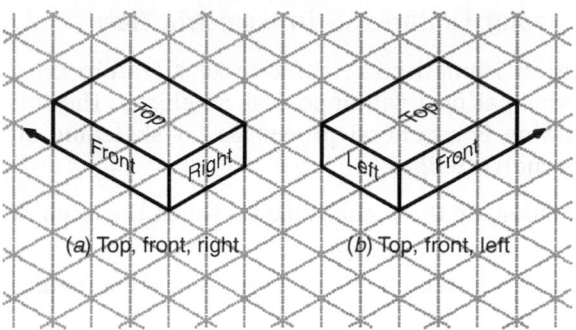

(a) Top, front, right   (b) Top, front, left

Figure 3-45 Two isometric views of a rectangular prism

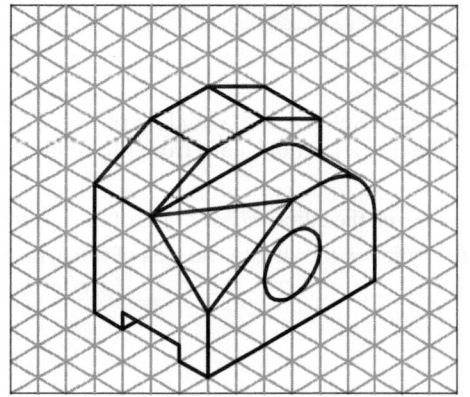

Figure 3-44 Isometric sketch of cut block using isometric grid paper

Object Orientation Guidelines

Generally speaking, an object's longest principal dimension should appear as a horizontal dimension on the front face of the object. Assuming this to be the case, then an isometric view showing the top, front, and right side of an object is obtained by laying out the longest dimension on the leftmost *up and to the left* axis (Figure 3-45a). Alternatively, to see a top, front, and left isometric view, place the longest dimension along the rightmost *up and to the right* axis (Figure 3-45b).

Step-by-step isometric sketch example for a cut block (see Figure 3-46)

1. PEB construction:
   a. Identify the front view of the object to be sketched.
   b. Using isometric axes (or isometric grid paper), lightly sketch a properly proportioned bounding box. To see the top, front, and right faces of the object, lay out the longest (horizontal) dimension along the *up and to the left* axis.
2. Still using construction lines, add feature details.
3. Go bold.

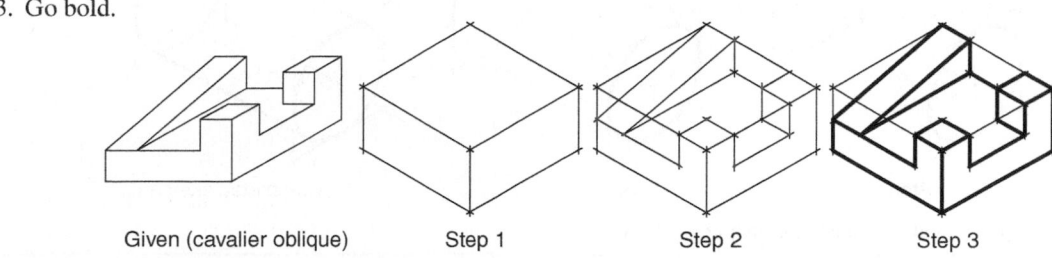

Given (cavalier oblique)   Step 1   Step 2   Step 3

Figure 3-46 Multiple steps for cut block isometric sketch

Circular Features in an Isometric View

In an isometric sketch, a circular feature on a principal face will appear as an ellipse. The following is a general procedure for sketching an isometric ellipse:

1. Identify the planar face on which the circular feature is to appear.
2. Lightly sketch the principal axes of the feature on the planar face.
3. Locate the quadrant points along these axes (equidistant from the intersection of the two axes).
4. If necessary, lightly sketch the bounding *rhombus* that just contains the circular edge.
5. Using bold lines, sketch in the elliptical shape. The ellipse should pass through the quadrant points and will be tangent to the rhombus sides.

Figure 3-47 demonstrates the use of this procedure in the construction of an isometric cylinder. In Figure 3-48, hole features are sketched on three faces of a box.

Step-by-step isometric sketch example for a cylinder (see Figure 3-47)

1. Sketch the axis, and locate the centers of the front and back faces of the cylinder. Note that in this example, the cylinder axis is horizontal.
2. Locate the quadrant points on the front face of the cylinder, and then sketch in the bounding rhombus. Note that the quadrant points are along the two axes in the plane of the face and equidistant from the center.
3. In bold, sketch the elliptical front face of the cylinder.
4. Locate quadrant points, sketch the bounding rhombus, and lightly sketch the elliptical back face of the cylinder.
5. Lightly sketch two lines to represent the *limiting elements* of the cylinder, parallel to the axis of the cylinder and tangent to the front and back elliptical faces.
6. Go bold, showing only the visible portion of the back edge.

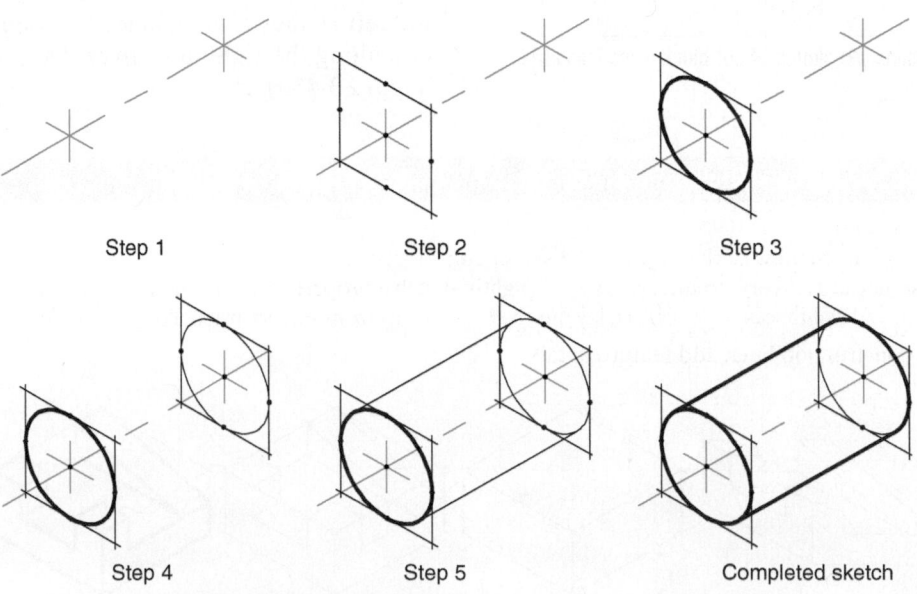

Step 1   Step 2   Step 3

Step 4   Step 5   Completed sketch

Figure 3-47 Construction of an isometric sketch of a cylinder

1. Locate the centers of the holes on each of the three visible faces of the box.
2. Locate the quadrant points of the holes, and then sketch in the bounding rhombus.
3. In bold, sketch the ellipses. A given ellipse will be tangent to its bounding rhombus.

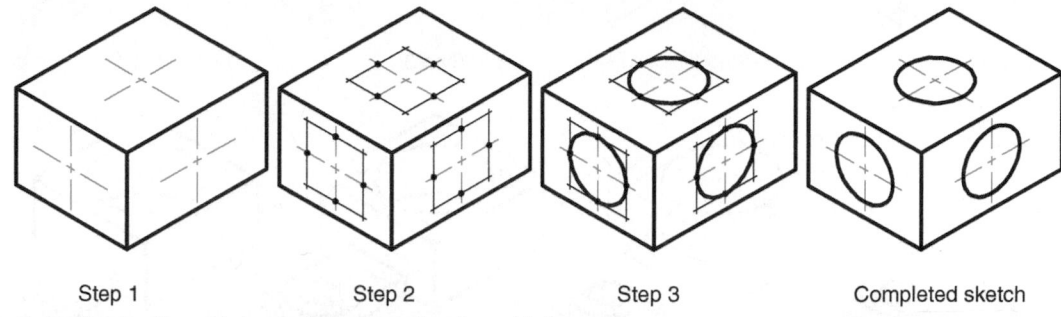

| Step 1 | Step 2 | Step 3 | Completed sketch |

Figure 3-48 Construction of holes on three faces of an isometric box

1. PEB construction:
   a. Construct two boxes.
2. Locate hole centers:
   a. Upper-left through hole.
   b. Lower-right arc and through hole.
3. Locate top-face quadrants, rhombus construction:
   a. Locate quadrants for upper-left through hole.
   b. Use construction lines to create rhombus.
   c. Locate quadrants for lower-right arc.
4. Top-face ellipse construction:
   a. Sketch upper-left full ellipse in bold.
   b. Sketch lower-right partial ellipse in bold.
   c. Locate quadrants for lower-right through hole.
   d. Use construction lines to create rhombus.
5. Top-face ellipse construction continued:
   a. Sketch lower-right full ellipse in bold.
6. Bottom-face ellipse construction:
   a. Upper-left full ellipse using light construction lines.
   b. Lower-right full ellipse using light construction lines.
   c. Lower-right partial ellipse using light construction lines.
7. Go bold.

ISOMETRIC SKETCHES

(Continues)

59

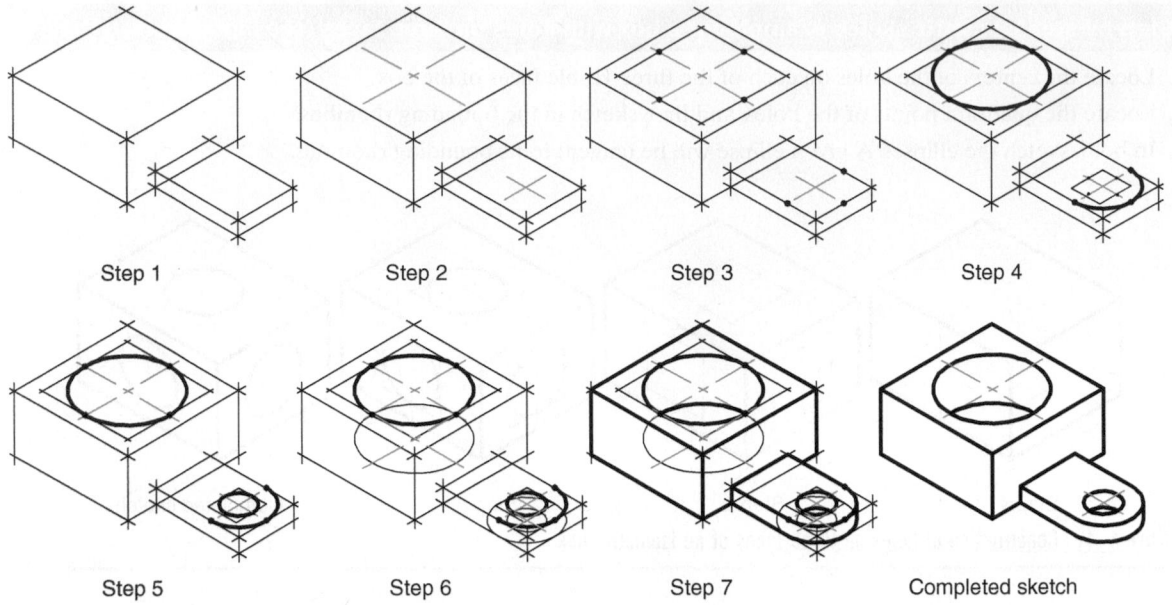

Figure 3-49 Multiple steps for isometric sketch of object with circular features

A drawing is said to be *scalable* if dimensional information can be derived from it, even if the drawing itself is not to scale. If, for example, the actual measure of a certain distance depicted on the drawing is known, then other dimensions on the drawing can be approximated by forming proportional relationships between the actual and the measured distances.

$$\frac{x_{\text{actual}}}{y_{\text{actual}}} = \frac{x_{\text{measured}}}{y_{\text{measured}}}$$

Multiview
- Parallel projection technique
- One PEB face is parallel to the projection plane

Therefore, all edges that are parallel to the projection plane are scalable.

Oblique
- Parallel projection technique
- One PEB face is parallel to the projection plane

Therefore, all edges that are parallel to the projection plane are scalable.
- For cavalier oblique, the receding axis is scaled the same as the other principal axes

Therefore, all edges parallel to the receding axis are scalable on a cavalier oblique.

Isometric
- Parallel projection technique
- The PEB is oriented such that all three principal axes are equally foreshortened

Therefore, all edges parallel to any principal axis are scalable.

Trimetric

- Parallel projection technique
- The PEB is oriented such that all three principal axes are foreshortened by different amounts

Therefore, all edges parallel to a single principal axis are scalable. In effect, there are three separate scales in a trimetric projection, one for each principal axis.

Note: For any planar projection technique, if an object (or feature) is parallel to the projection plane, the feature will be projected true shape. For a parallel projection, these features will also be projected true size.

▌ QUESTIONS

TRUE OR FALSE

1. Parallel object edges always appear parallel in a parallel projection.

2. In a trimetric projection, none of the angles between the projected principal axes is equal.

3. Assuming that a projection plane is offset to a new parallel position, an axonometric projection of an object onto the projection plane will be identical (i.e., same size and shape) in either location.

4. In an isometric drawing, each principal axis is foreshortened by approximately 82% of its true length.

5. The receding axis angle of an oblique projection is governed by the out-of-plane angle α.

MULTIPLE CHOICE

6. Which of the following is not considered to be a main element of a projection system?
   a. 3D object
   b. 2D cutting plane
   c. Projectors
   d. 2D projection plane
   e. 2D projected image

7. For the cavalier oblique drawing of the cut block shown in Figure P3-1, which edges are scalable (i.e., directly measurable)?
   a. BC and AC
   b. EF and DJ
   c. FH and HJ

   d. GJ and DJ
   e. All of the above
   f. None of the above

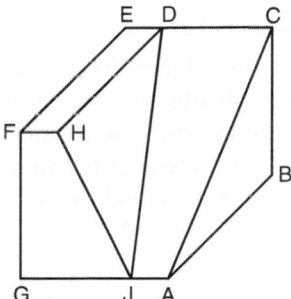

Figure P3-1 Cavalier oblique drawing of a cut block

8. For the isometric drawing of the cut block shown in Figure P3-2, which edges are scalable (i.e., directly measurable)?
   a. AB
   b. BF
   c. BD
   d. DF
   e. All of the above
   f. None of the above

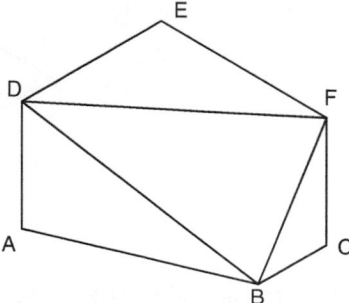

Figure P3-2 Isometric pictorial view of a cut block

9. Referring to Figure P3-3, if the sight lines are parallel to each other and also perpendicular to the infinite projection plane, $\beta = 15°$ and $\phi = 30°$, what is the type of projection of the block that will appear on the projection plane?

   a. One-point perspective
   b. Two-point perspective
   c. Trimetric
   d. Isometric
   e. Dimetric
   f. Cabinet oblique
   g. Cavalier oblique
   h. General oblique
   i. None of the above

10. Referring to Figure P3-3, if the sight lines are perpendicular to the infinite projection plane, $\beta = 30°$ and $\phi = 20°$ initially, which of the following is true of the ratio between the projected length a' and the actual length a as ϕ is increased to 70°?

    a. Increases
    b. Decreases
    c. Does not change
    d. Cannot determine without additional information

11. Given the isometric view of the cut block objects appearing in Figures P3-4 through P3-65, use oblique right grid paper in the back of the book (or download worksheet from the book website) to sketch a cavalier oblique view of the object.

12. Given the isometric view of the cut block objects appearing in Figures P3-4 through P3-65, use oblique right grid paper in the back of the book (or download worksheet from the book website) to sketch a cabinet oblique view of the object.

13. Given the isometric view of the cut block objects appearing in Figures P3-4 through P3-65, use oblique left grid paper in the back of the book (or download worksheet from the book website) to sketch a cavalier oblique view of the object.

14. Given the isometric view of the cut block objects appearing in Figures P3-4 through P3-65, use oblique left grid paper in the back of the book (or download worksheet from the book website) to sketch a cabinet oblique view of the object.

15. Given the isometric view of the cut block objects appearing in Figures P3-4 through P3-65, use a blank sheet of paper to create a

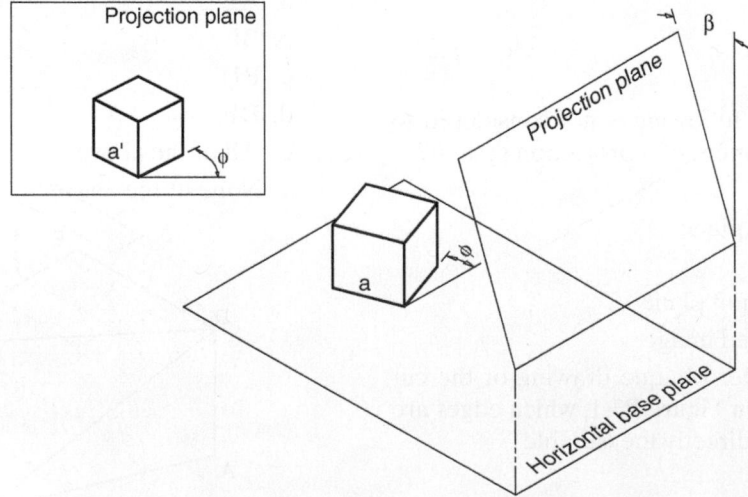

Figure P3-3 Pictorial projection where lengths a and a' represent the actual and projected lengths of the block, respectively (Figure adapted from the work of Michael H. Pleck)

well-proportioned cavalier oblique sketch of the object.

16. Given the isometric view of the cut block objects appearing in Figures P3-4 through P3-65, use a blank sheet of paper to create a well-proportioned cabinet oblique sketch of the object.

17. Given the cavalier oblique view of the cut block objects appearing in Figures P3-66 through P3-95, use isometric grid paper in the back of the book (or download worksheet from the book website) to sketch an isometric view of the object.

18. Given the cabinet oblique view of the cut block objects appearing in Figures P3-66

through P3-95, use isometric grid paper in the back of the book (or download worksheet from the book website) to sketch an isometric view of the object.

19. Given the cavalier oblique view of the cut block objects appearing in Figures P3-66 through P3-95, use a blank sheet of paper to create a well-proportioned isometric sketch of the object.

20. Given the cabinet oblique view of the cut block objects appearing in Figures P3-66 through P3-95, use a blank sheet of paper to create a well-proportioned isometric sketch of the object.

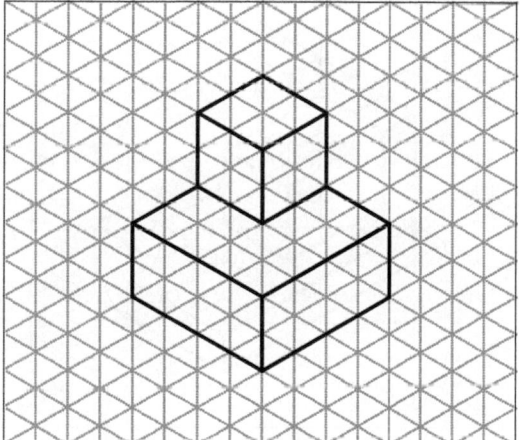

Figure P3-4

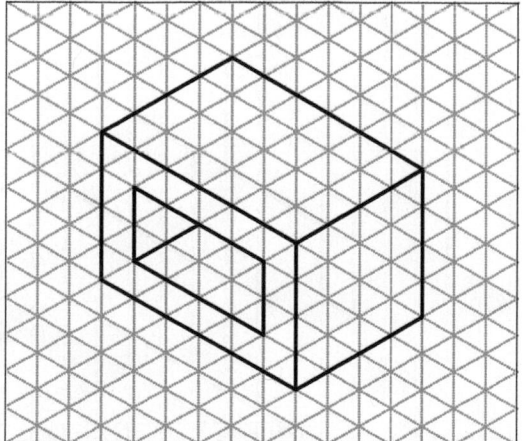

Figure P3-5

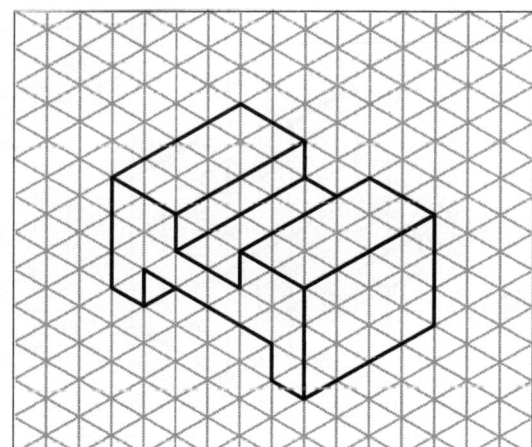

Figure P3-6

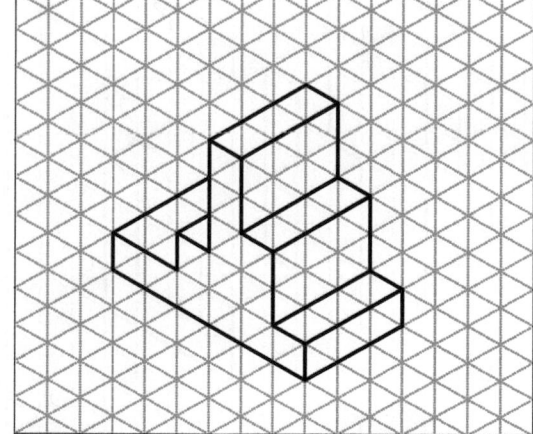

Figure P3-7

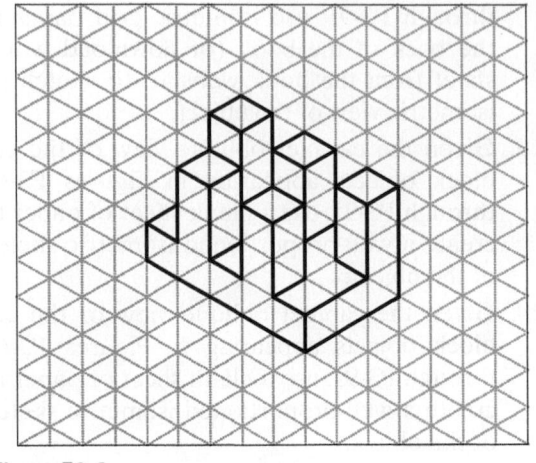

Figure P3-8

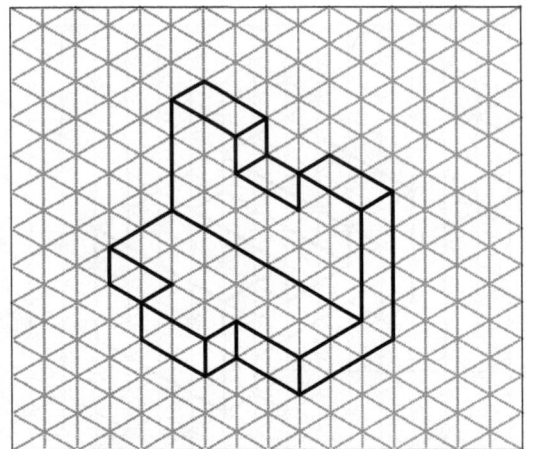

Figure P3-9

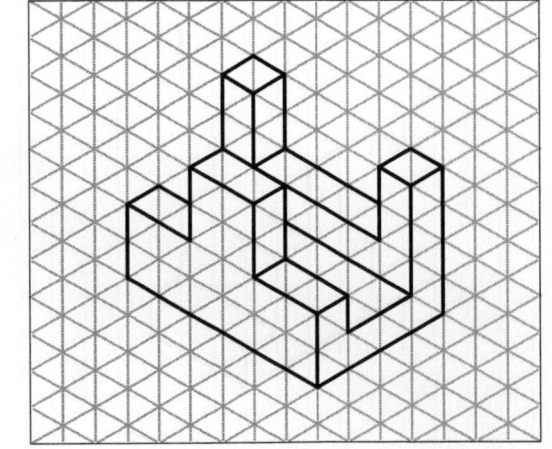

Figure P3-10

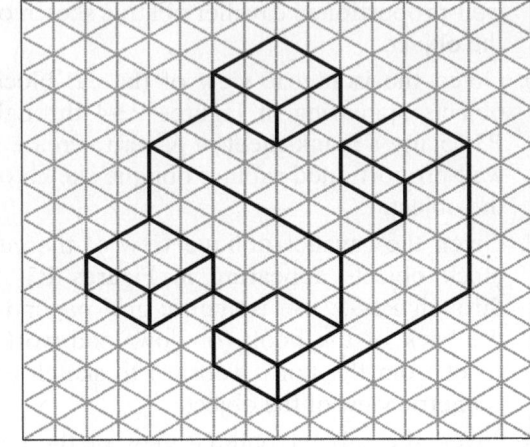

Figure P3-11

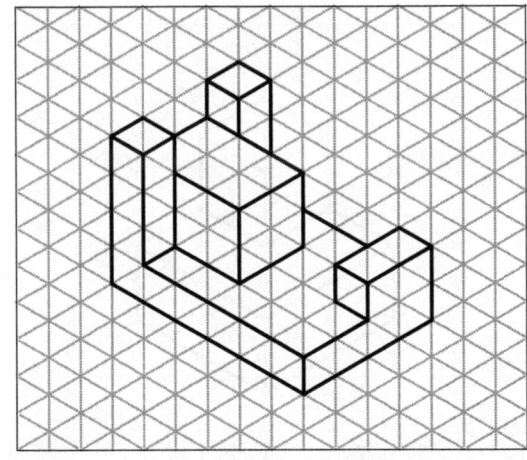

Figure P3-12

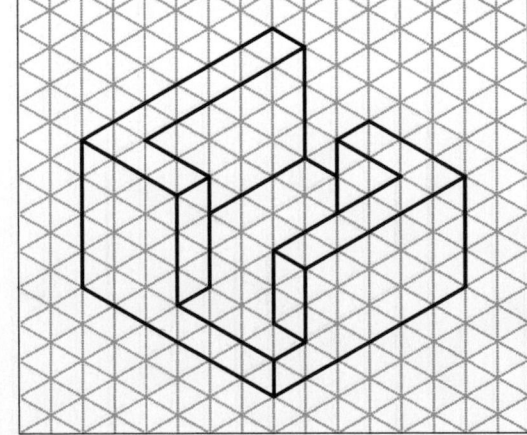

Figure P3-13

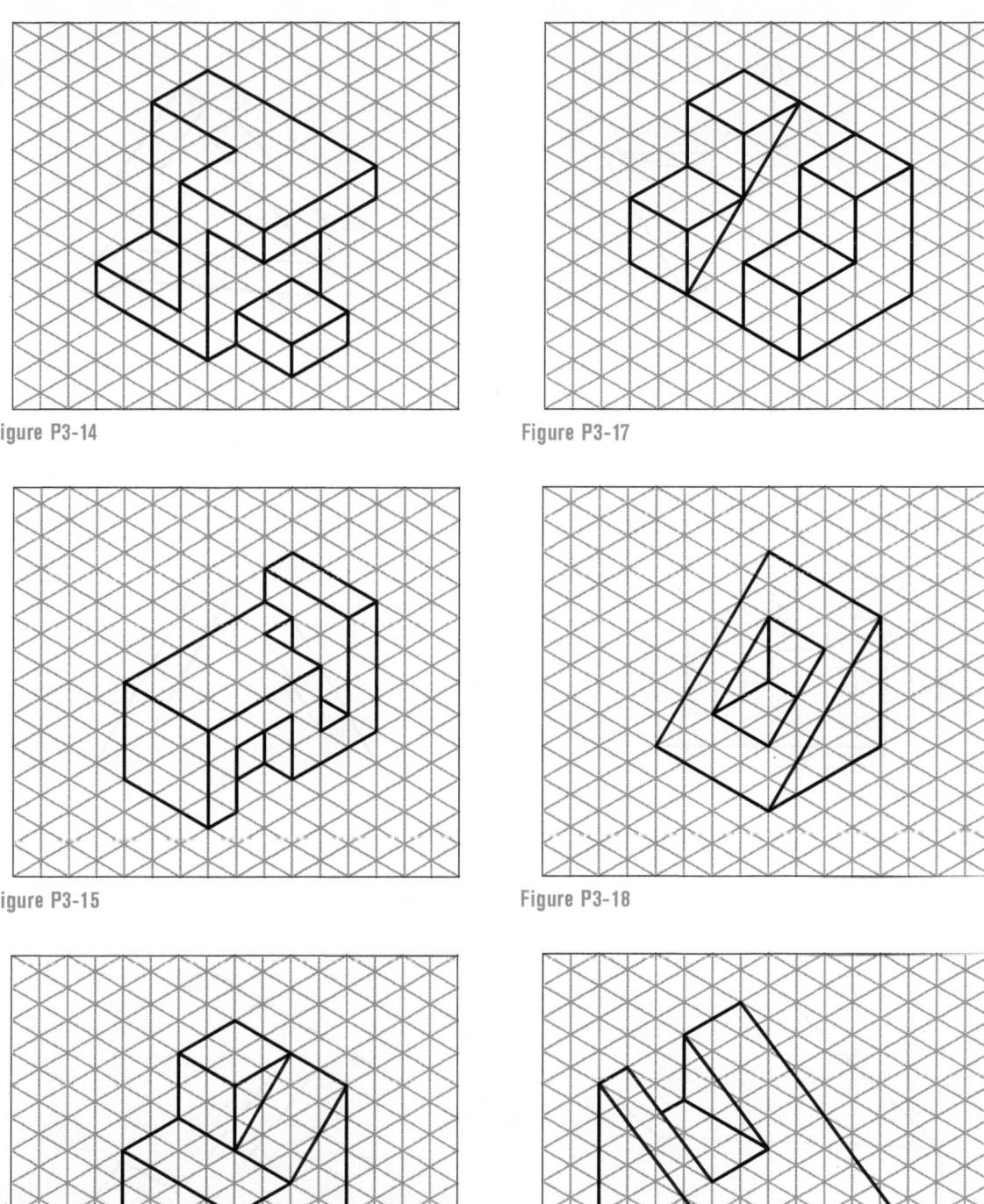

Figure P3-14

Figure P3-17

Figure P3-15

Figure P3-18

Figure P3-16

Figure P3-19

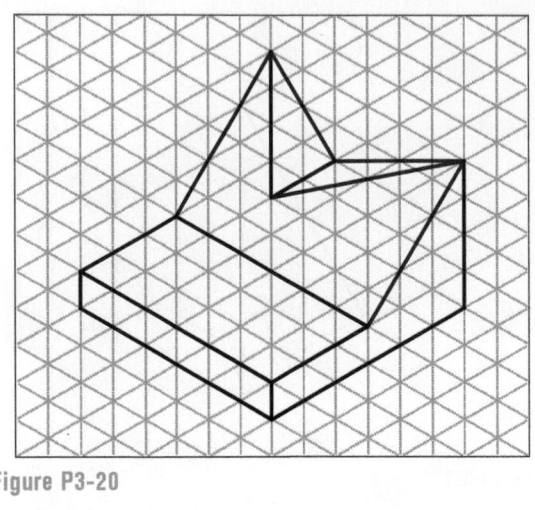

Figure P3-20

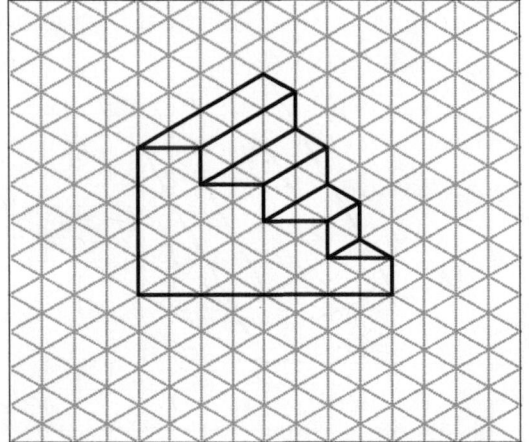

Figure P3-21

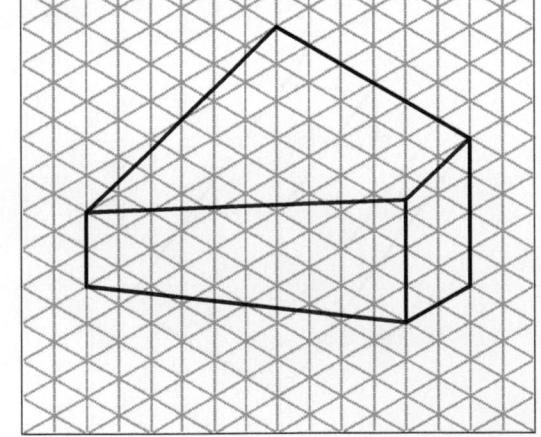

Figure P3-22

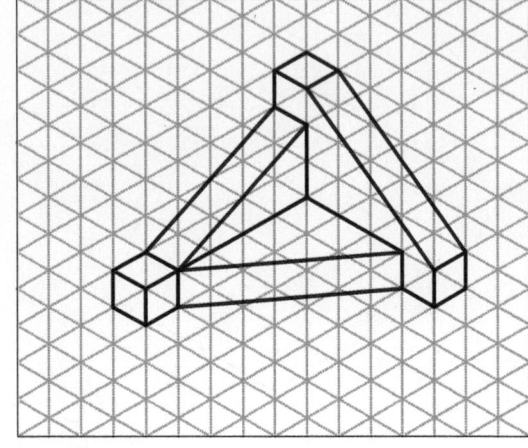

Figure P3-23

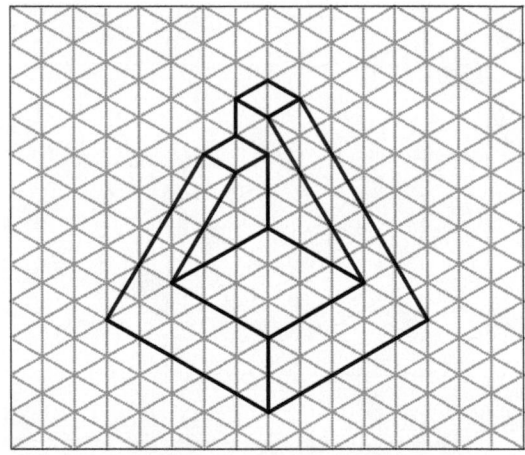

Figure P3-24

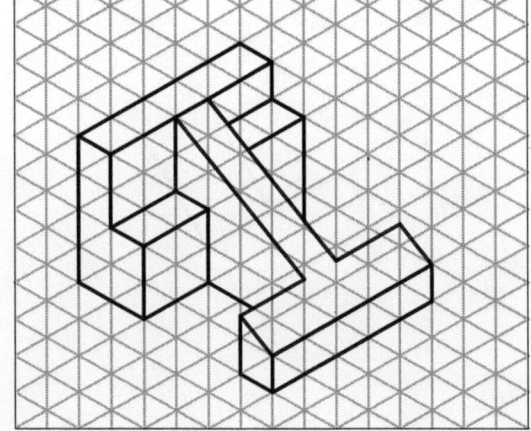

Figure P3-25

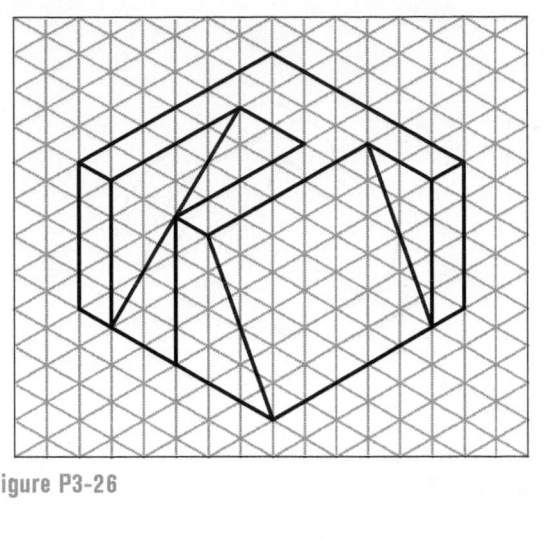

Figure P3-26

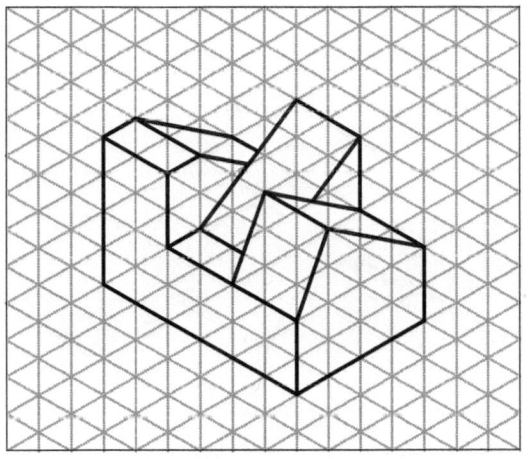

Figure P3-27

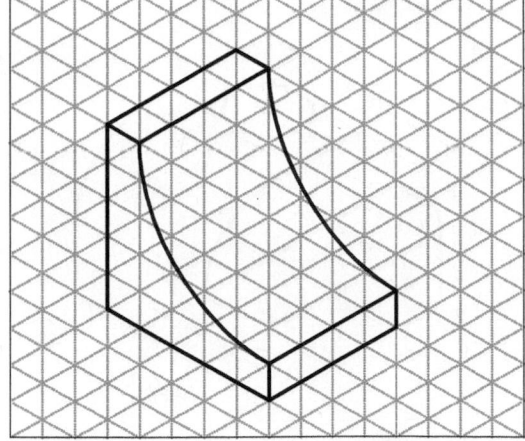

Figure P3-28

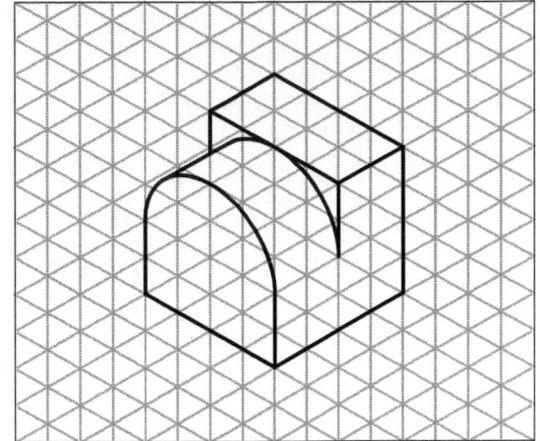

Figure P3-29

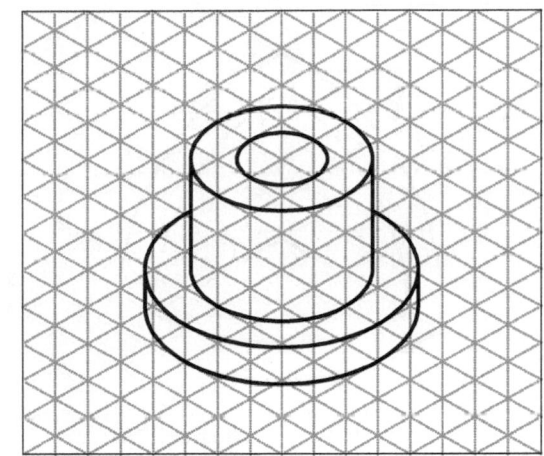

Figure P3-30

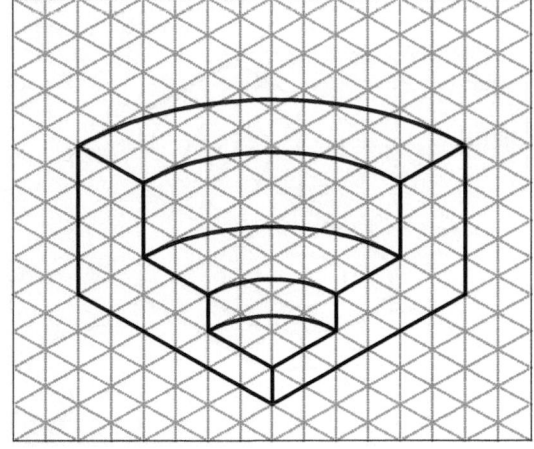

Figure P3-31

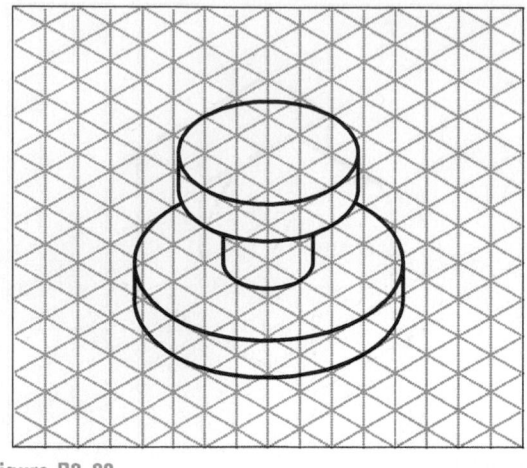

Figure P3-32

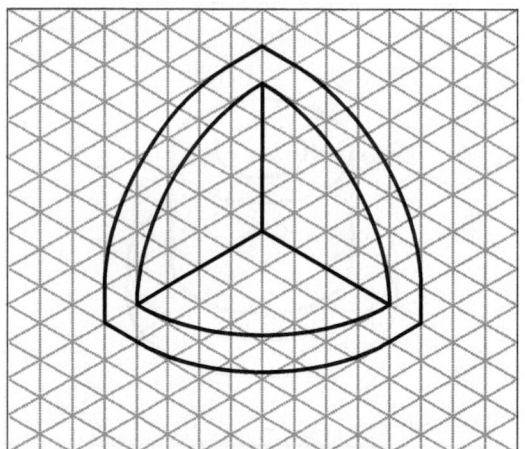

Figure P3-33

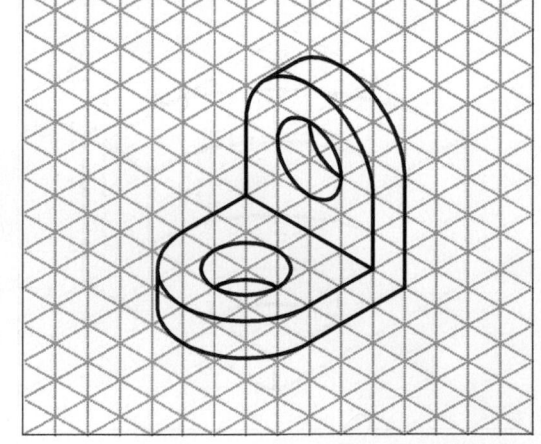

Figure P3-34

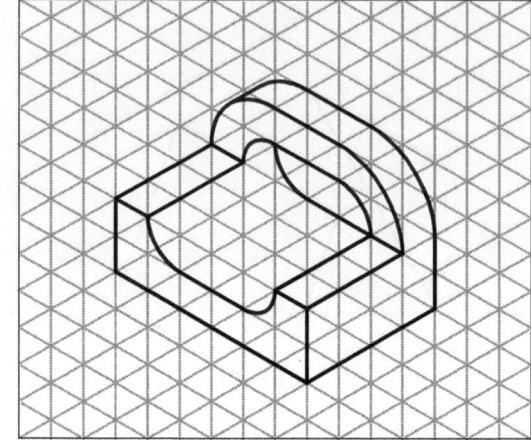

Figure P3-35

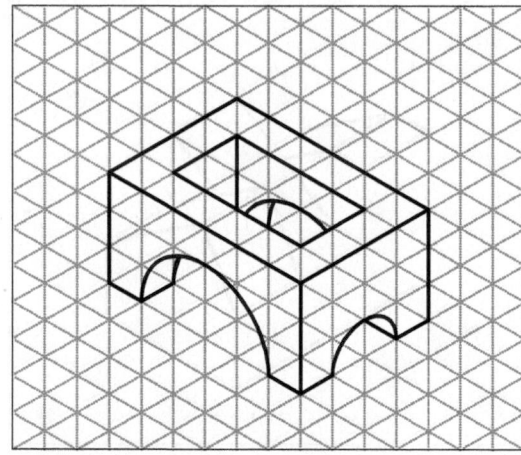

Figure P3-36

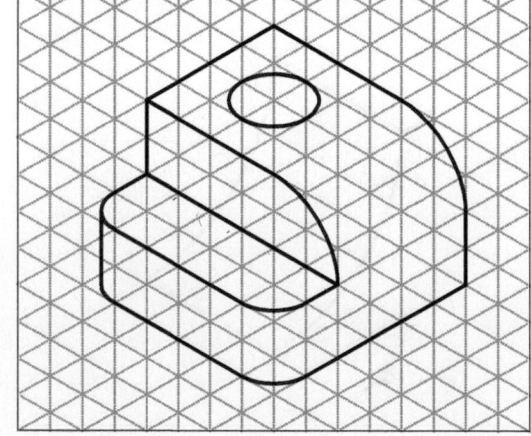

Figure P3-37

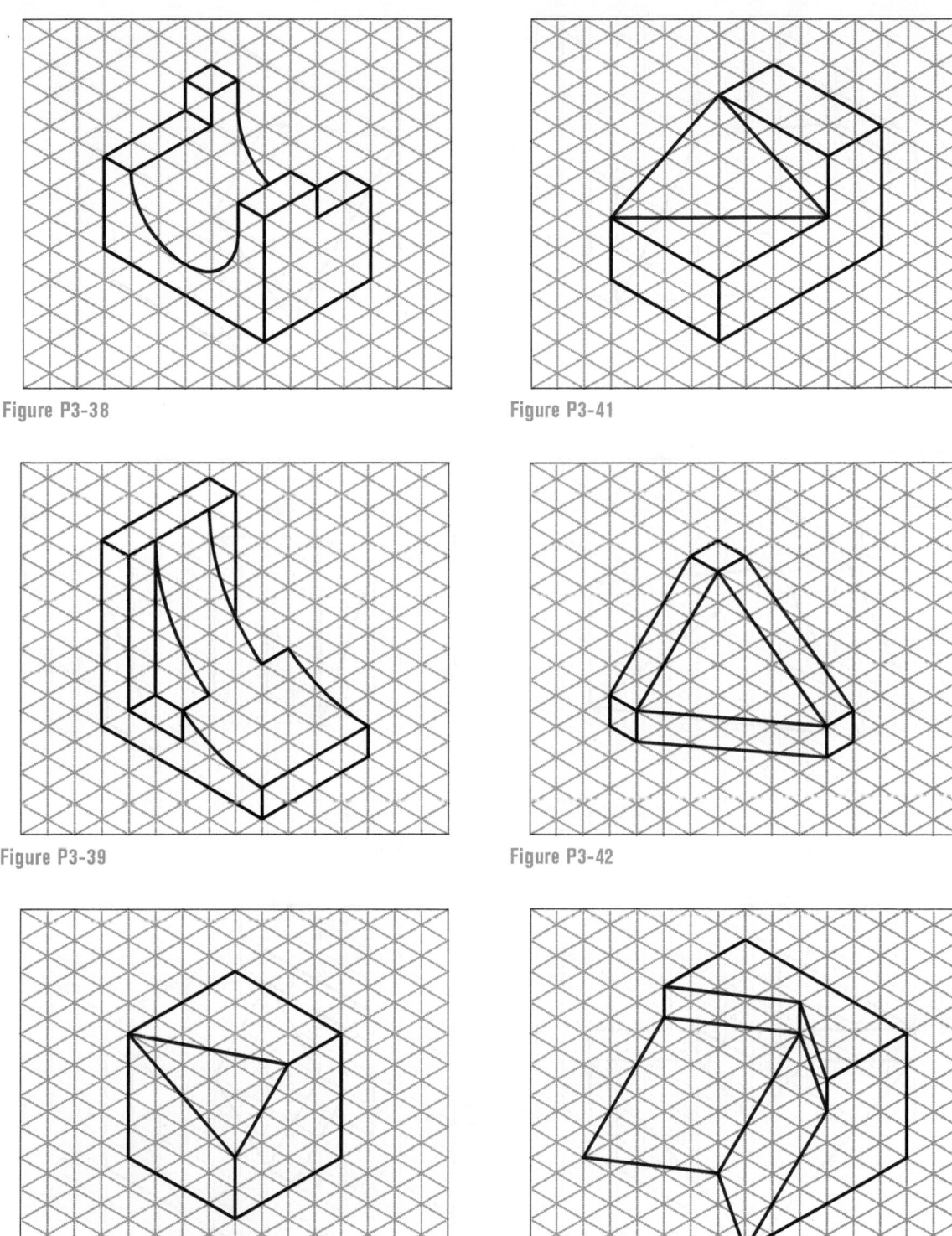

Figure P3-38

Figure P3-41

Figure P3-39

Figure P3-42

Figure P3-40

Figure P3-43

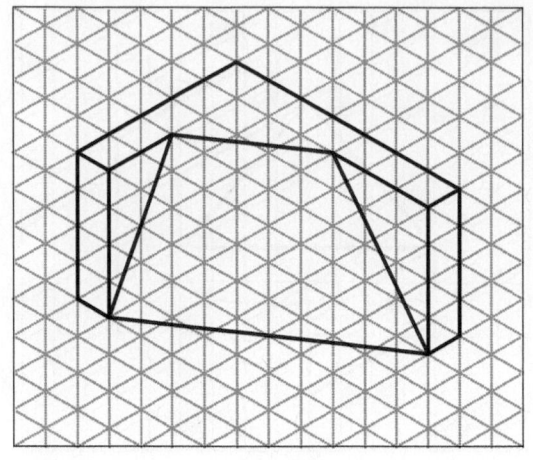

Figure P3-44

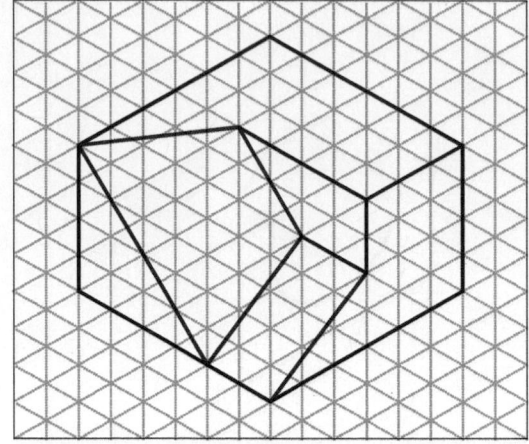

Figure P3-47

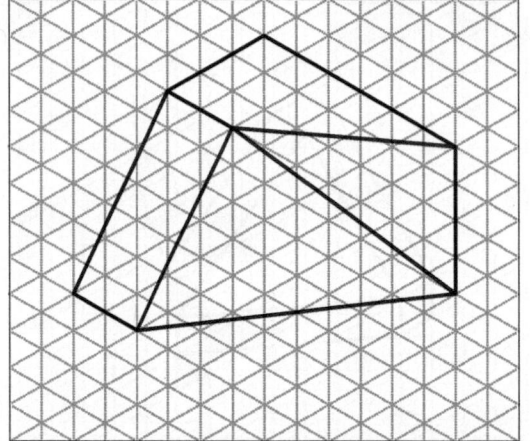

Figure P3-45

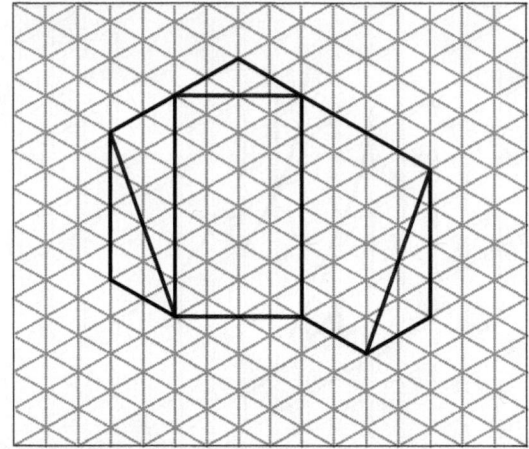

Figure P3-48

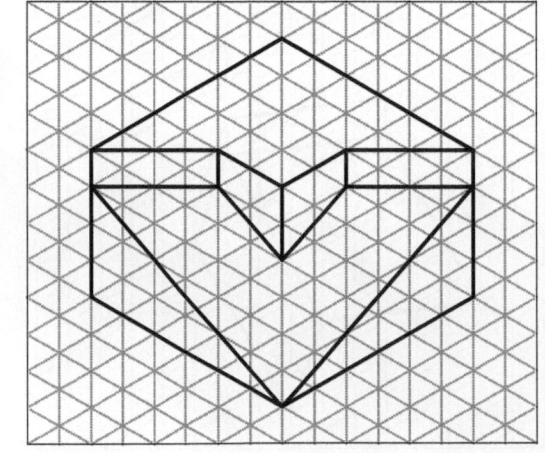

Figure P3-46

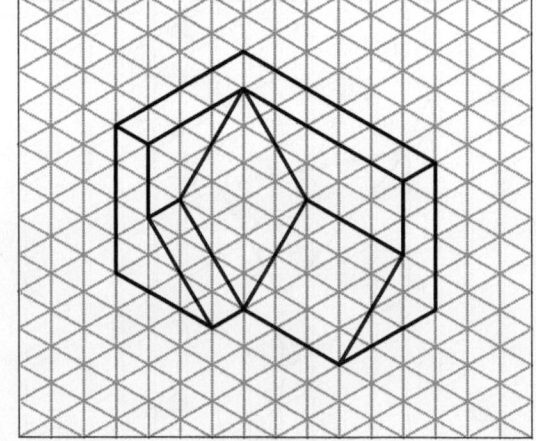

Figure P3-49

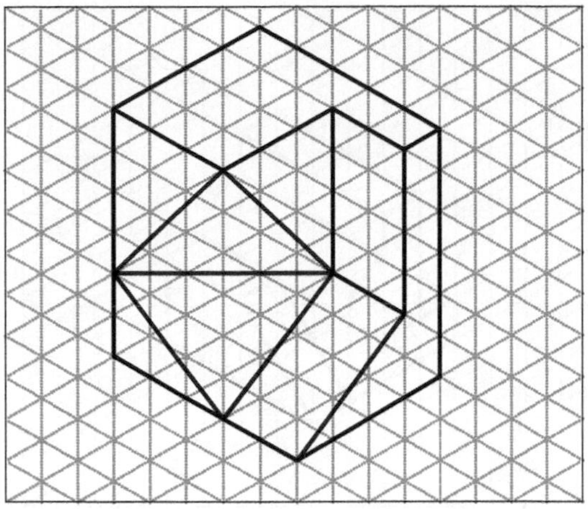

Figure P3-50

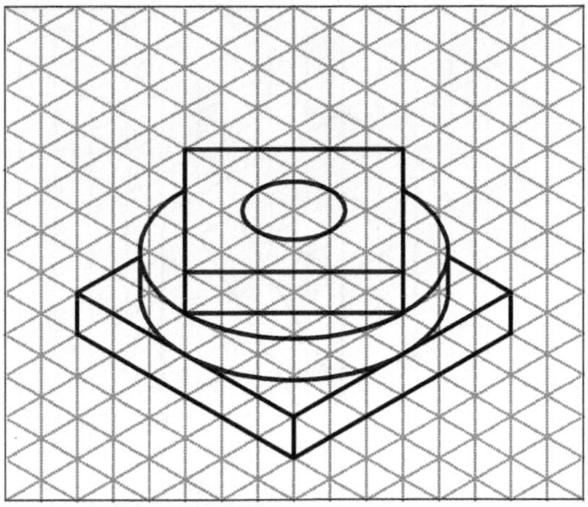

Figure P3-53

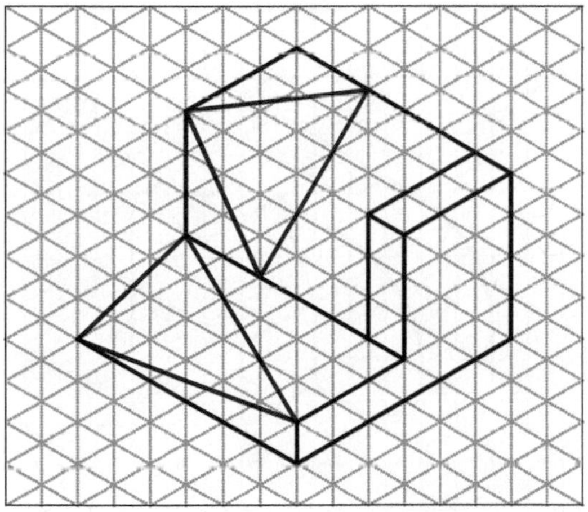

Figure P3-51

Figure P3-54

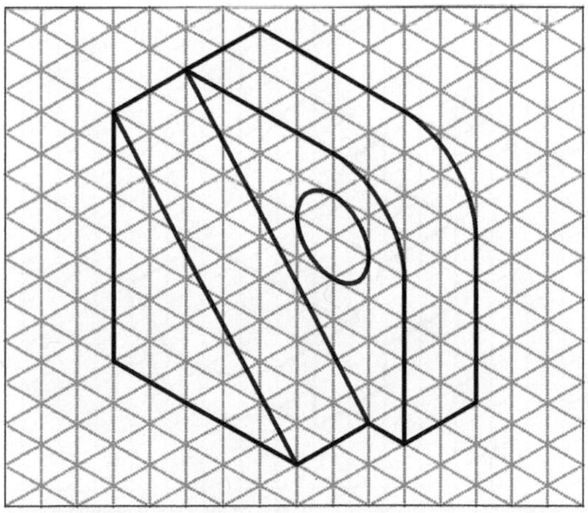

Figure P3-52

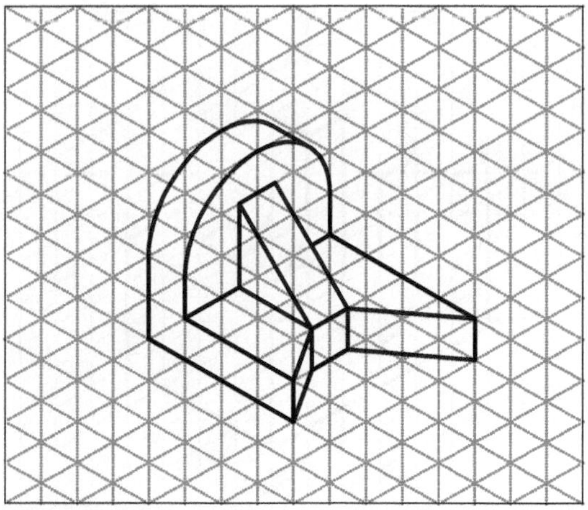

Figure P3-55

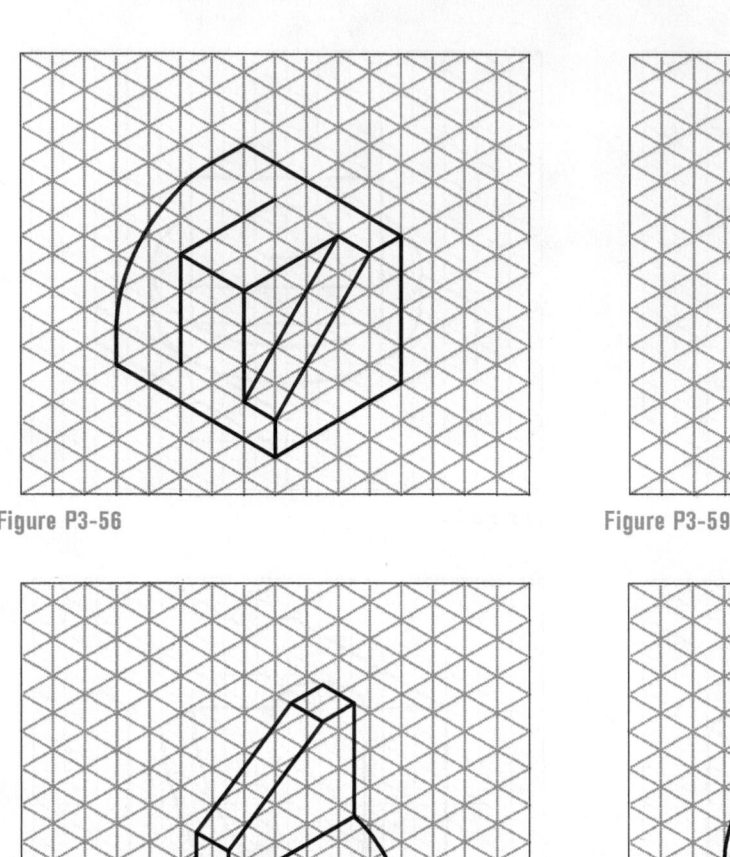

Figure P3-56

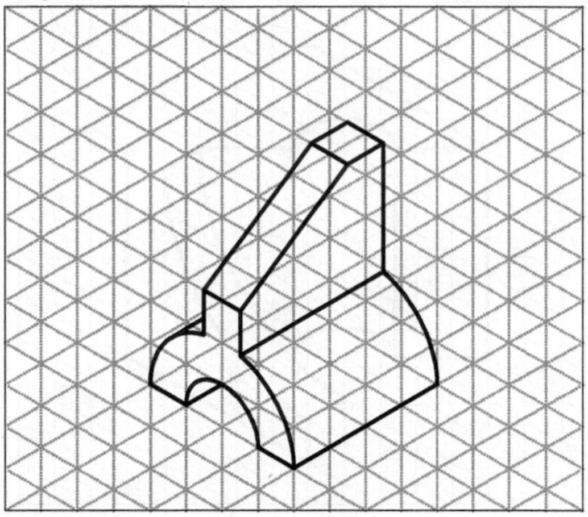

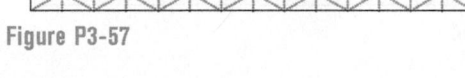

Figure P3-59

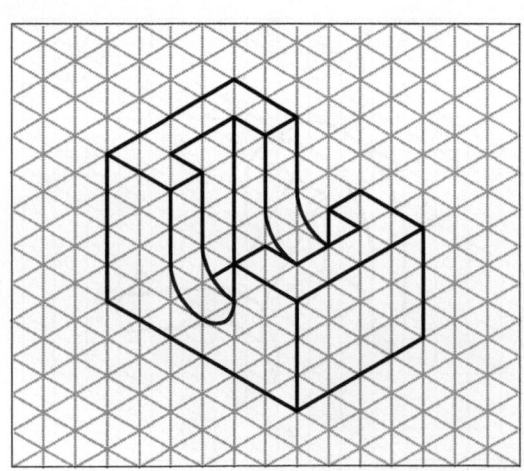

Figure P3-57

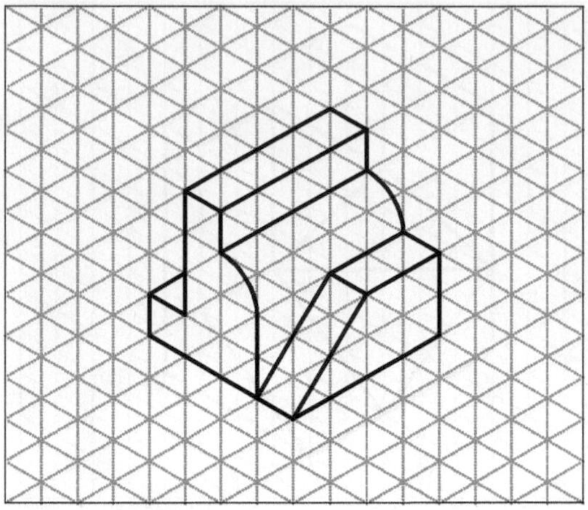

Figure P3-60

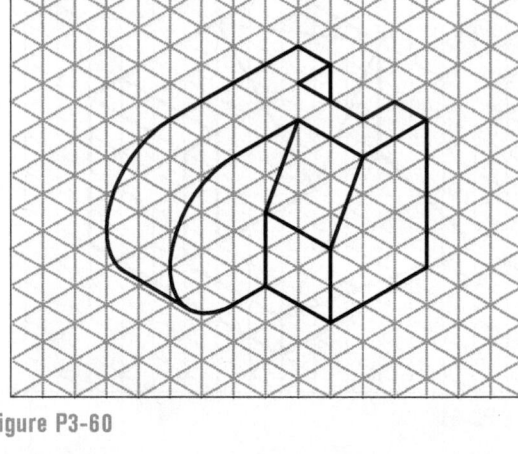

Figure P3-58

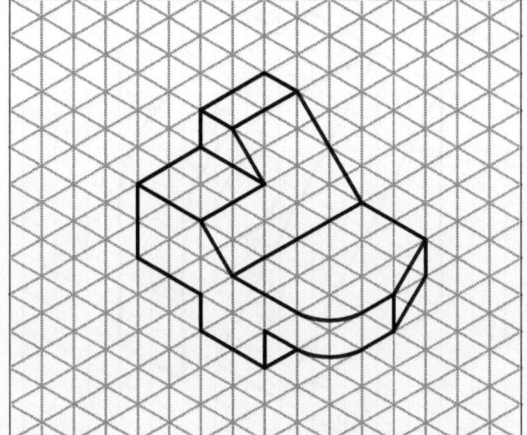

Figure P3-61

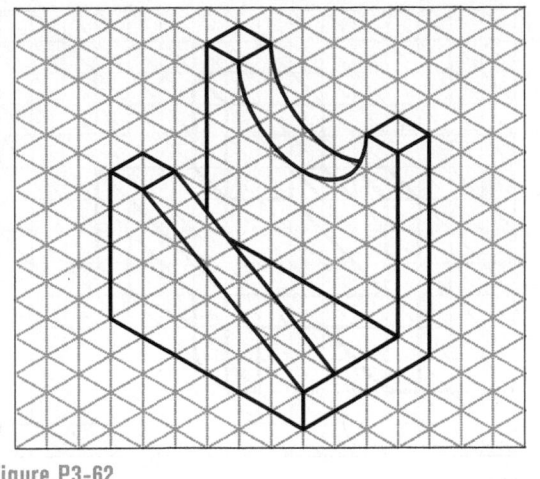

Figure P3-62

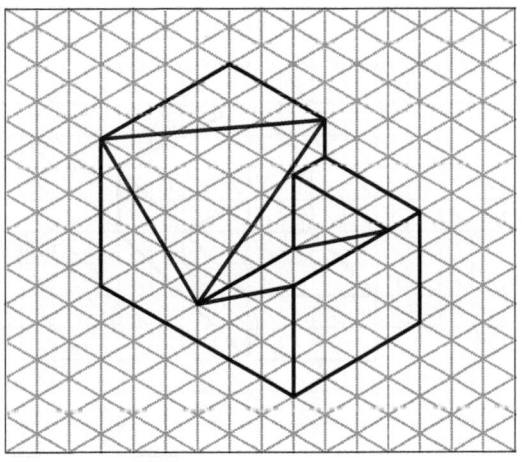

Figure P3-63

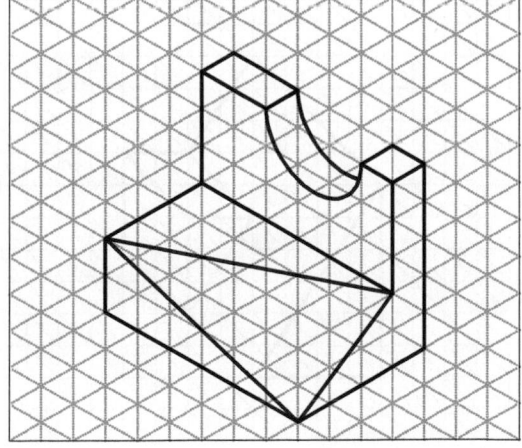

Figure P3-64

Figure P3-65

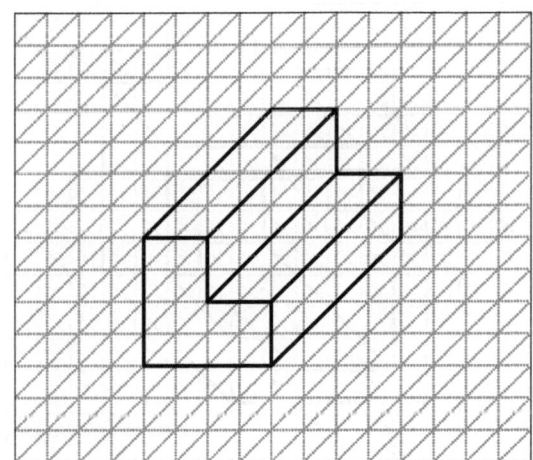

Figure P3-66

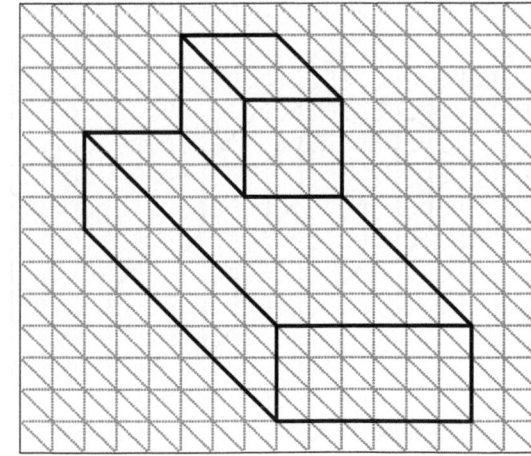

Figure P3-67

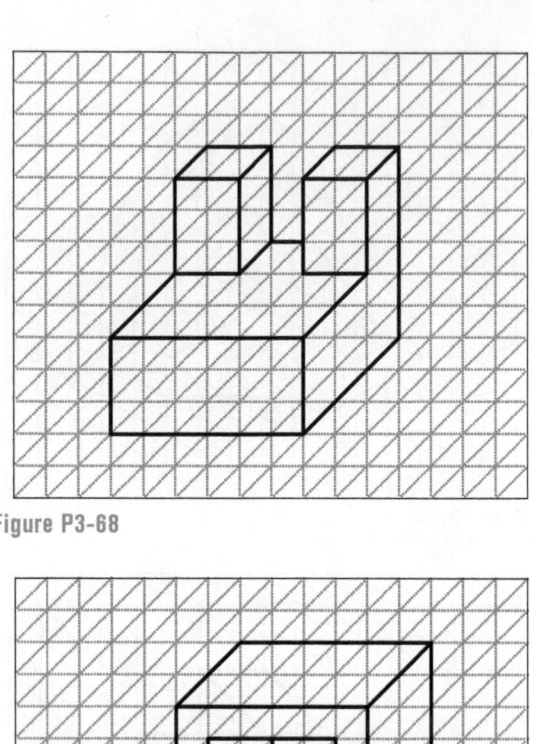

Figure P3-68

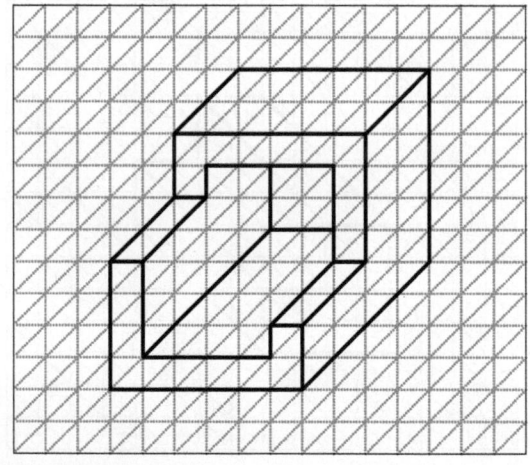

Figure P3-69

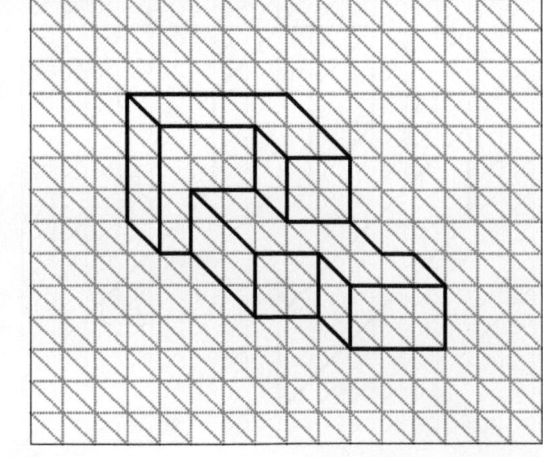

Figure P3-70

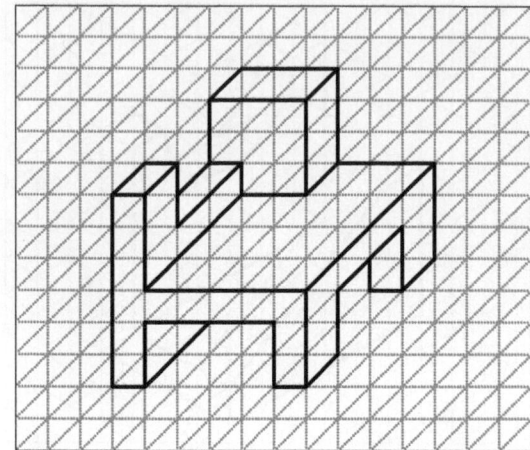

Figure P3-71

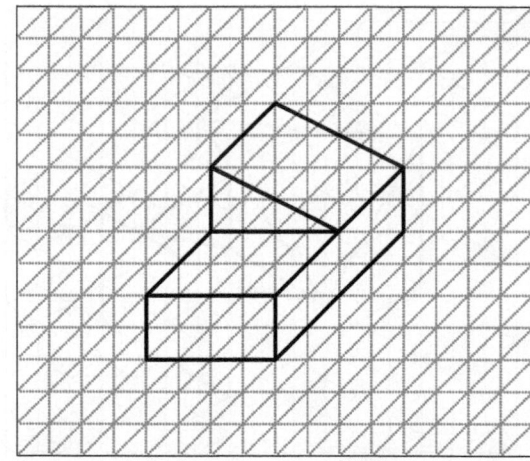

Figure P3-72

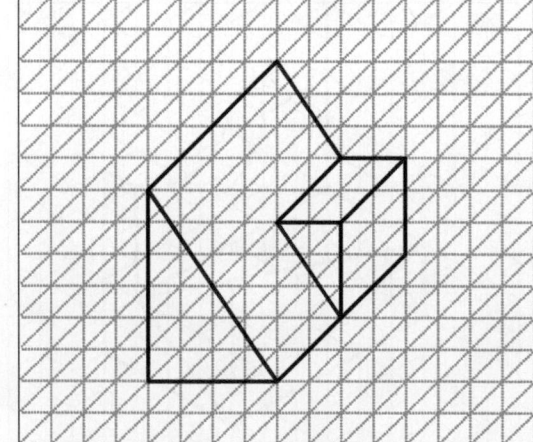

Figure P3-73

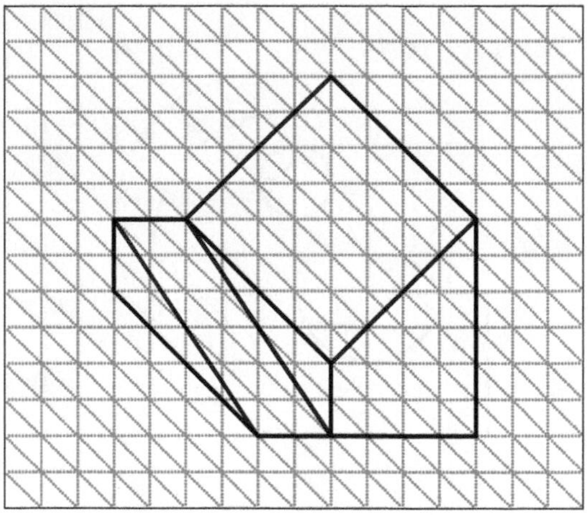

Figure P3-74

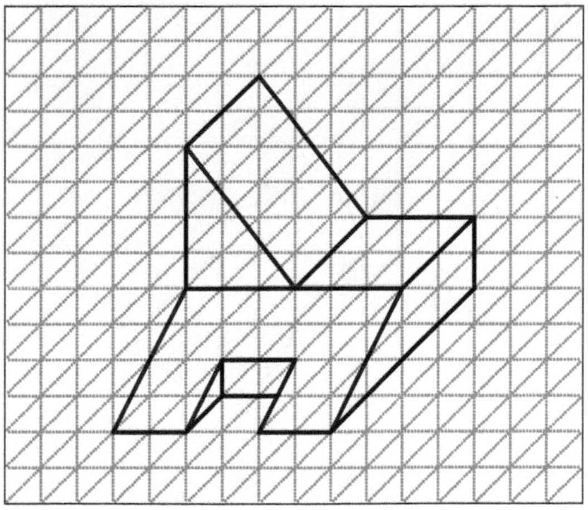

Figure P3-77

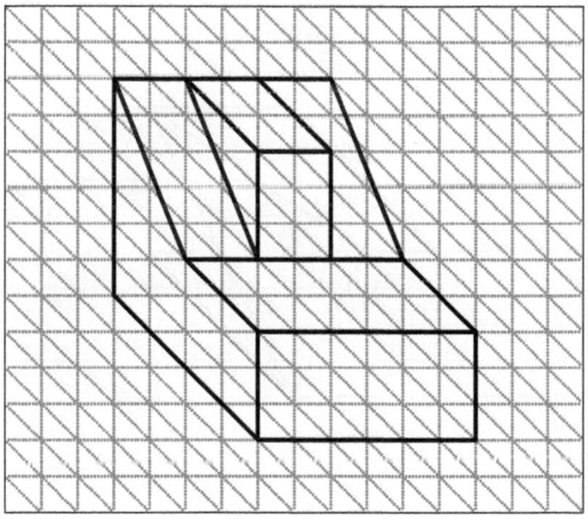

Figure P3-75

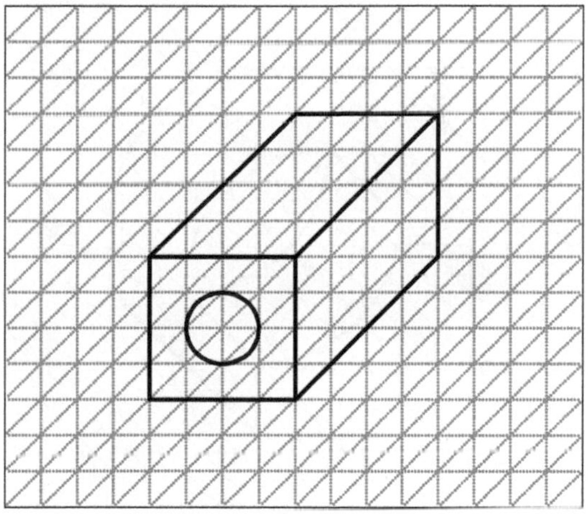

Figure P3-78

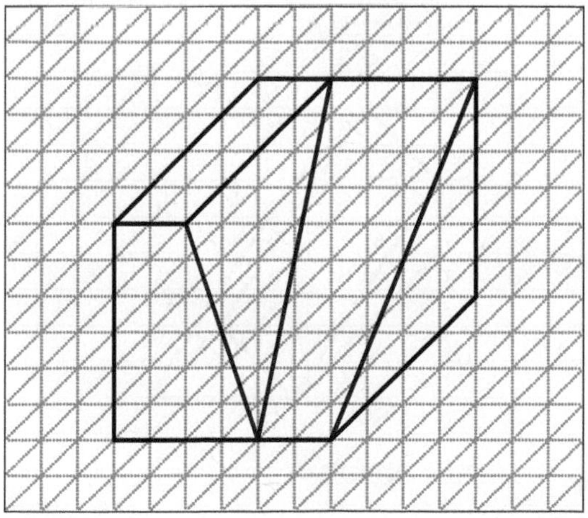

Figure P3-76

Figure P3-79

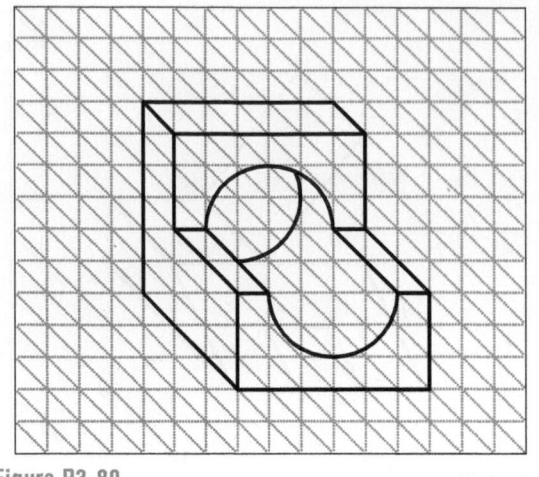

Figure P3-80

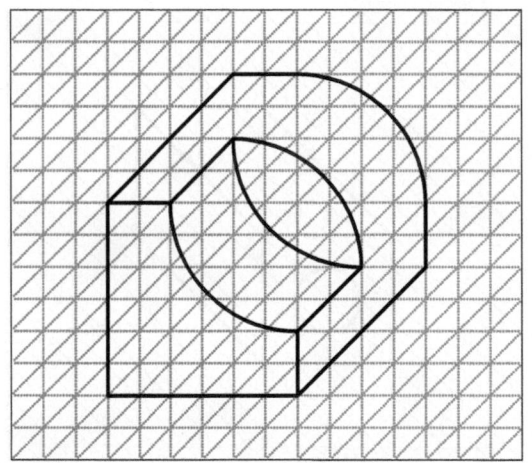

Figure P3-81

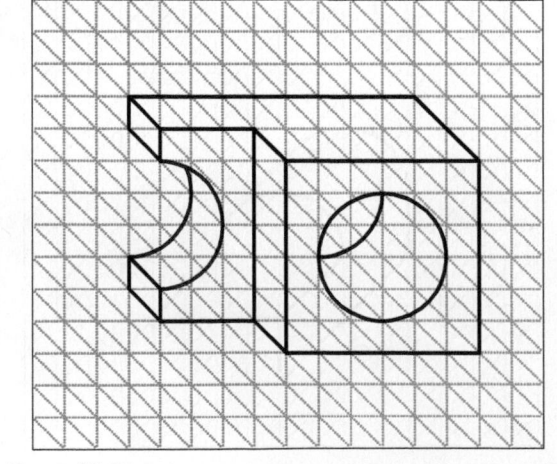

Figure P3-82

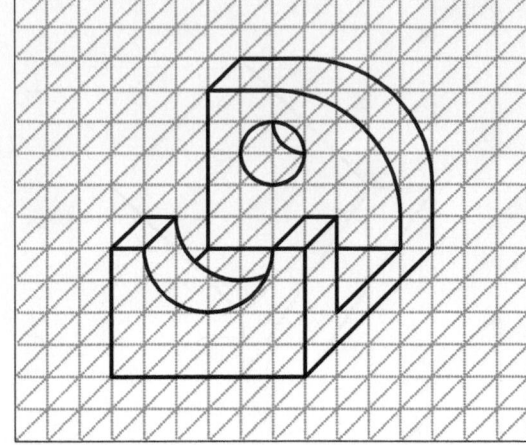

Figure P3-83

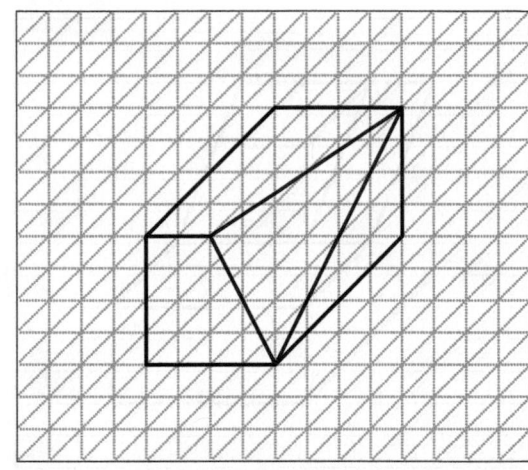

Figure P3-84

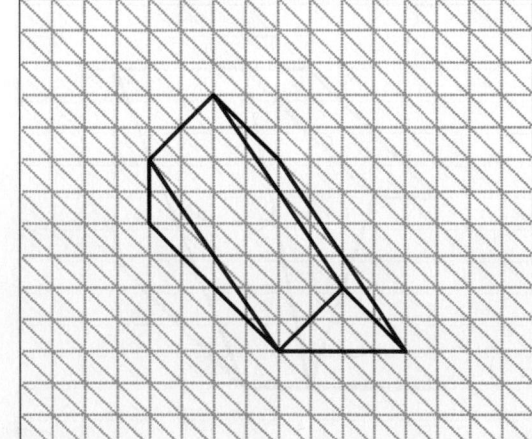

Figure P3-85

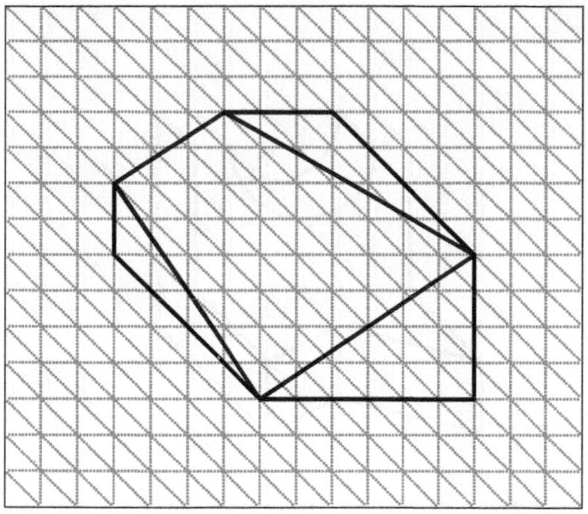

Figure P3-86

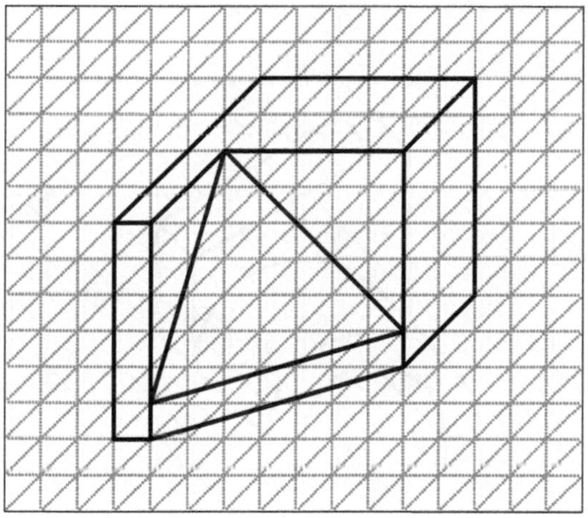

Figure P3-87

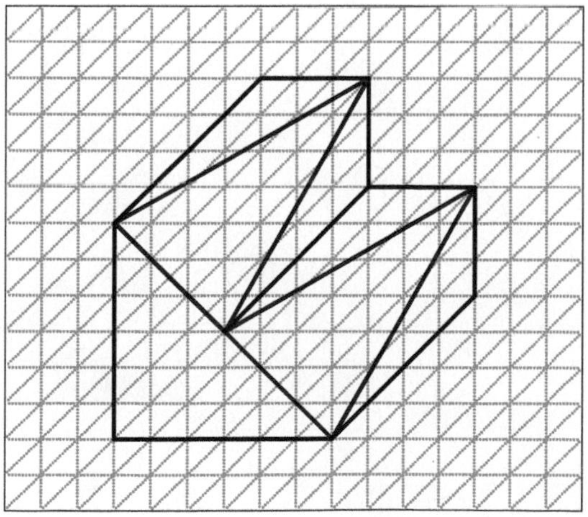

Figure P3-88

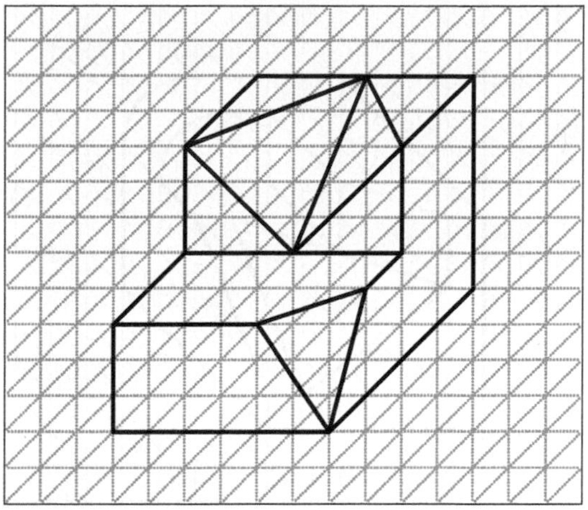

Figure P3-89

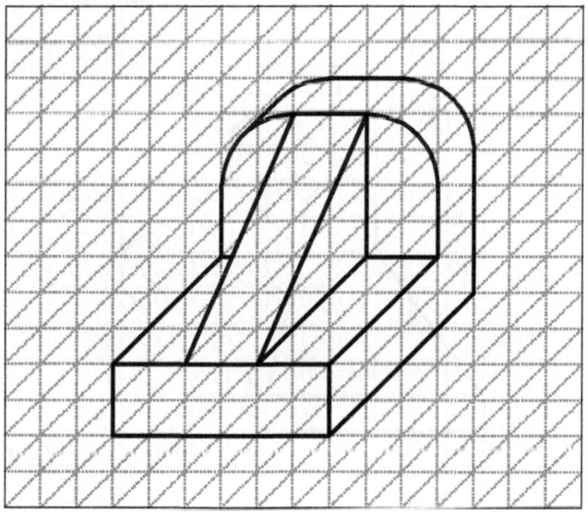

Figure P3-90

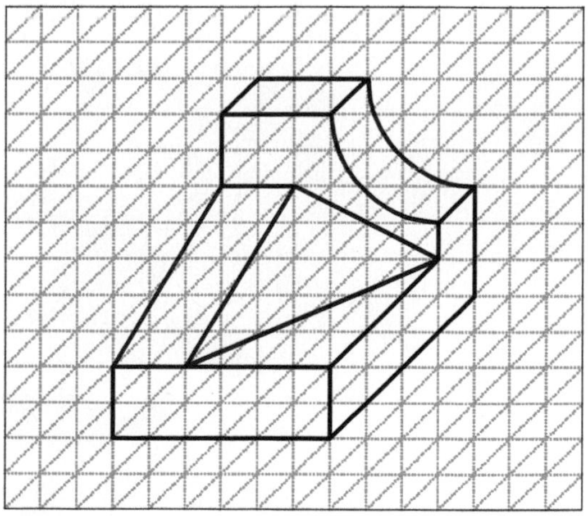

Figure P3-91

QUESTIONS

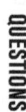

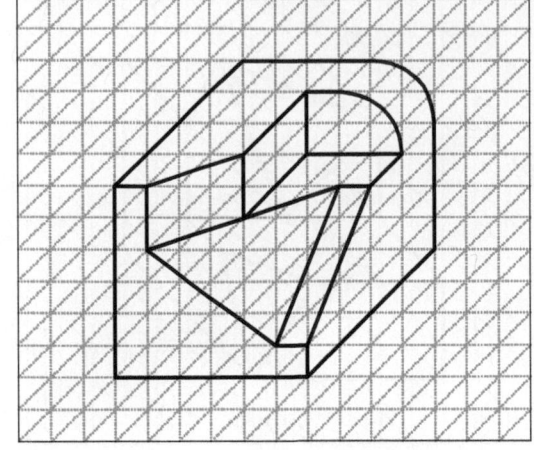

Figure P3-92

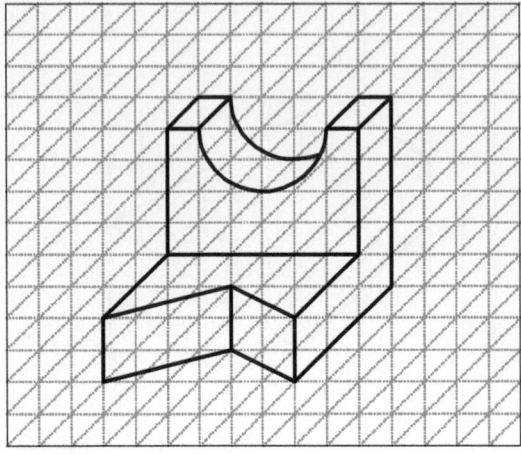

Figure P3-94

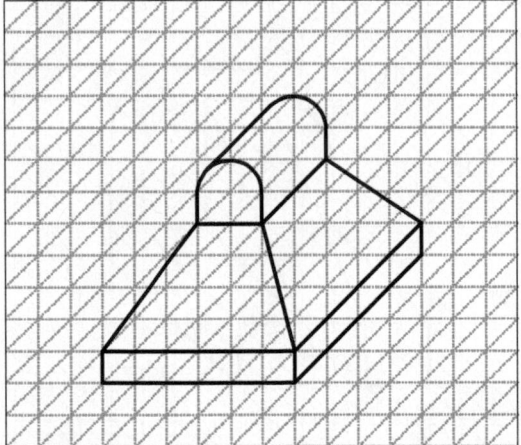

Figure P3-93

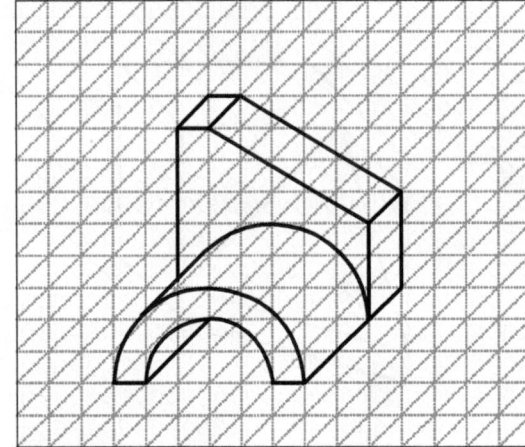

Figure P3-95

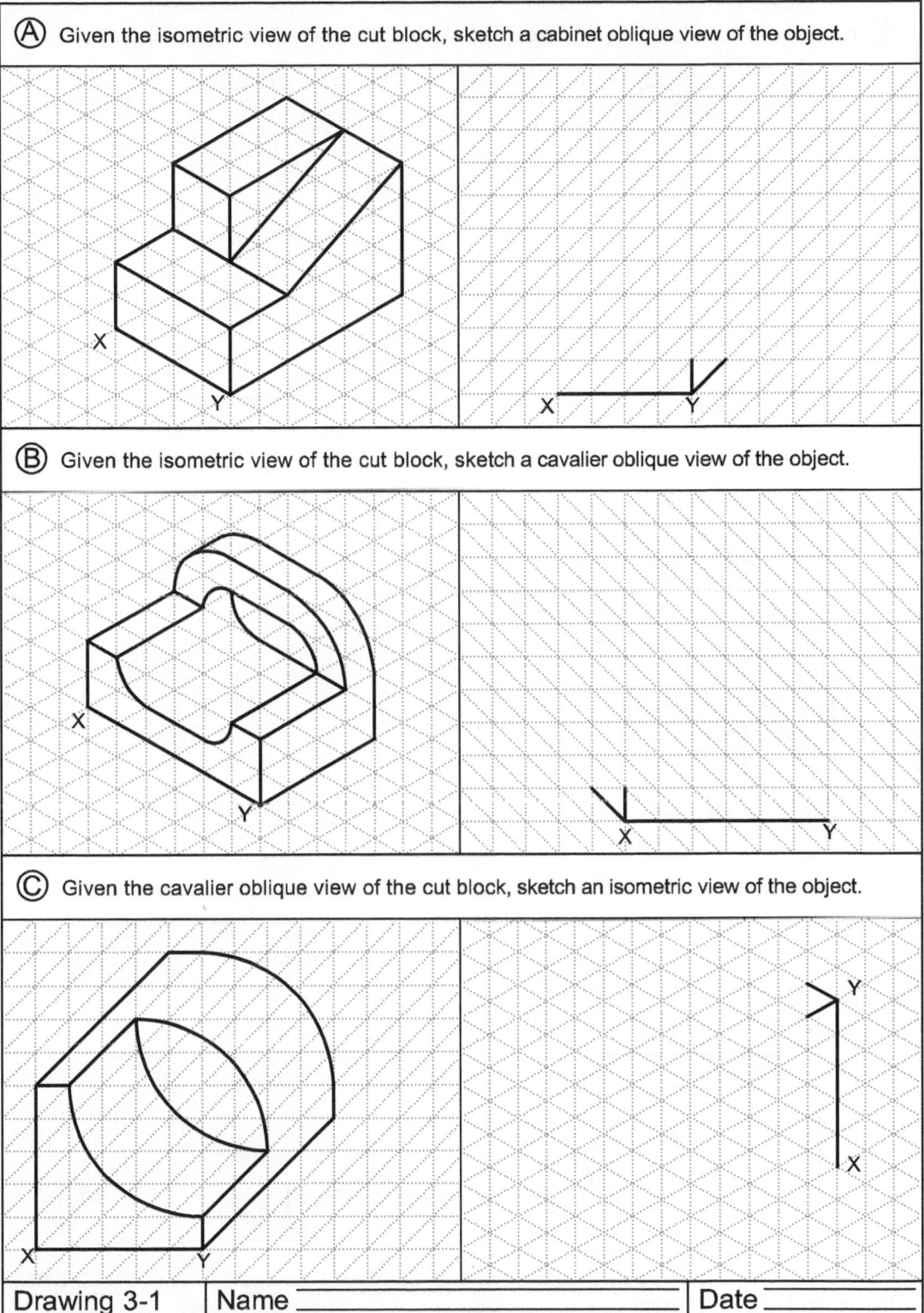

(A) Given the isometric view of the cut block, sketch a cabinet oblique view of the object.

(B) Given the isometric view of the cut block, sketch a cavalier oblique view of the object.

(C) Given the cavalier oblique view of the cut block, sketch an isometric view of the object.

| Drawing 3-1 | Name | Date |

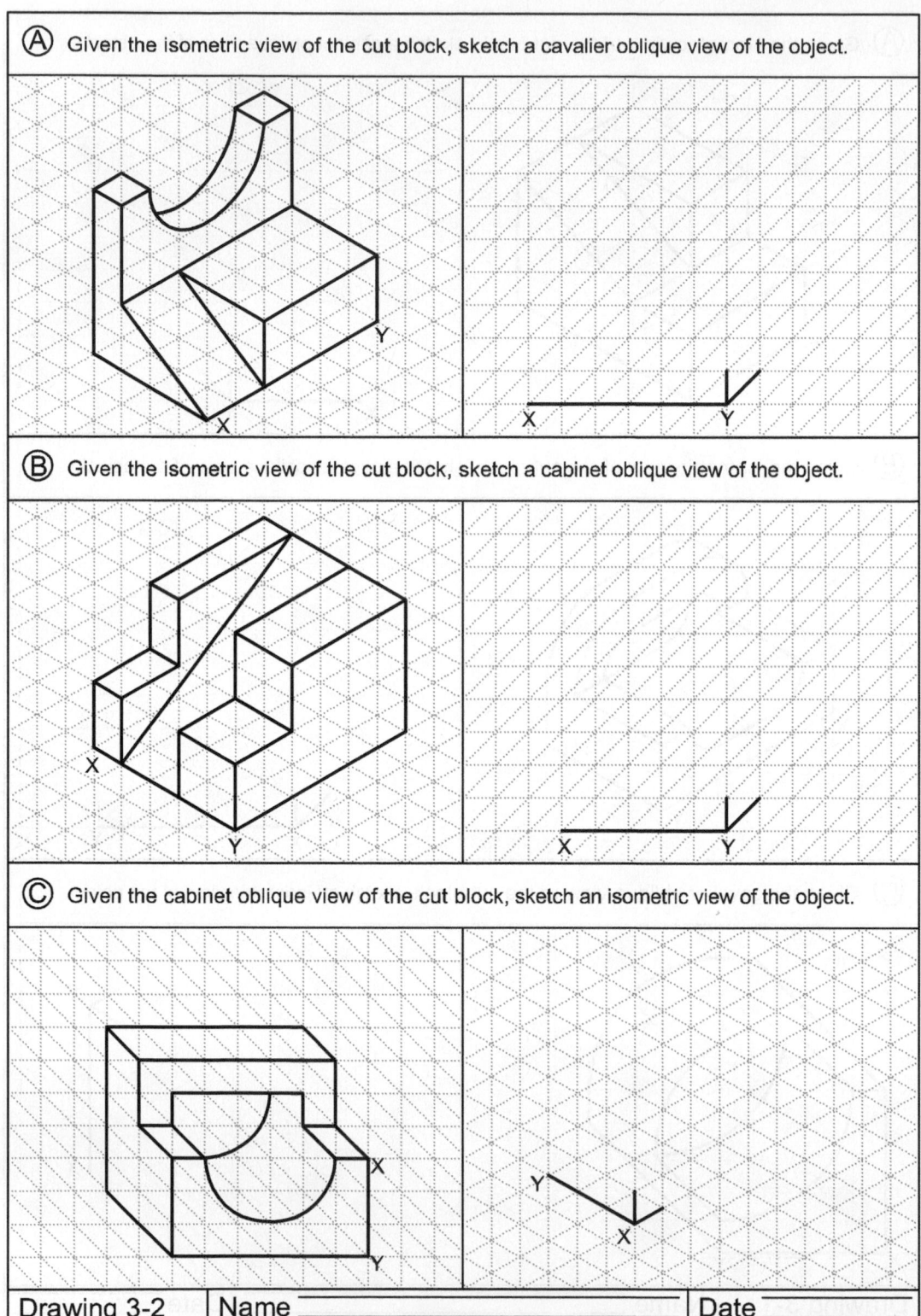

Ⓐ Given the isometric view of the cut block, sketch a cavalier oblique view of the object.

Ⓑ Given the isometric view of the cut block, sketch a cabinet oblique view of the object.

Ⓒ Given the cabinet oblique view of the cut block, sketch an isometric view of the object.

Drawing 3-2	Name	Date

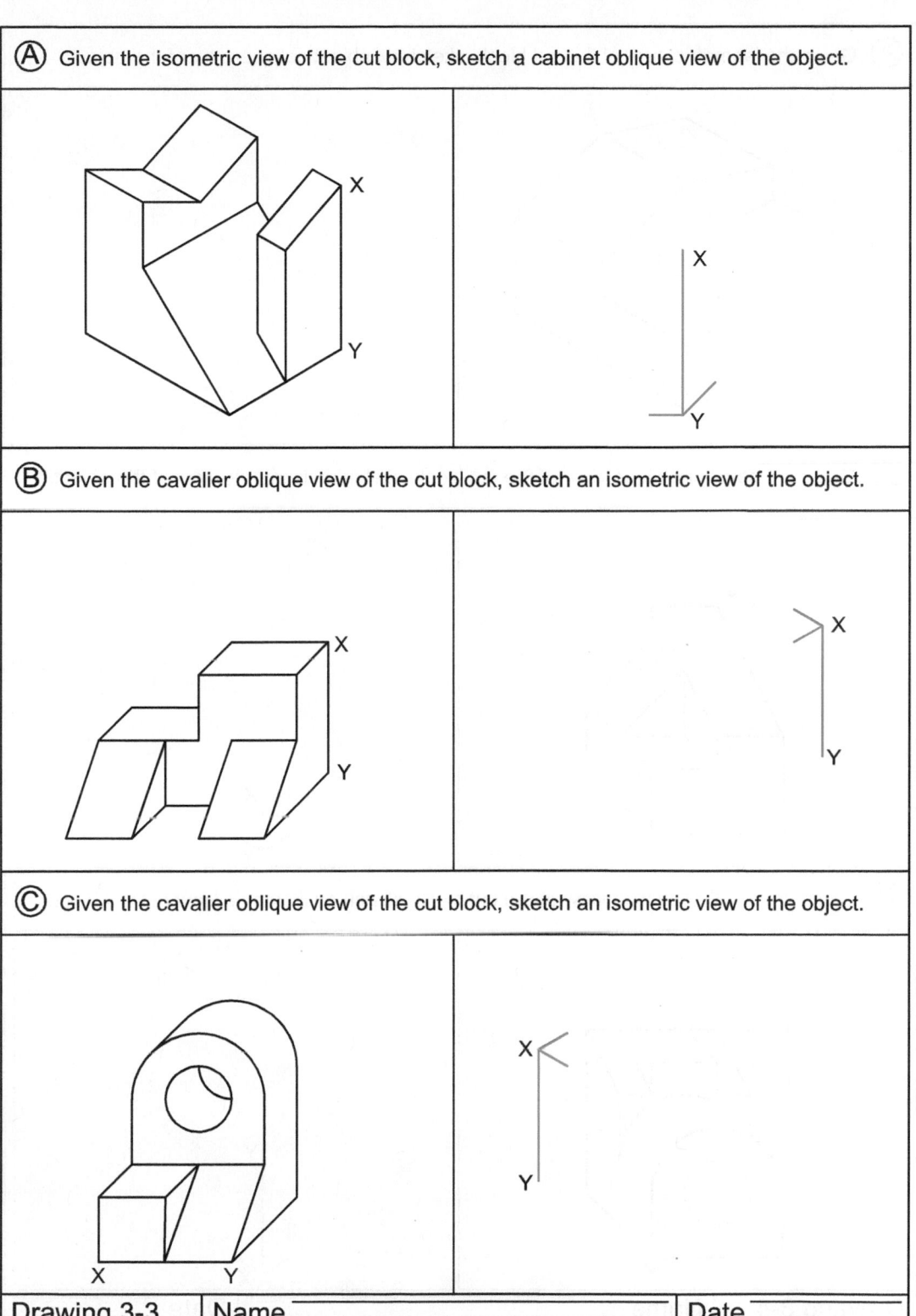

Ⓐ Given the isometric view of the cut block, sketch a cabinet oblique view of the object.

X

Y

Ⓑ Given the cavalier oblique view of the cut block, sketch an isometric view of the object.

X

Y

Ⓒ Given the cavalier oblique view of the cut block, sketch an isometric view of the object.

X

Y

Drawing 3-3 | Name _____ | Date _____

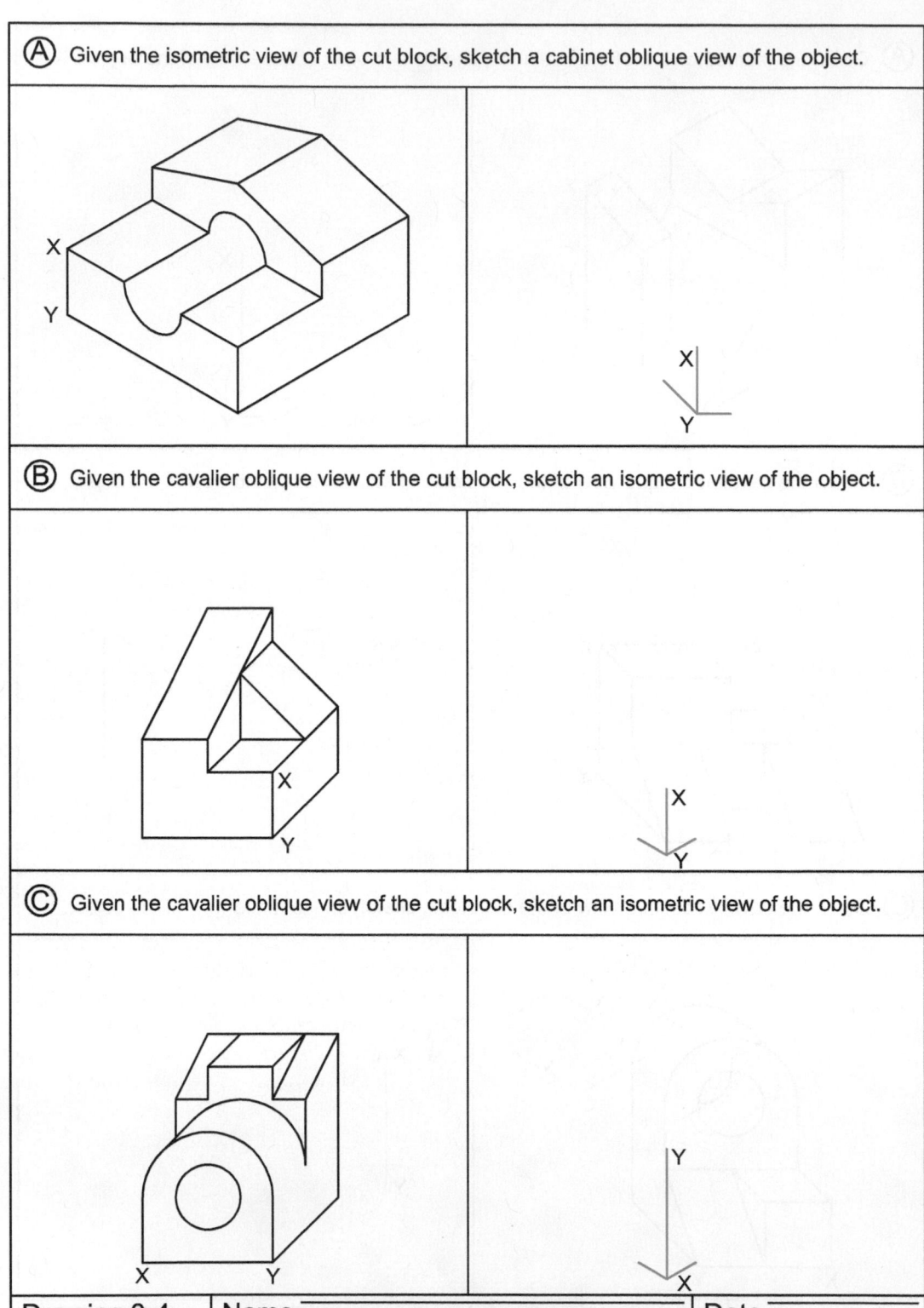

Ⓐ Given the isometric view of the cut block, sketch a cabinet oblique view of the object.

Ⓑ Given the cavalier oblique view of the cut block, sketch an isometric view of the object.

Ⓒ Given the cavalier oblique view of the cut block, sketch an isometric view of the object.

| Drawing 3-4 | Name _____ | Date _____ |

4 MULTIVIEWS

▌MULTIVIEW SKETCHING

Introduction–Justification and Some Characteristics

Multiview drawings are at the core of what has traditionally been thought of as engineering graphics. The purpose of a multiview drawing is to fully represent the size and shape of an object using one or more views. Along with notes and dimensions, these views provide the information needed to fabricate the part.

Chapter 3 included a brief discussion of the characteristics of multiview projection. These characteristics, as seen in Figure 4-1, include

(1) parallel projectors normal to the projection plane and (2) the object positioned so that one principal face is parallel to the projection plane.

As a consequence of this geometry, a multiview drawing can show only one object face. This means that in most cases, more than one view is needed to fully describe the object. It is for this reason that this orthographic projection technique is called multiview projection.

Although only two of the three sets of linear dimensions (i.e., width, depth, height) are projected in any one view, all of this projected information parallel to the projection plane is directly scalable.

Glass Box Theory

Because more than one view is typically needed to document an object using multiview projection, **glass box theory** is used to describe the arrangement of the different multiviews with respect to one another. Imagine that the object to be documented is placed inside a glass box, as shown in Figure 4-2 on page 84. The object is positioned so that its sides are orthogonal to the sides of the glass box. Note also that the principal dimensions of the object (width, depth, and height) are also indicated in this figure.

In Figure 4-3 on page 84, the six sides of the glass box are used as projection planes, upon which the six principal views (top and bottom, front and back, right and left) of the object are projected.

Imagine further that some of the glass box sides are hinged to one another. These hinges, or

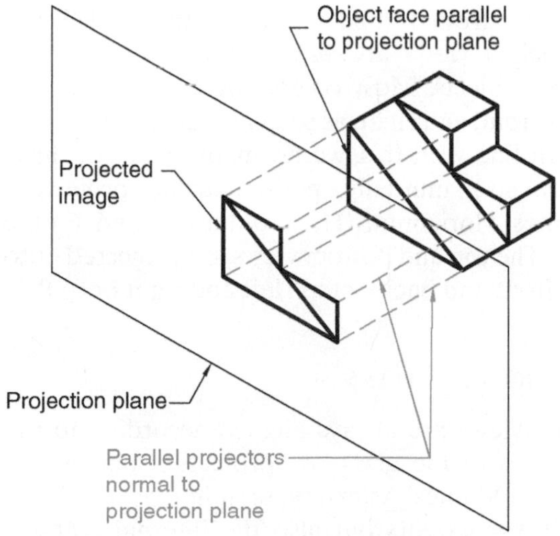

Figure 4-1 Pictorial view of the multiview projection process

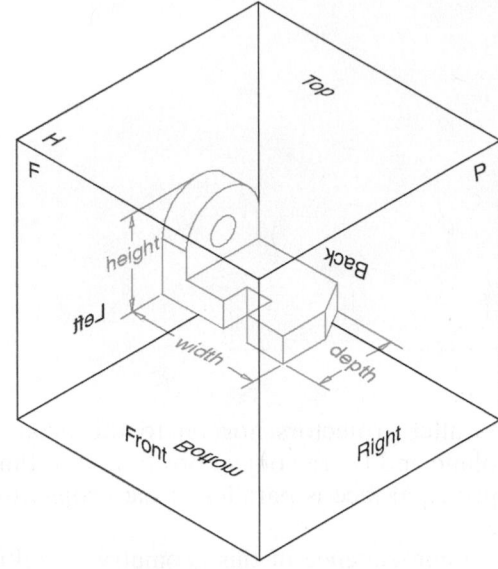

Figure 4-2 Object placed inside glass box

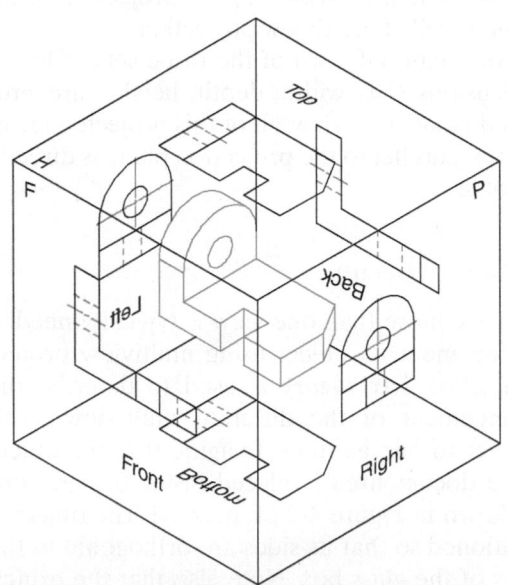

Figure 4-3 Object projected onto box sides

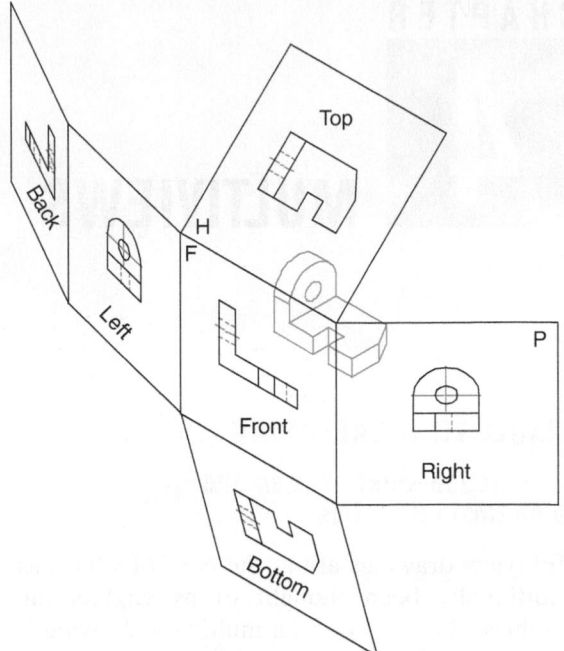

Figure 4-4 Glass box being opened

fold lines, when opened (as shown in Figure 4-4) and then laid flat, result in the six views being arranged as shown in Figure 4-5.

Note in Figure 4-5 that four of the views are hinged to the front view, which is traditionally treated as the primary view in multiview projection. Also note that the top, front, and bottom views are all aligned vertically, whereas the back, left, front, and right views are horizontally aligned.

All six principal views are not normally required to completely document an object. Note in Figure 4-5 the similarities between top and bottom, front and back, and left and right. Three principal views are sufficient to fully describe most objects. Most commonly these views are top, front, and right, as seen in Figure 4-6.

In line with this, we normally speak of three (not six) mutually perpendicular projection planes: Horizontal (H), Frontal (F), and Profile (P). The top and bottom views are projected onto H, front and back onto F, left and right onto P.

Alignment of Views

Multiviews are always aligned according to the dictates of the glass box projection planes and their fold lines. As can be seen in Figure 4-7, not only the extents but also the internal features should be aligned.

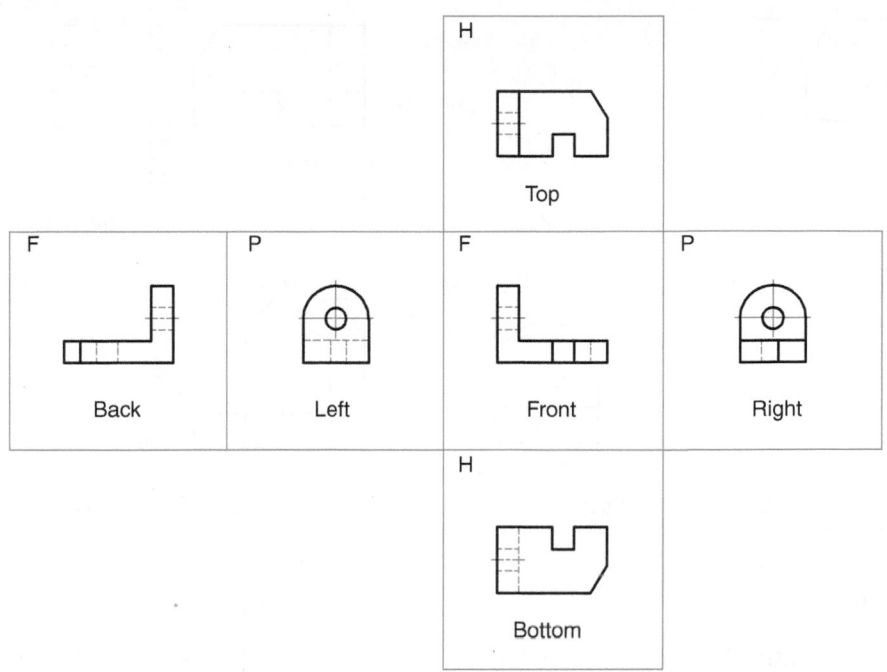

Figure 4-5 Glass box opened and laid out flat

Figure 4-8 on page 86 shows that the top and front views are vertically aligned and share the width dimension, whereas the front and right views are horizontally aligned and share the height dimension. Aligned views that share a common dimension are said to be **adjacent**. The top and right views, though not aligned, share the depth dimension. Views that share a common dimension, but are not aligned, are said to be **related**.

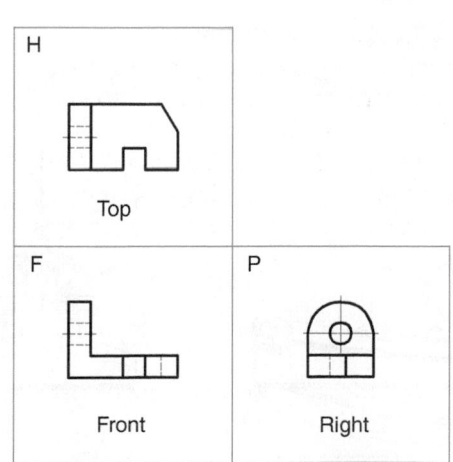

Figure 4-6 Top, front, right multiview arrangement

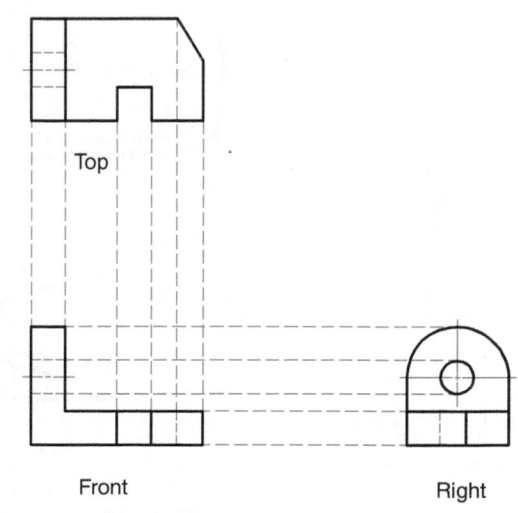

Figure 4-7 Feature alignment

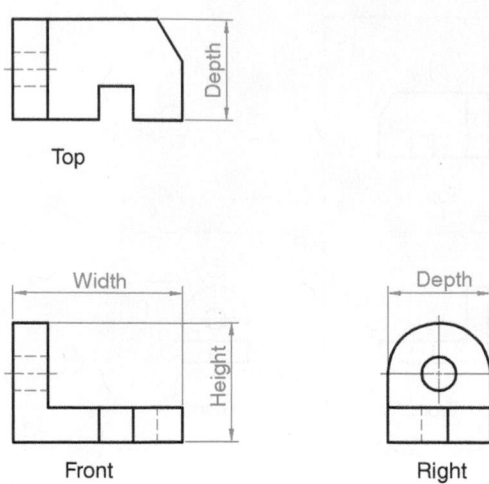

Figure 4-8 Three views with shared dimensions

Figure 4-9 Transfer of depth using a miter line

Transfer of Depth

Every point or feature appearing in one view must be aligned along parallel projectors in their adjacent and related views. Between adjacent views, feature information can be transferred directly along parallel projectors (see Figure 4-7). Either a trammel or a 45-degree miter line can be used to transfer depth information between related views (see Figure 4-9).

View Selection

The most descriptive view should be selected as the front view. In addition, the longest principal dimension should appear as a horizontal dimension in the front view. For example, in the multiview drawing of the boat in Figure 4-10, the side of the boat appears in the front view. This is because it is the most descriptive view and because the longest dimension appears in the front view as horizontal.

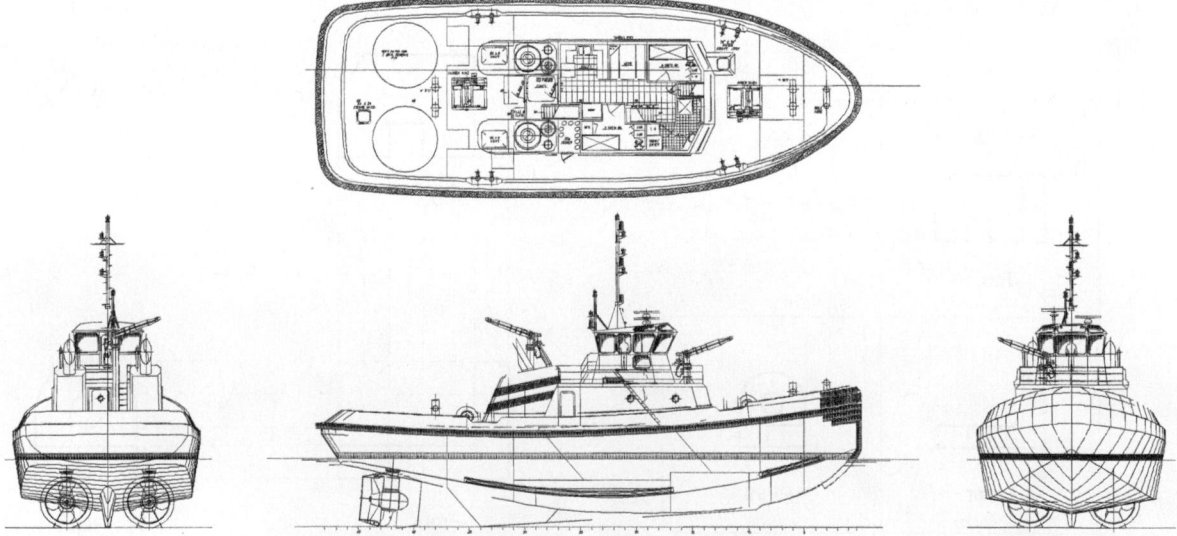

Figure 4-10 Multiview drawing of a boat (Courtesy of Jensen Maritime Consultants, Inc.)

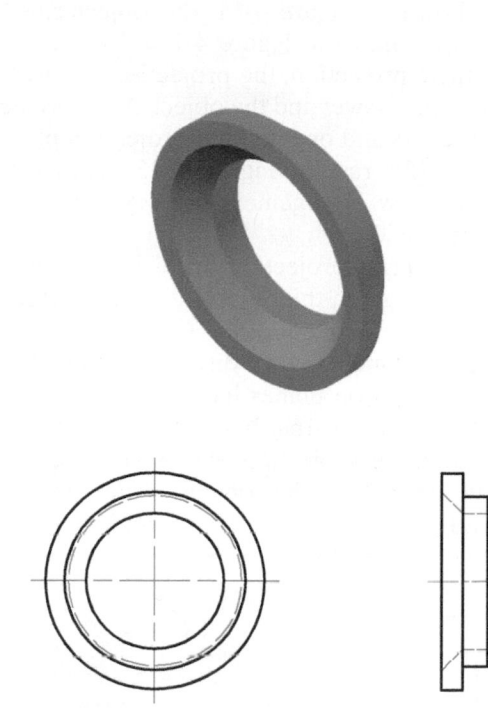

Figure 4-11 Two-view drawing

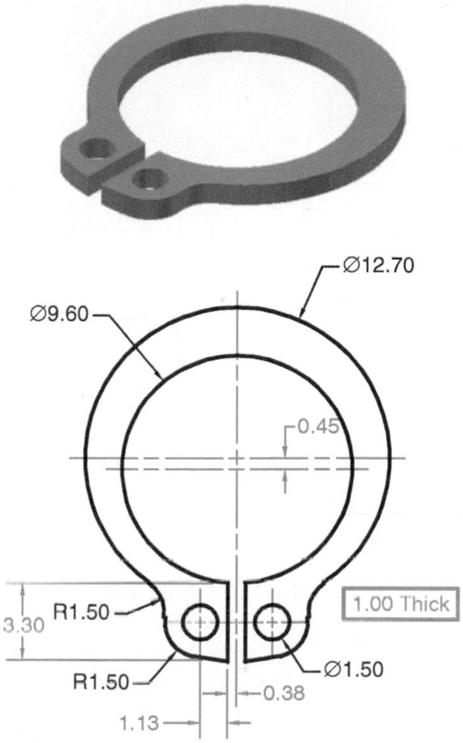

Ø12.70
Ø9.60
0.45
R1.50
3.30
R1.50
1.00 Thick
Ø1.50
0.38
1.13

Figure 4-12 One-view drawing

Another guideline related to view selection is to include the minimum number of views that allows for a complete, unambiguous representation of the object. For most objects, three views are required to fully document the part. In some cases, however, only two views are needed (see Figure 4-11). Simple extruded parts (e.g., washers, bushings) may require only a single view, along with a dimensional callout (see Figure 4-12).

In the event that two views provide the same information, choose the view that has the least number of hidden lines. In Figure 4-13 the right view is preferable to the left view, because the right view has fewer hidden lines.

Third-Angle and First-Angle Projection

Both third-angle and first-angle projection are used to determine how the principal views are arranged with respect to one another. To explain the difference between the two, use a horizontal plane and a (vertically oriented) frontal plane to divide three-dimensional space into quadrants, numbered as shown in Figure 4-14 on page 88.

Third-angle projection, used in the United States and Great Britain, assumes that an object to be projected resides in the third quadrant.

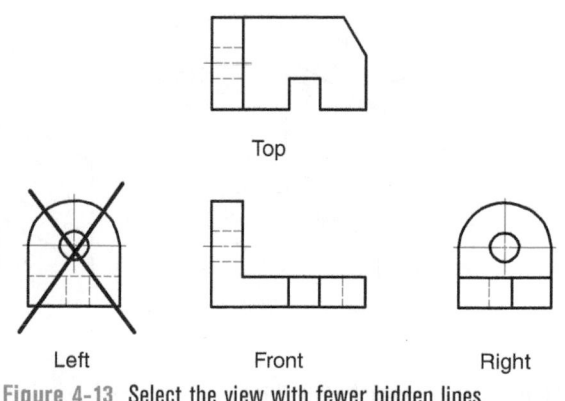

Top

Left   Front   Right

Figure 4-13 Select the view with fewer hidden lines

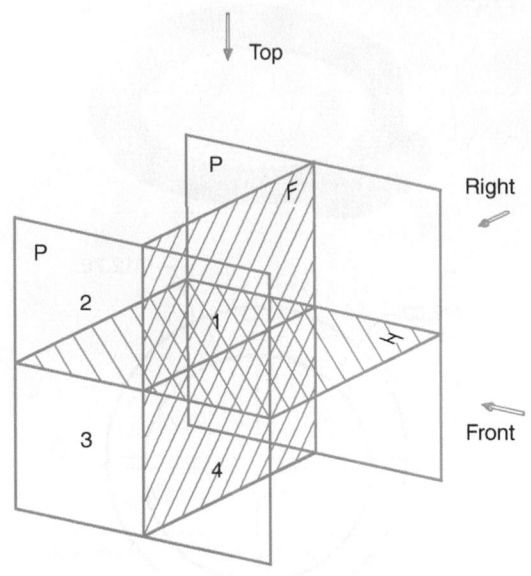

Figure 4-14 Horizontal and front plane, two profile planes

Using the viewing directions for top, front, and right shown in Figure 4-14, the object can be isolated as shown in Figure 4-15*a*. Note that in third-angle projection, the projection planes are between the viewer and the object. After projecting the views and opening the projection planes, Figure 4-15*b* results. Figure 4-15*c* shows the resulting view arrangement employed in third-angle projection.

In first-angle projection, used in the rest of Europe and Asia, the object to be projected is placed in the first quadrant (see Figure 4-14). Using the same viewing directions, the object and its projection planes have been isolated as shown in Figure 4-16*a*. Note that in first-angle projection, the projection planes are all placed behind or below the object. After projecting and then unfolding the projection planes, Figure 4-16*b* results.

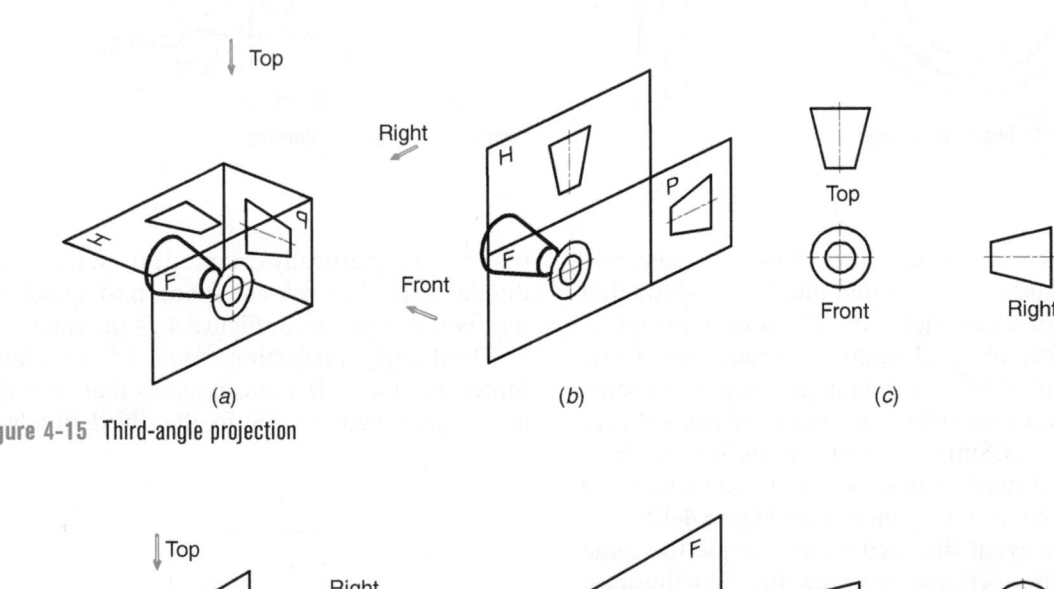

Figure 4-15 Third-angle projection

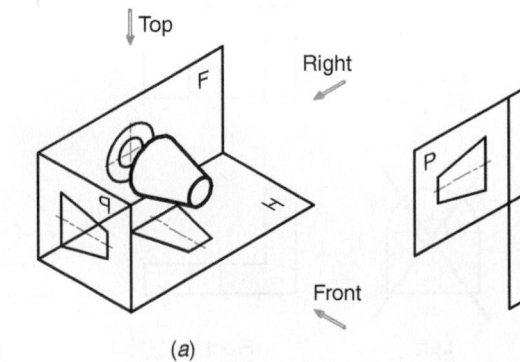

Figure 4-16 First-angle projection

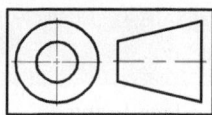
Third-angle projection
symbol

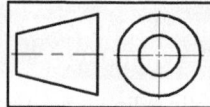
First-angle projection
symbol

Figure 4-17 Third- and first-angle projection symbols

To indicate whether third- or first-angle projection has been used, the international symbol of a truncated cone shown in Figure 4-17 is used.

Line Conventions

In a drawing view, dark thick *continuous lines* are used to represent the:

1. Edge view of a surface
2. Edge between two intersecting surfaces
3. Extent of a contoured surface

Examples of each of these are shown in Figure 4-18.

Dark thin dashed lines, called *hidden lines*, are used to represent features that are hidden in a particular view. Similar to visible continuous lines, hidden lines are used to represent:

1. A hidden edge of a surface
2. A hidden change of planes
3. The hidden extents or limiting elements of a hole

See Figure 4-19 for examples of each of these occurrences.

Centerlines are used in a variety of situations. These thin dark lines typically extend about 5 millimeters beyond the feature being represented. Centerlines are commonly used to represent the axis of a cylinder or hole. In a circular view, crossing centerlines are used. These centerlines should extend beyond the largest-diameter (or largest-radius) concentric

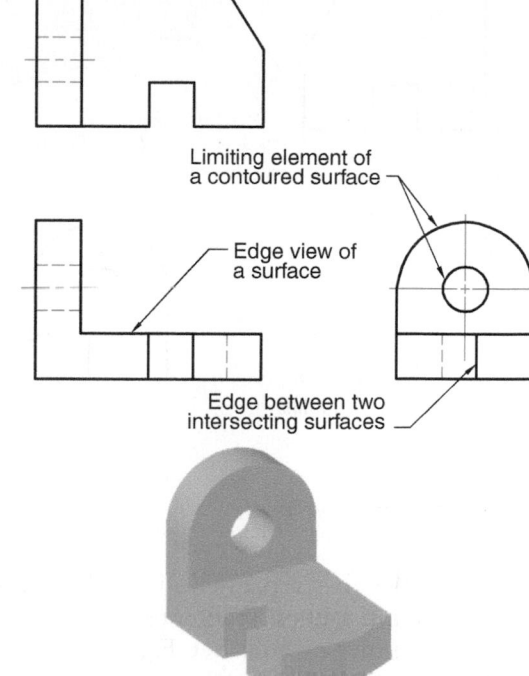

Limiting element of a contoured surface

Edge view of a surface

Edge between two intersecting surfaces

Figure 4-18 Uses of visible continuous line

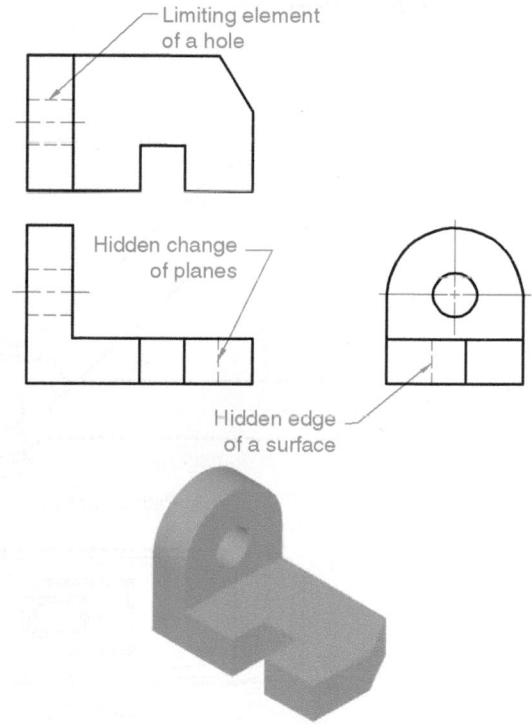

Limiting element of a hole

Hidden change of planes

Hidden edge of a surface

Figure 4-19 Uses of hidden line

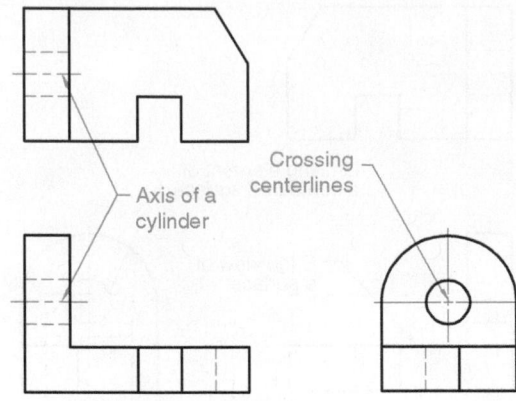

Figure 4-20 Uses of centerline

circle (or arc) being represented. In the rectangular view, a single centerline represents the axis of the cylinder or hole. See Figure 4-20 for an example.

Centerlines are also used to indicate symmetry, to show a path of motion, or to represent bolt circles. See Figure 4-22.

Multiview drawing of a cylinder (see Figure 4-21)

A solid cylinder has two circular edges. In the rectangular view these edges project as straight lines. In order to complete the representation, the *limiting elements*, or extents, of the cylinder are also represented as continuous lines.

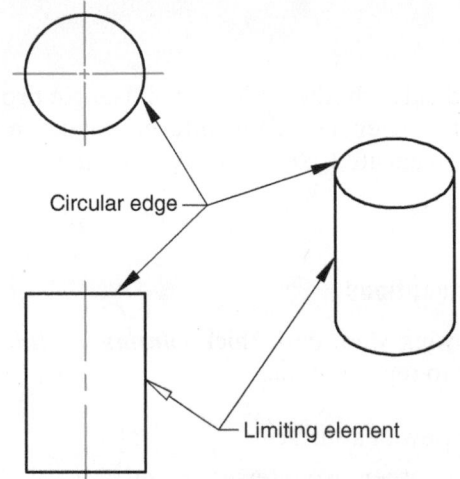

Figure 4-21 Multiview drawing of a cylinder

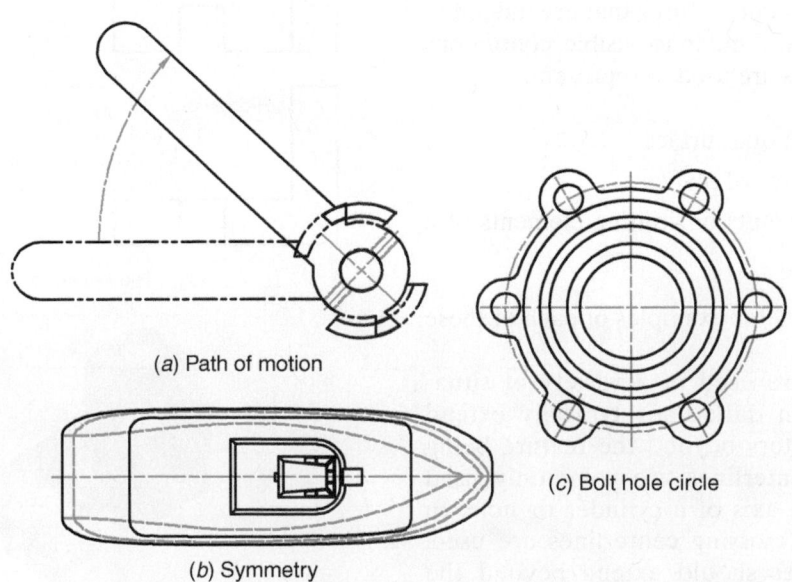

(a) Path of motion

(b) Symmetry

(c) Bolt hole circle

Figure 4-22 Other uses of centerline

Line Precedence

Different object features may sometimes coincide in a multiview sketch. When this occurs, the following order of line precedence is used to determine which lines are represented and which are not: (1) visible, (2) hidden, (3) center. Figure 4-23 illustrates three different collinear line combinations.

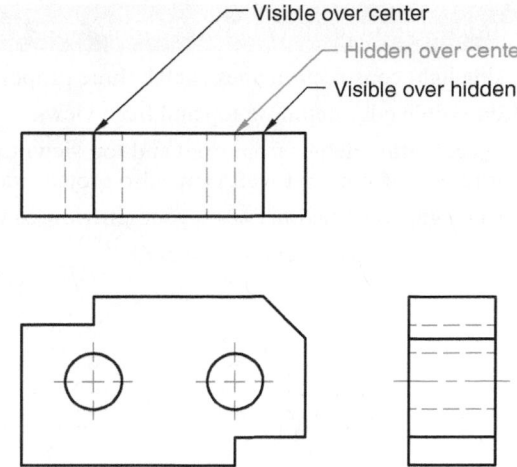

Figure 4-23 Line precedence

1. Using light construction lines, sketch three properly proportioned bounding box views.
2. Add front view feature details.
3. Project feature details from front view to adjacent views.
4. Starting with curved features, go bold.

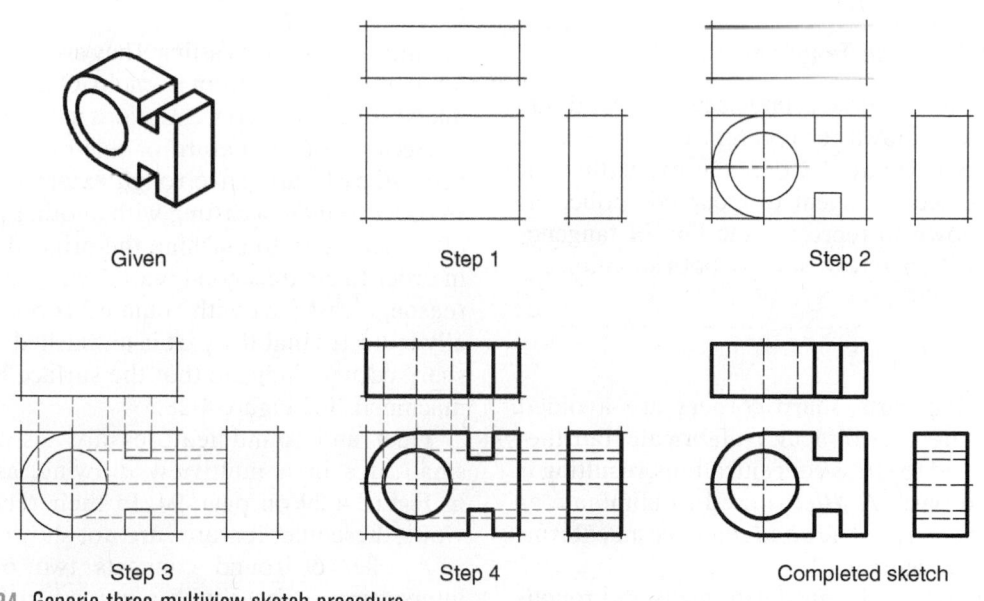

Figure 4-24 Generic three multiview sketch procedure

MULTIVIEW SKETCHING

91

1. Using light construction lines, sketch three properly proportioned bounding box views.

2. Add visible edge details in top and front views.

3. Project feature details from front and top views to left view; use a miter line or trammel to transfer depth information from top to left view. Also project additional feature details between top and front views.

4. Using light construction lines, layout left view, as well as hidden lines in other views.

5. Starting with curved features, go bold.

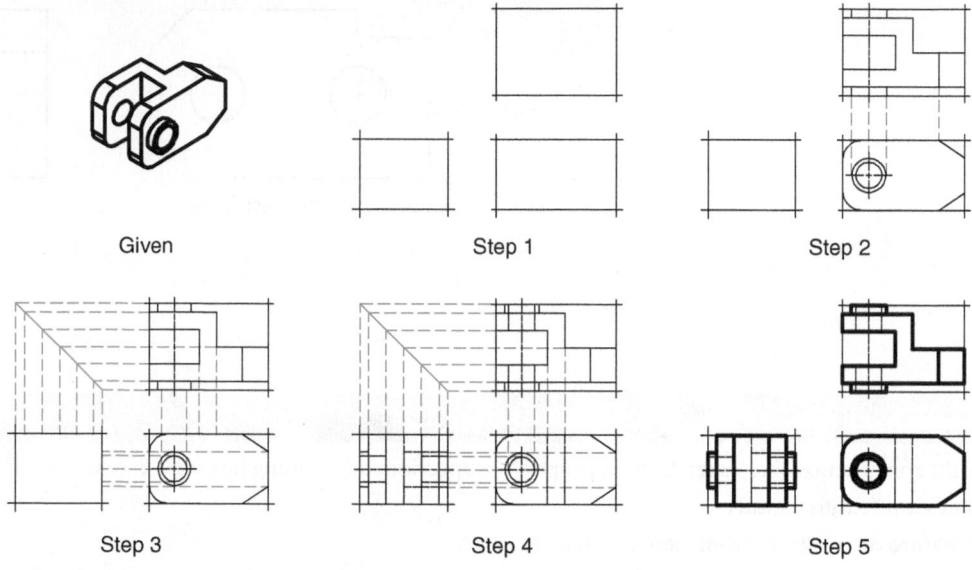

Given   Step 1   Step 2

Step 3   Step 4   Step 5

Figure 4-25 Step-by-step multiview sketch example

Intersections and Tangency

When a planar surface intersects a curved surface, a line is drawn to represent the intersecting surfaces. See Figure 4-26. In the event that the planar surface is tangent to a curved surface, no edge is shown to represent the line of tangency. See Figure 4-26 for examples of both situations.

Fillets and Rounds

In designing parts, sharp corners are avoided. Not only are they difficult to fabricate, but they can also lead to stress concentrations, resulting in weakened parts. A *fillet* is used to eliminate an internal corner, and a *round* removes an external corner. See Figure 4-27.

Cast parts are designed with fillets and rounds. Fillets prevent cracks, tears, and shrinkage at re-entry angles when casting. They also make it easier to extract a part from its mold. Once in service, fillets reduce stress concentrations in the cast part.

Because of the nature of this manufacturing process, cast parts have rough external surfaces. In order to mate a casting with another part, it is often necessary to machine the original surfaces in order to create a good mating surface. For this reason, a cast part with rounded corners generally indicates that the part is unfinished, whereas sharp corners indicate that the surface has been machined. See Figure 4-28.

Fillet and round features are displayed as small arcs in a multiview drawing, as shown in Figure 4-29 on page 94. In their rectangular views, these fillet features are not shown.

A fillet or round connects two otherwise intersecting surfaces with a curved surface tangent to the original surfaces. Because there is no

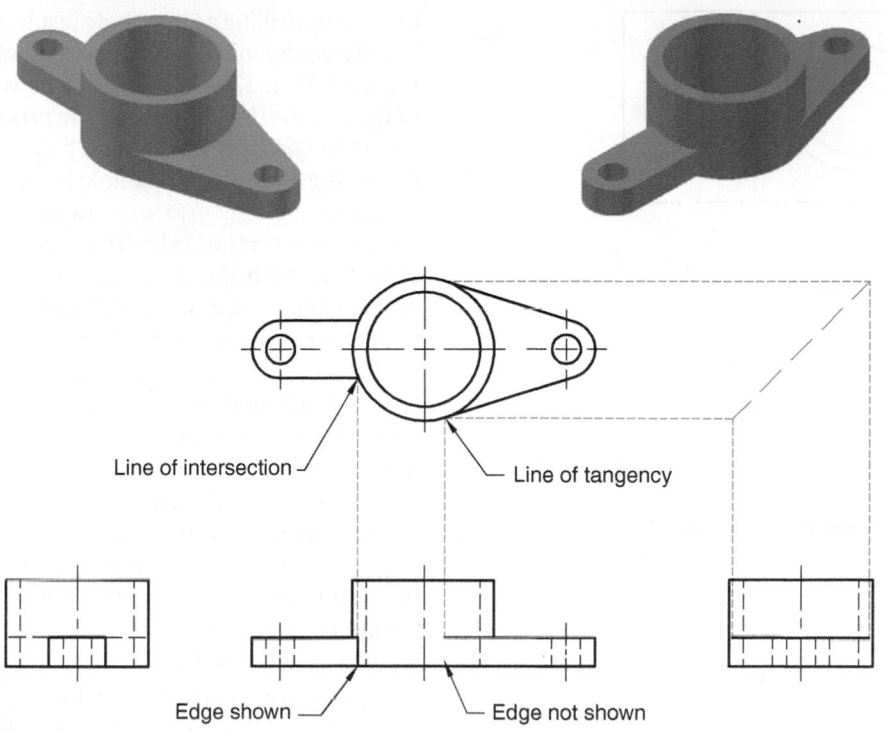

Line of intersection

Line of tangency

Edge shown

Edge not shown

Figure 4-26 Intersection versus tangency

real change in planes, the top view of the object shown in Figure 4-30 on page 94 would normally be shown as a single surface (see Figure 4-30*a*). In order to provide a clearer representation, however, fillets and rounds are sometimes ignored in the rectangular view; edges are then drawn at the imaginary intersection of the two planes, as shown in Figure 4-30*b*.

Machined Holes

Machined holes are formed by various machining operations. These include drilling, boring, and reaming. The specific machining operation used to create the hole is not specified on the drawing, leaving this

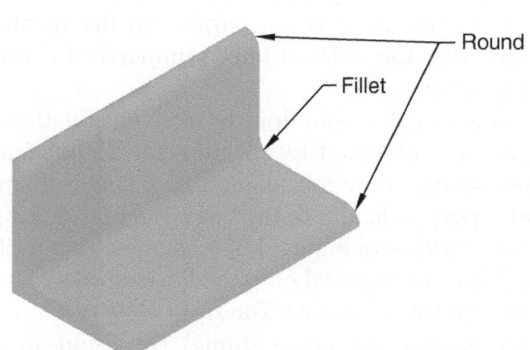

Round

Fillet

Figure 4-27 Fillets and rounds

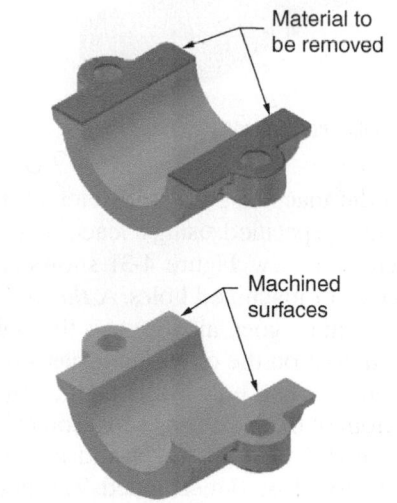

Material to be removed

Machined surfaces

Figure 4-28 Cast part surfaces before and after machining

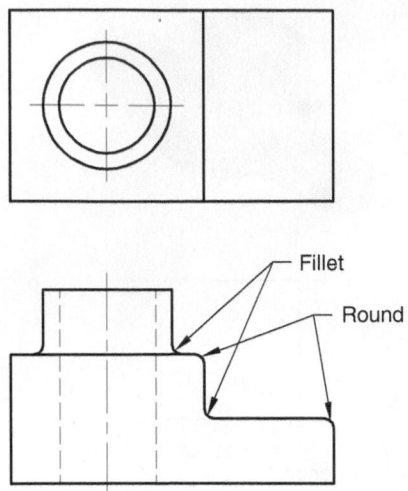

Figure 4-29 Fillet features display as arcs

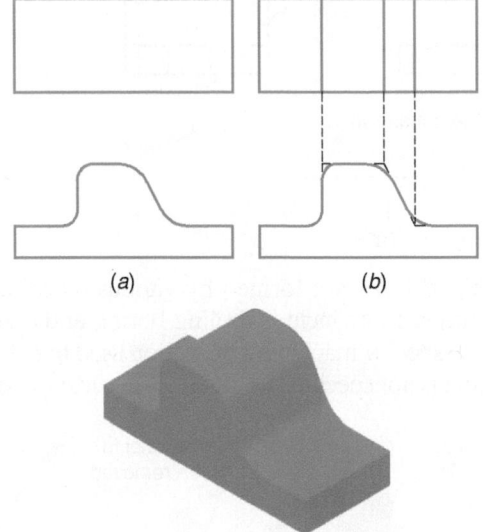

Figure 4-30 Fillet conventions

formed by drilling a larger hole inside a smaller hole to enlarge the initial portion of the hole. As seen in Figure 4-31, a 120-degree shoulder is a byproduct of the counterdrill operation. The process of drilling and then conically enlarging a hole is called *countersinking*. A countersunk hole is used for flat head fasteners and may also serve as a chamfered guide for shafts and other cylindrical parts. In a countersunk hole, both the diameter and the angle of the countersink are specified. Although the angle of the countersink is typically 82 degrees, by convention it is often drawn as 90 degrees. *Spotfacing* is the process of machining the surface around a drilled hole, typically on a cast part, in order to provide a smooth mating surface for washers, bolt heads, nuts, and so on. The cylindrical diameter created by the spotfacing operation is specified; the required depth is left to the machinist. *Counterboring* is the process of cylindrically enlarging the initial portion of a drilled hole. The counterbore operation results in an enlarged hole with a flat bottom. A counterbore hole permits a bolt head to be flush with or recessed below the surface of the part. In a *threaded* hole, an internal thread is made by drilling a hole with a tap drill.

Conventional Representations: Rotated Features

A true orthographic projection of a part with radially distributed features like ribs, holes, and spokes can be confusing to visualize and difficult to construct. The front view of Figure 4-32*a*, for example, shows the true projection of a part with radially distributed ribs and holes. Note the lack of symmetry about the centerline. The rib(s) on the left side of Figure 4-32*a* will not be easy to draw, because it is not parallel to the frontal plane. Also, the holes are not symmetrical about the centerline.

To avoid this situation, by convention these views are simplified by rotating the radial features so that they are aligned in a single plane that is perpendicular to the line of sight. Looking at the top view of Figure 4-32*b*, imagine that a rib and a hole are rotated onto the horizontal axis, as indicated by the arrows. The front view of Figure 4-32*b* shows this conventional representation, with the revolved features now aligned.

decision to the machinist. The diameter of a hole, not the radius, is specified, using a leader extending from the circular view. Figure 4-31 shows several different kinds of machined holes. A *through* hole, formed by drilling, goes all the way through the part. A *blind* hole, on the other hand, has a specific depth. Because a blind hole is also formed by drilling, the bottom of the hole comes to a conical point formed by the drill bit. Only the cylindrical portion of the hole should be dimensioned. The angle of the drill bit is 30 degrees. A *counterdrilled* hole is

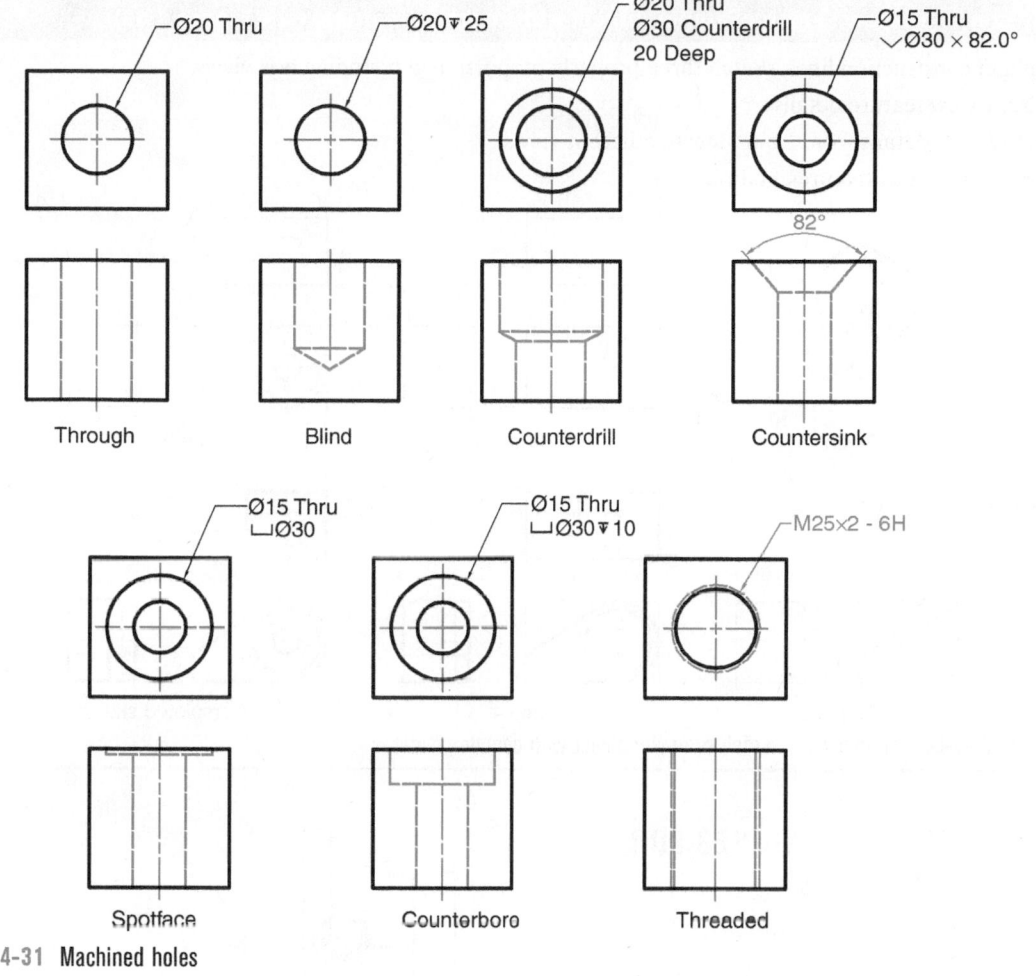

Figure 4-31 Machined holes

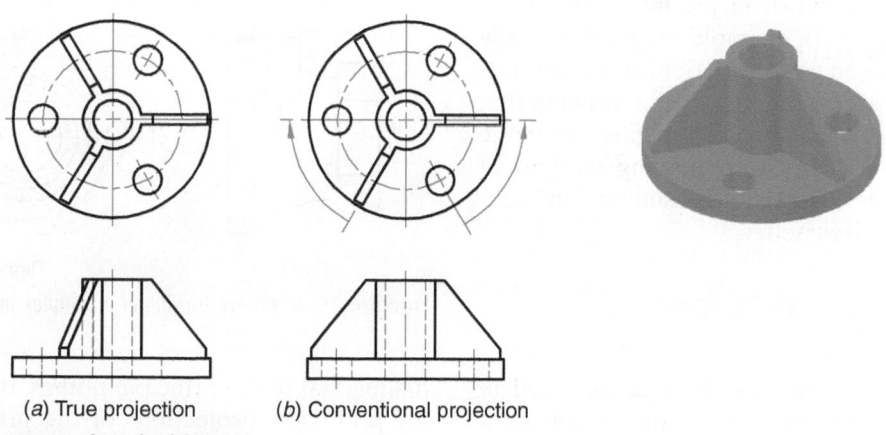

Figure 4-32 Treatment of revolved features

1. Using light construction lines, sketch three properly proportioned bounding box views.
2. Add front-view feature details.
3. Project feature details from front view to adjacent views.
4. Starting with curved features, go bold.

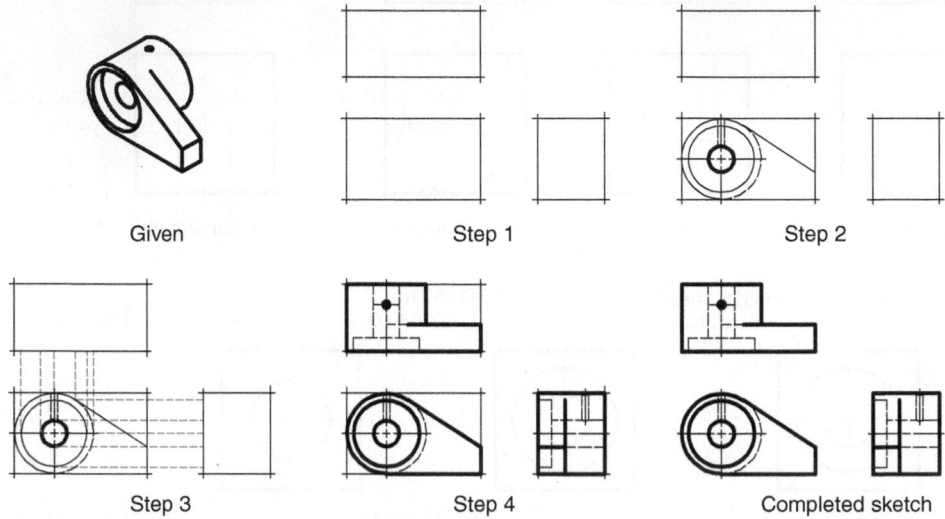

Given   Step 1   Step 2

Step 3   Step 4   Completed sketch

Figure 4-33 Step-by-step multiview sketch example: object with complex features

■ VISUALIZATION TECHNIQUES FOR MULTIVIEW DRAWINGS

Introduction and Motivation

Visualization is a process by which shape information on a drawing is translated to give the viewer an understanding of the object represented. The part shown in the multiview drawing in Figure 4-34, for example, may not be easily recognizable, even to an experienced engineer. It is only after careful reading of the drawing that a mental image of the product begins to emerge (see Figure 4-35). In the remaining sections of this chapter, various spatial visualization techniques will be discussed.

Treatment of Common Surfaces

NORMAL SURFACES

A rectangular prism like the one depicted in Figure 4-36 contains only normal surfaces. A *normal surface* is a planar surface that is

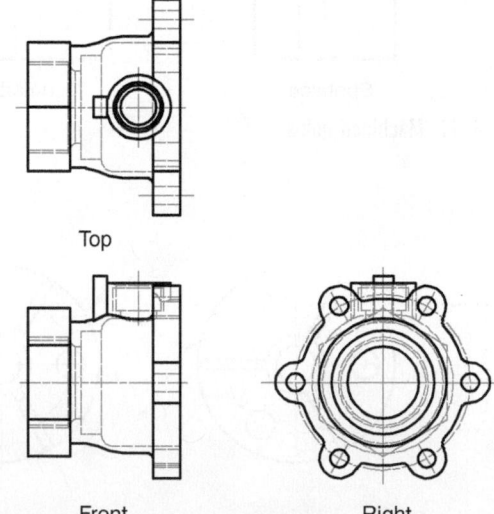

Top

Front   Right

Figure 4-34 Multiview drawing of a complex part

orthogonal to the principal planes. If we look at the multiview projections of the prism, we see that surface A appears as an area in the top view,

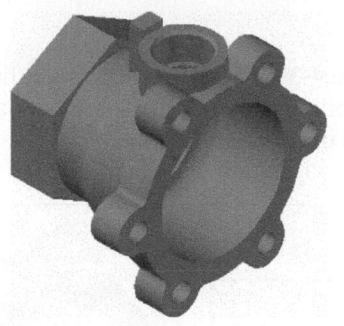

Figure 4-35 Rendered view of the complex part in Figure 4-34

whereas in the other views, surface A appears as an edge. Also, note that because surface A is parallel to the horizontal projection plane, it is shown true size (TS) in the top view. In a three view drawing, a normal surface appears as a true size surface in one view and on edge in the other two views.

INCLINED SURFACES

Surface B in Figure 4-37 is called an inclined surface. An *inclined surface* can be described as a normal surface that has been rotated about a line parallel to a principal axis. An inclined surface is perpendicular to one principal plane and is inclined (i.e., neither parallel nor perpendicular) to the others. An inclined surface appears as a foreshortened (i.e., not true size) surface in two views, whereas in the third view the inclined surface appears on edge. This edge length is a true length.

OBLIQUE SURFACES

An *oblique surface* is a planar surface that has been rotated about two principal axes. An oblique surface is inclined to all three principal projection planes. Surface C in Figure 4-38 is an oblique surface. Note that surface C appears as an area in all three multiviews; in none of these views is the oblique surface true size.

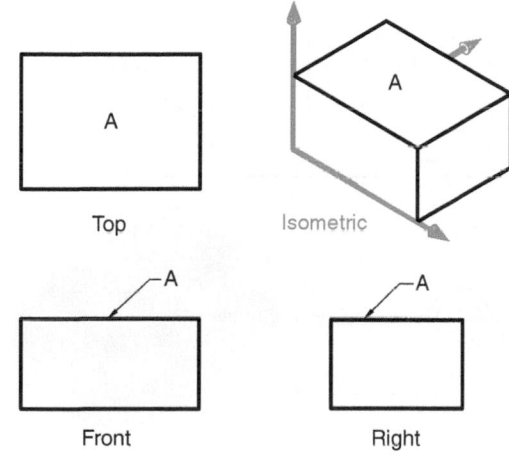

Figure 4-36 Normal surfaces

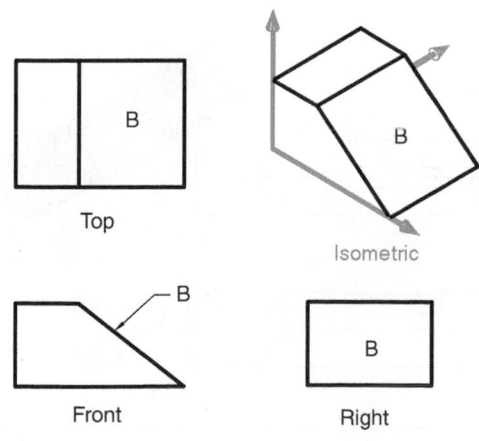

Figure 4-37 Inclined surfaces

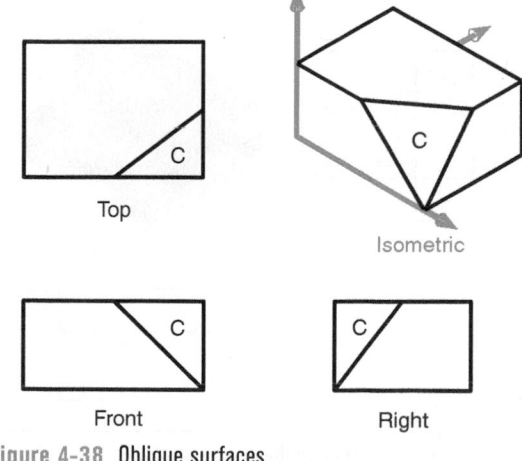

Figure 4-38 Oblique surfaces

Projection Studies

One way to improve visualization skills is to study four views (three multiviews and one pictorial view) of simple objects like those appearing in Figure 4-39. These projection studies improve one's ability to recognize common shapes and features in combination.

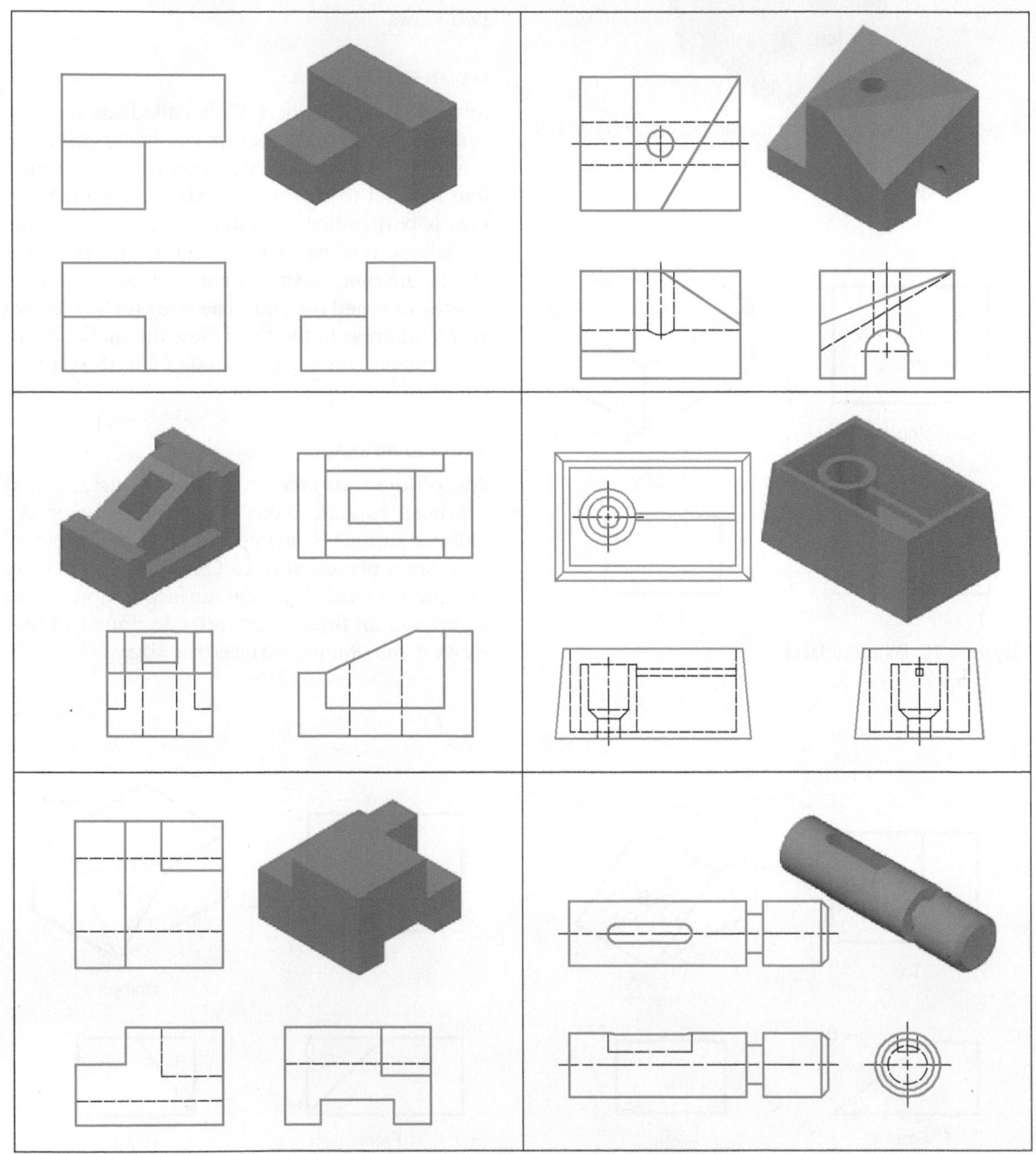

Figure 4-39 Projection studies

Adjacent Areas

The top view in Figure 4-40 on page 100 shows three distinct areas. Because no two adjacent areas can lie in the same plane, these areas must represent different surfaces. Some possible objects matching this view are also shown in the figure.

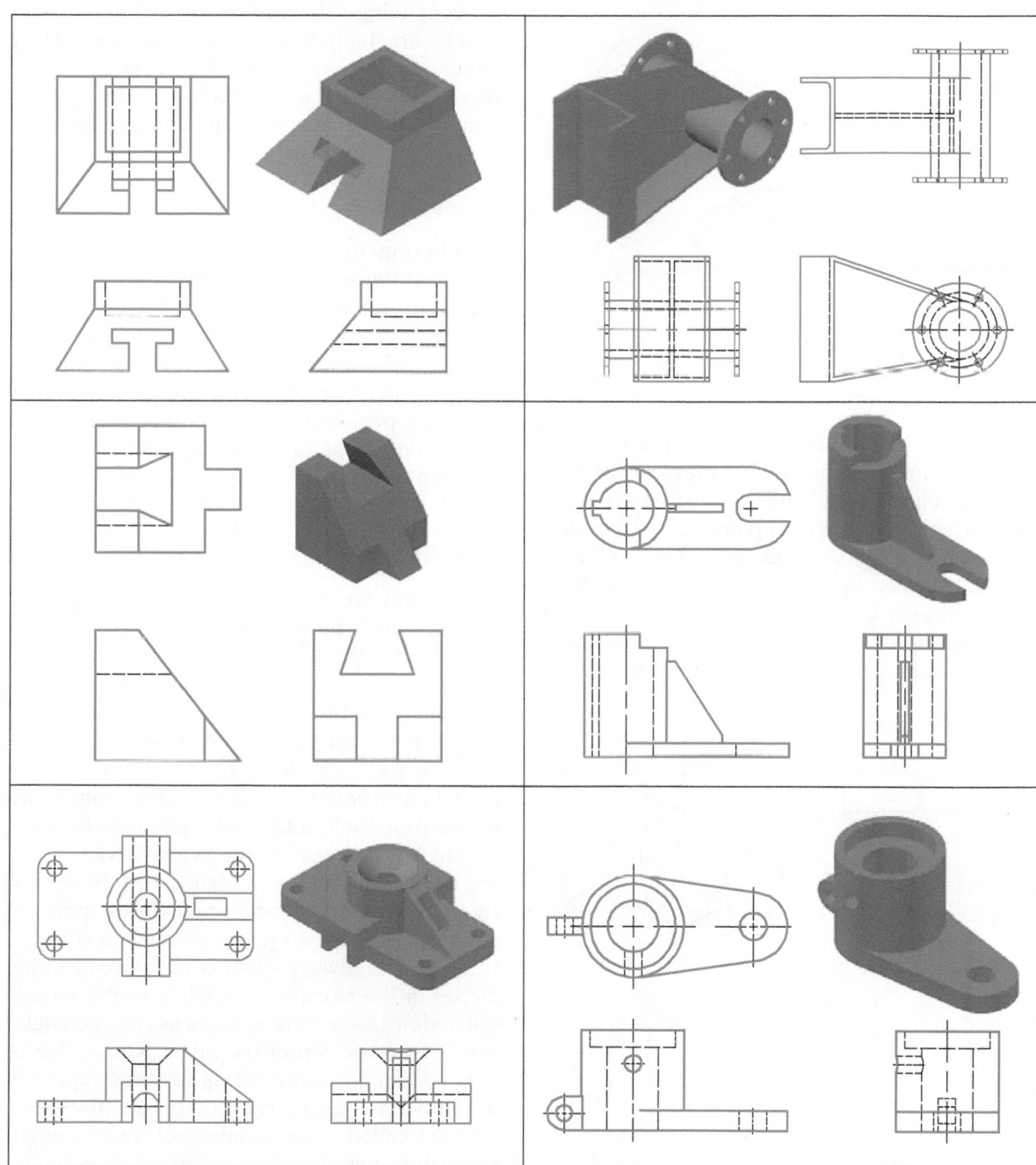

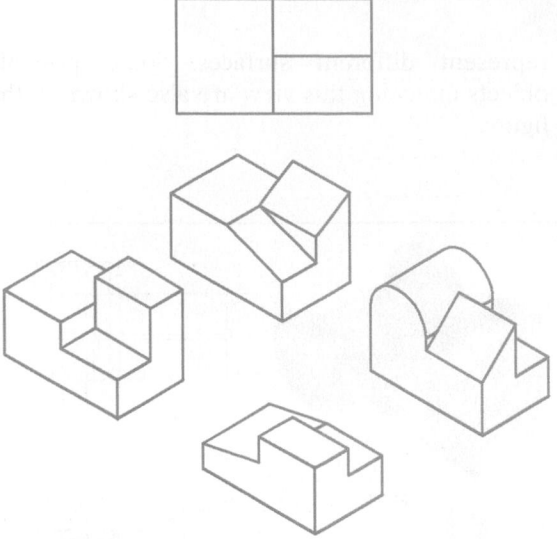

Figure 4-40 Adjacent areas

Surface Labeling

In order to help interpret a multiview drawing, it is sometimes useful to label these surfaces, as shown in Figure 4-41. Note that surface 1 is a normal surface. It appears as an area in the top view and as an edge in the other views. Surface 3 is an inclined surface, appearing on edge in the front view and as a foreshortened area in the other views. Surface 5 is an oblique surface and appears as a foreshortened area in all three views.

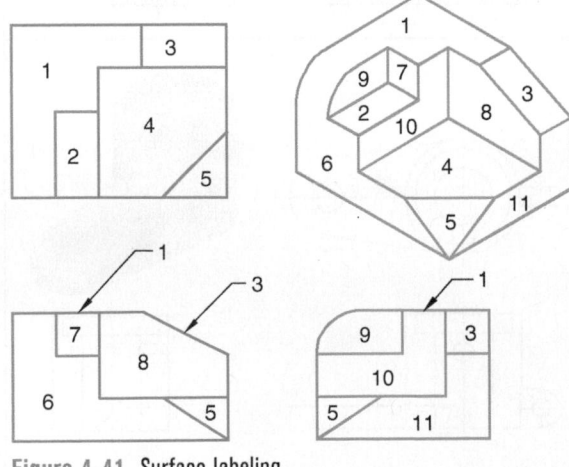

Figure 4-41 Surface labeling

Similar Shapes

Unless viewed on edge, a planar face will always be projected with the same number of vertices. In addition, these vertices will always connect in the same sequence, no matter what the view. These facts are useful when reading a multiview drawing containing either inclined or oblique surfaces (see Figure 4-42). Recall that an inclined surface appears as a foreshortened surface in two of three views and that an oblique surface appears as a foreshortened surface in all three views.

Vertex Labeling

In addition to labeling surfaces, it is also at times useful to label the vertices of a complicated surface in one view, and then to project these points into adjacent or related views. See Figure 4-43 for an example, where the vertices of an oblique surface are labeled in one view and then projected into the other views. Note the similar shape of this oblique surface in the different views.

Analysis by Feature

As we will see in greater detail in Chapter 7, parts are built up from features. These features include such three-dimensional shapes as extrusions, revolutions, holes, ribs, and chamfers. By combining features we arrive at a completed part. See, for example, Figure 4-44 on page 102, where a part is built up from various features, including an extrusion, a boss (raised cylinder), a counterbore hole, a rib, and fillets and chamfers. Figure 4-45 on page 102 shows a multiview drawing of the part shown in Figure 4-44. Note how these different manufacturing features appear in the multiview drawing. For example, a counterbore feature always appears as two concentric circles in the circular view, while in the rectangular view this feature appears as two rectangles stacked one on top of the other. Knowledge of how common manufacturing features appear in a multiview drawing can be very helpful when you are called upon to interpret more complicated drawings.

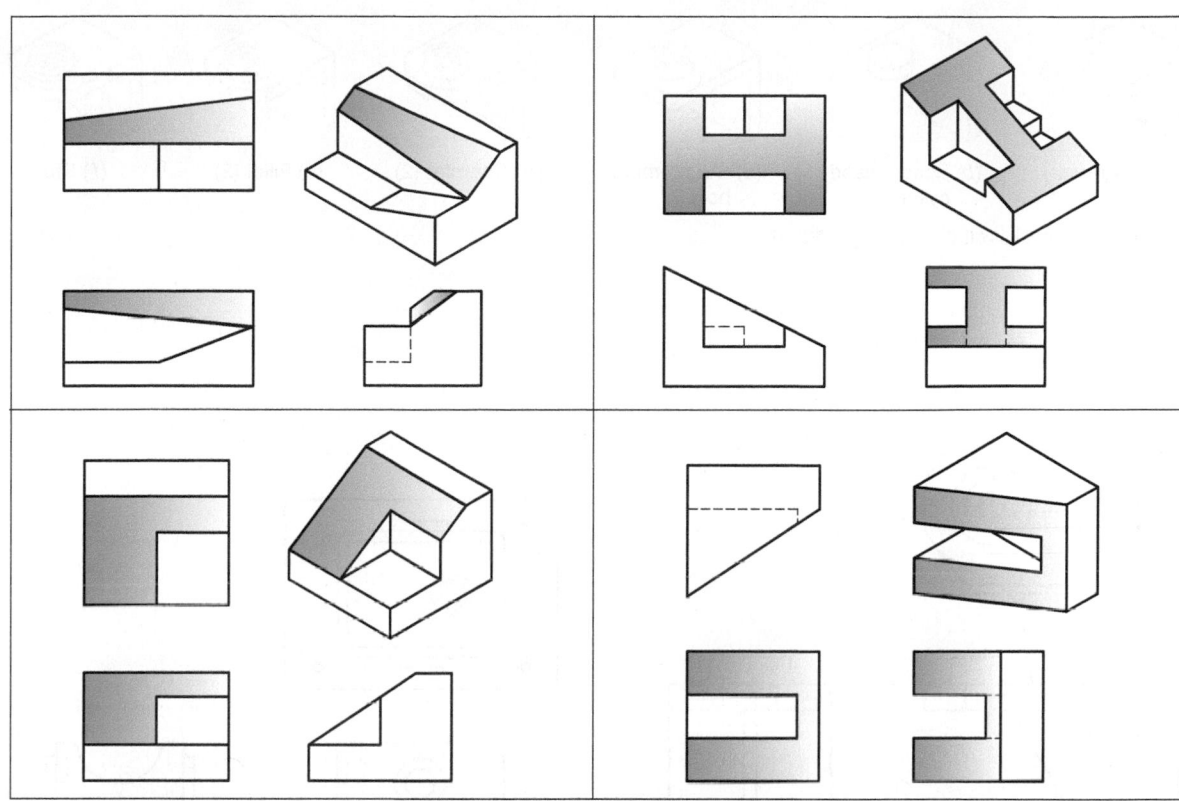

Figure 4-42 Similar shapes

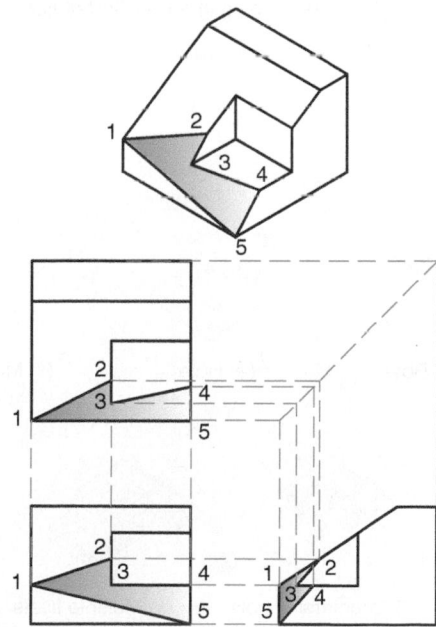

Figure 4-43 Vertex labeling

Now look at the multiview drawing shown in Figure 4-46 on page 102. Without the benefit of a pictorial view, this object is difficult to visualize. However, if we break the part down into recognizable features, this task becomes more manageable. See Figure 4-47 on page 102 for a breakdown of the features of the object depicted in Figure 4-46.

Missing-Line and Missing-View Problems

Two additional tools, or rather exercises, that are very useful in developing spatial reasoning are missing-line and missing-view problems. In a missing-line problem three views are given, but some lines are missing from the views. The objective is to identify the missing lines. This can be accomplished by identifying edges in one view that do not appear in an adjacent or related view. By projecting the location of these edges into the adjacent and related views, the location of the missing lines can be identified. Additional visual

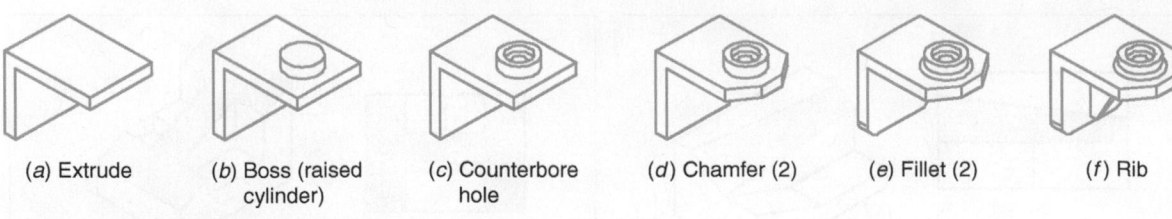

(a) Extrude   (b) Boss (raised cylinder)   (c) Counterbore hole   (d) Chamfer (2)   (e) Fillet (2)   (f) Rib

Figure 4-44 Breakdown of a part by features

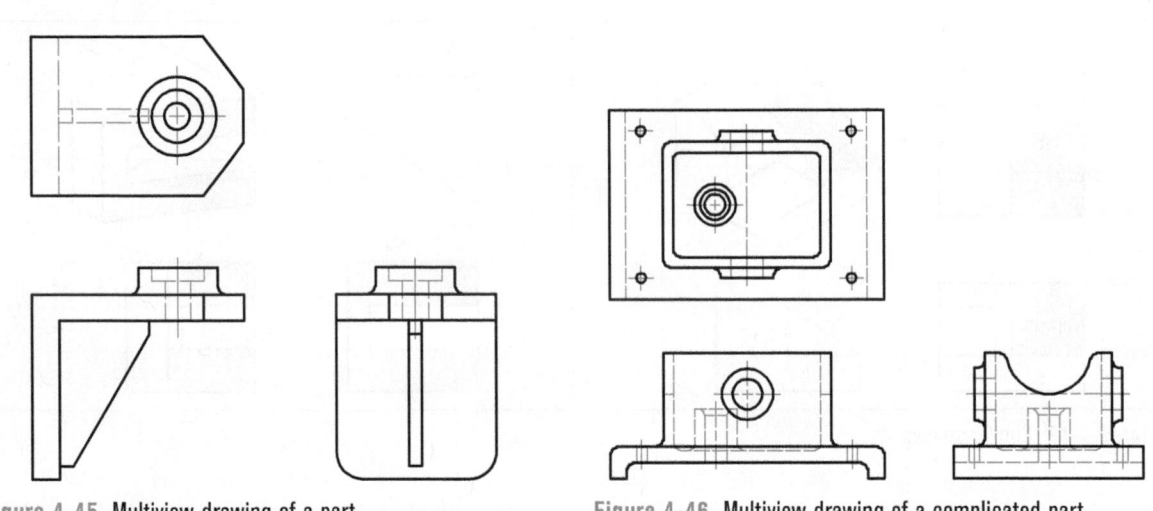

Figure 4-45 Multiview drawing of a part   **Figure 4-46** Multiview drawing of a complicated part

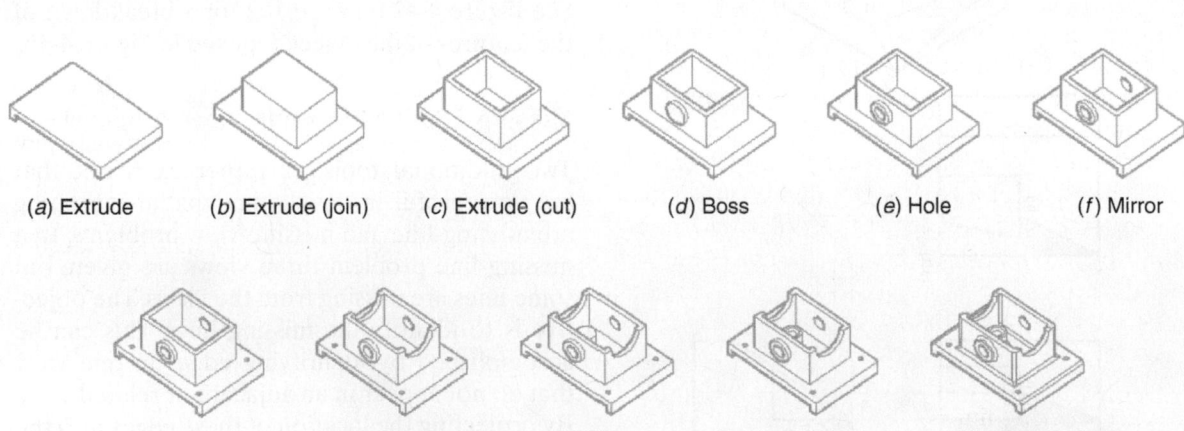

(a) Extrude   (b) Extrude (join)   (c) Extrude (cut)   (d) Boss   (e) Hole   (f) Mirror

(g) Hole (4)   (h) Extrude (cut)   (i) Boss   (j) Countersink hole   (k) Multiple fillets

Figure 4-47 Breakdown of a complicated part by features

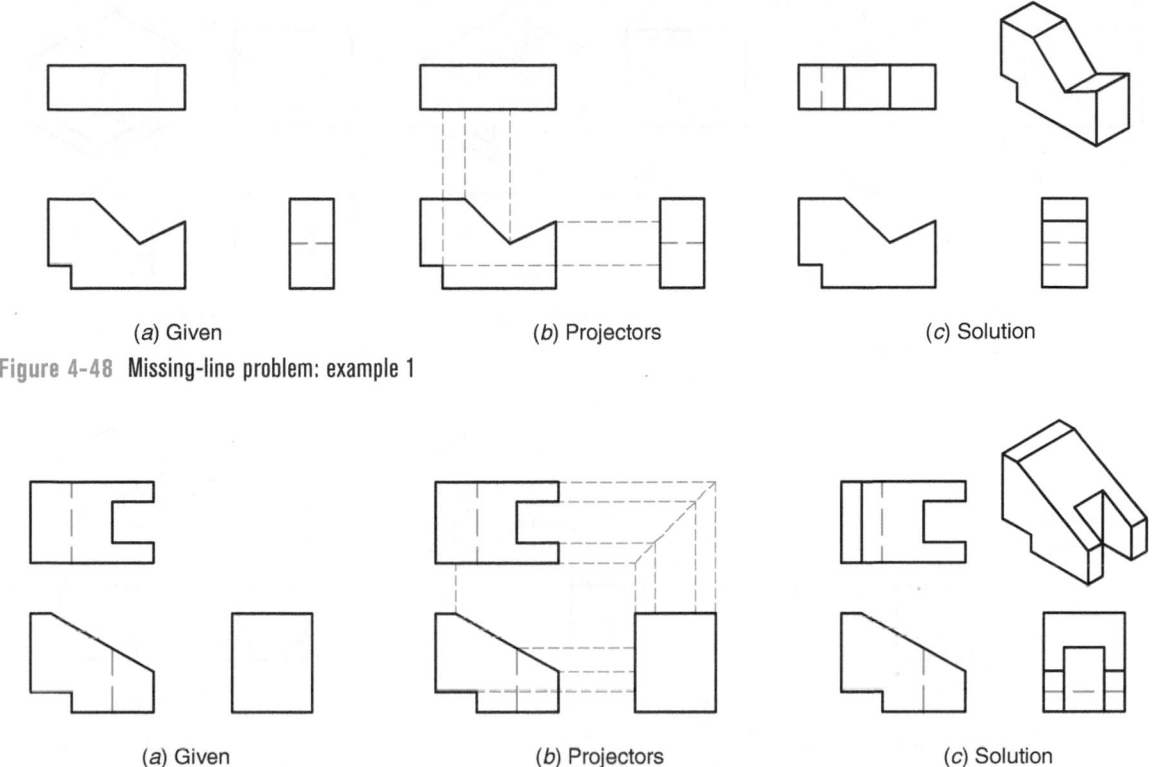

(a) Given   (b) Projectors   (c) Solution

Figure 4-48 Missing-line problem: example 1

(a) Given   (b) Projectors   (c) Solution

Figure 4-49 Missing-line problem: example 2

reasoning is required to identify the type (e.g., visible, continuous) and extent of the missing lines. Figures 4-48 and 4-49 provide two examples of missing-line problems.

More challenging still are missing-view problems. Here two of three views are given. The objective is to find the missing third view. As with missing-line problems, edge features in the given views can be projected to help identify the lines in the missing view. This technique is employed, for example, in Figures 4-50 and 4-51 on page 104. In nearly all missing-view problems, however, it is even more helpful to sketch a well-proportioned pictorial view of the object. Start by sketching the object's bounding box. Next use the given views to identify prominent object features. The right view of the object in Figure 4-51, for example, suggests a backwards "C" shape extrusion. Add this feature to the pictorial sketch. The vertical hidden line in the right view still needs to be accounted for. By employing visual reasoning and some trial and error, one discovers

that a wedge-shaped vertical cut can account for this hidden line, as well as for the internal vertical lines in the front view. The missing object is consequently composed of a "C" shape extrusion and the symmetrical wedge cut.

▌ QUESTIONS

TRUE OR FALSE

1. In a multiview drawing, it is always necessary to include at least three principal views to completely define the object.

2. In a multiview drawing, the right view should be used, even if it has more hidden lines than the left view.

3. In a three view drawing, an inclined planar surface will appear as a line in two of the principal views.

4. The angle at which the line-of-sight pierces the projection plane is the same for both multiview and axonometric projections.

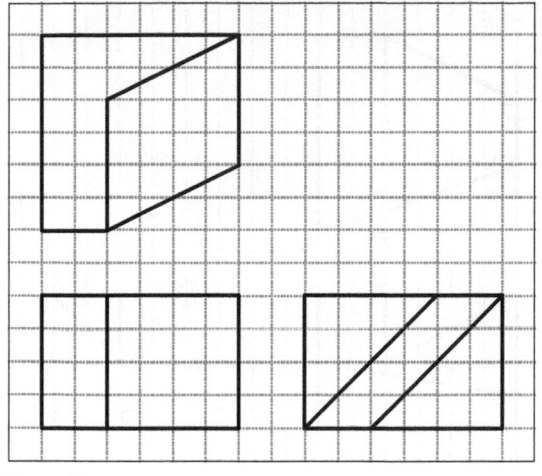

Figure P4-49

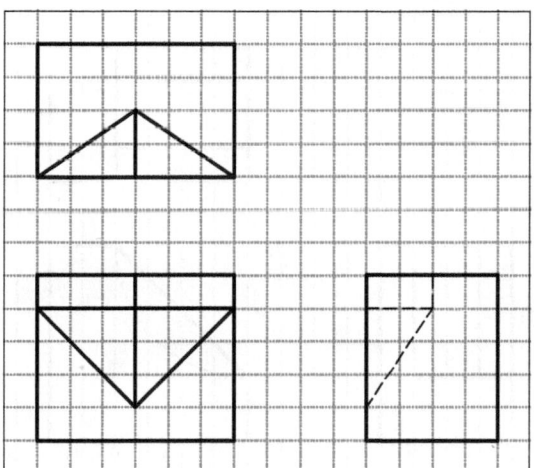

Figure P4-50

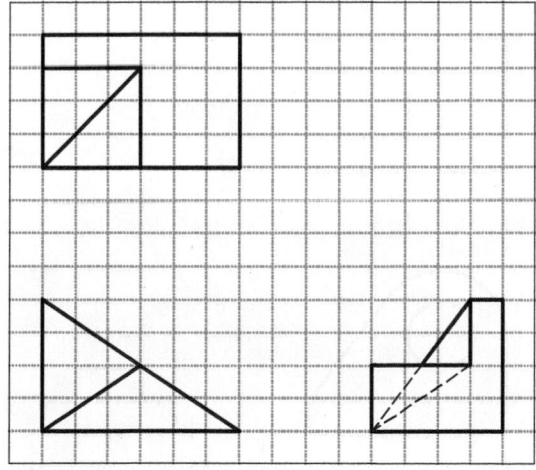

Figure P4-51

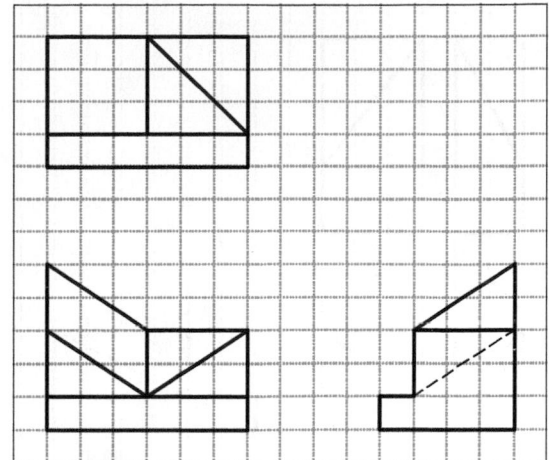

Figure P4-52

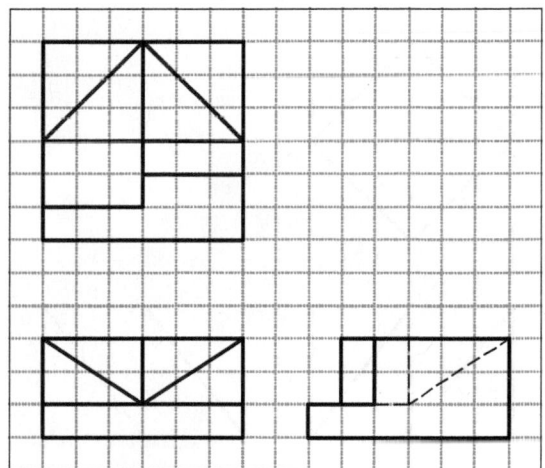

Figure P4-53

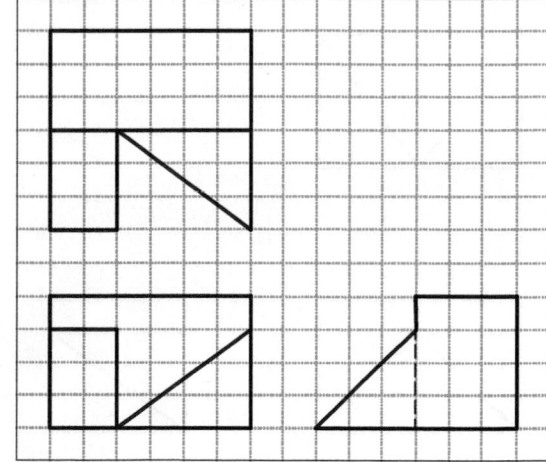

Figure P4-54

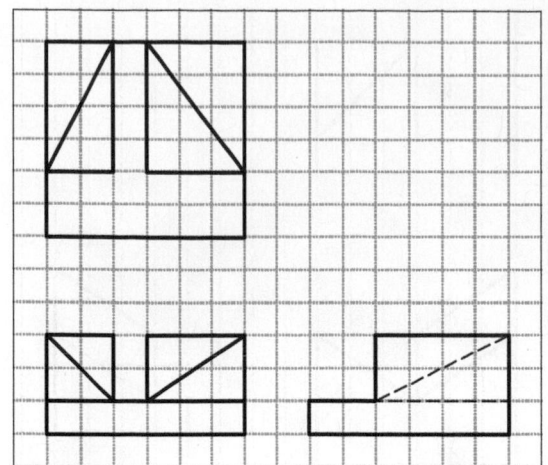

Figure P4-55

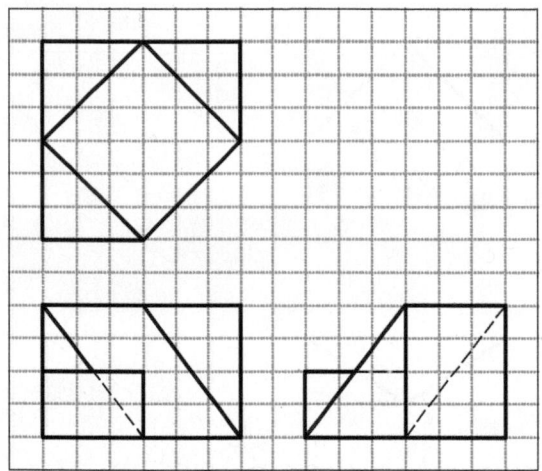

Figure P4-56

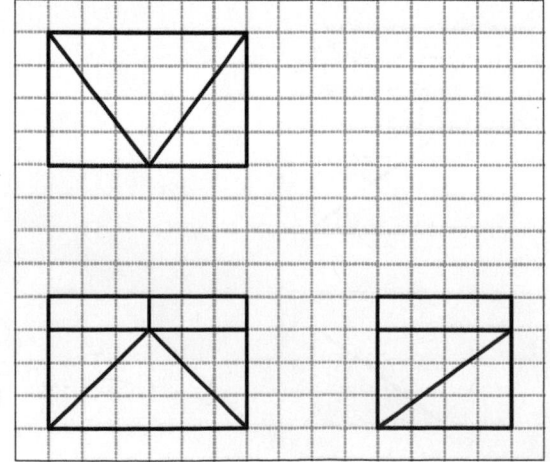

Figure P4-57

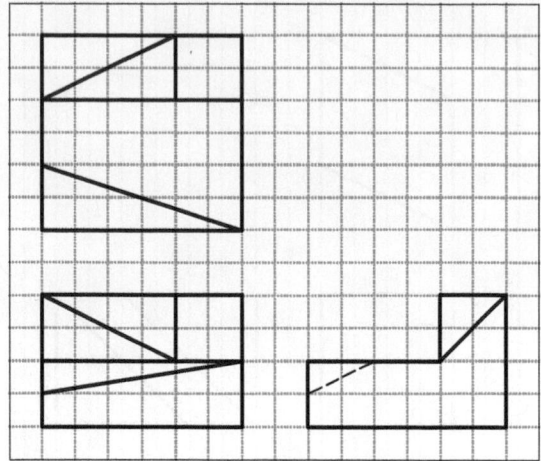

Figure P4-58

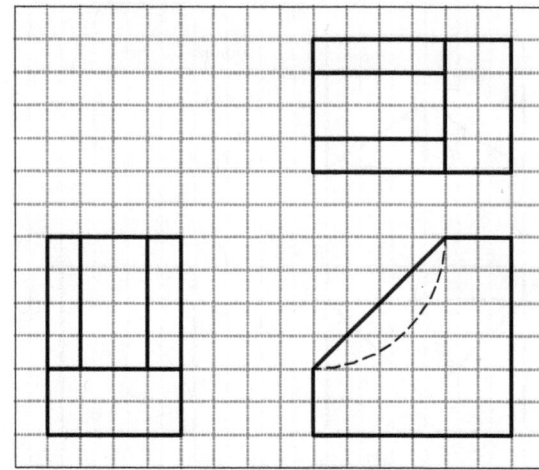

Figure P4-59

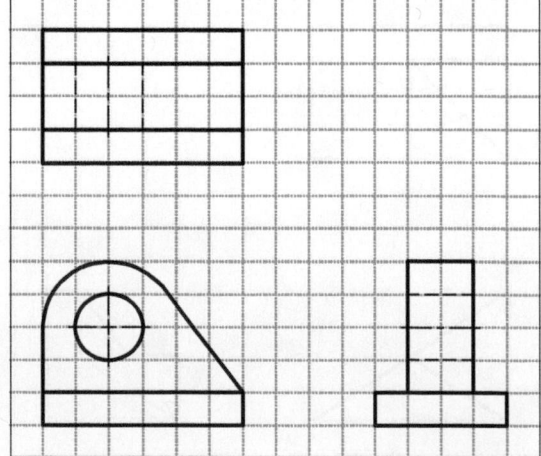

Figure P4-60

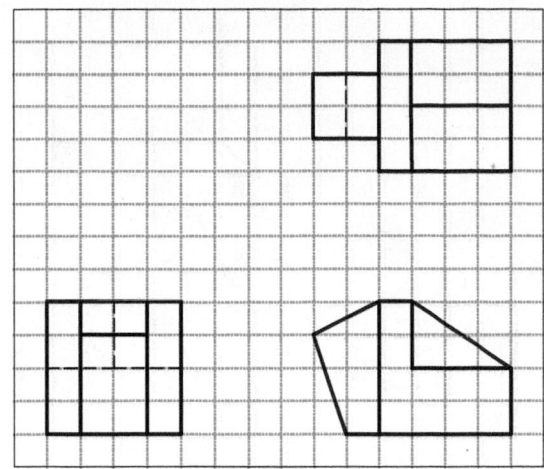

Figure P4-61

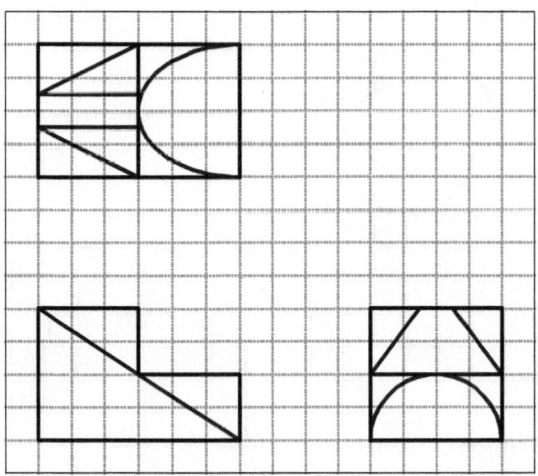

Figure P4-62

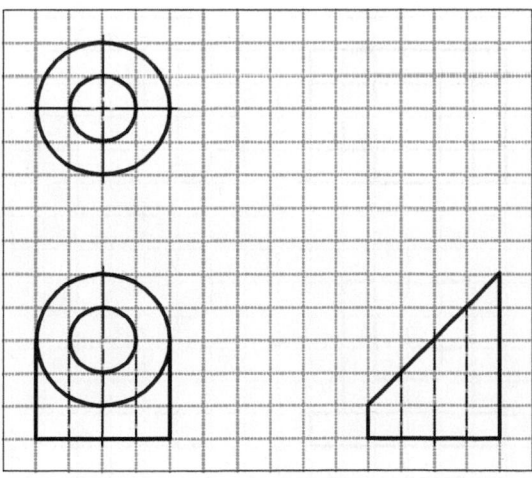

Figure P4-63

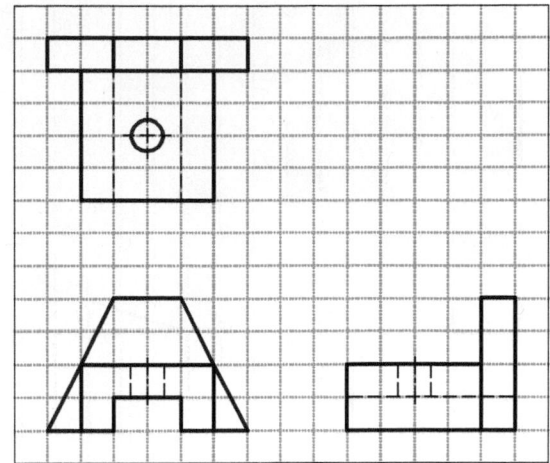

Figure P4-64

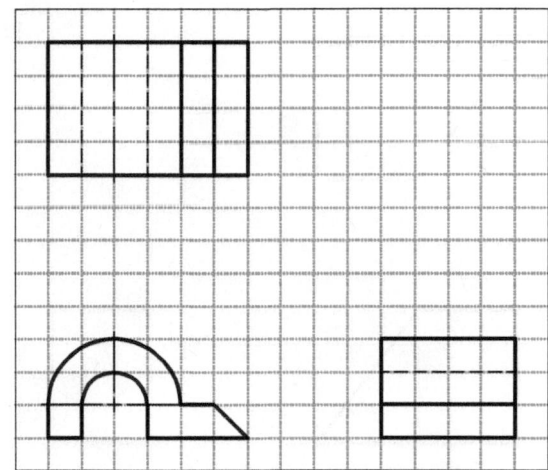

Figure P4-65

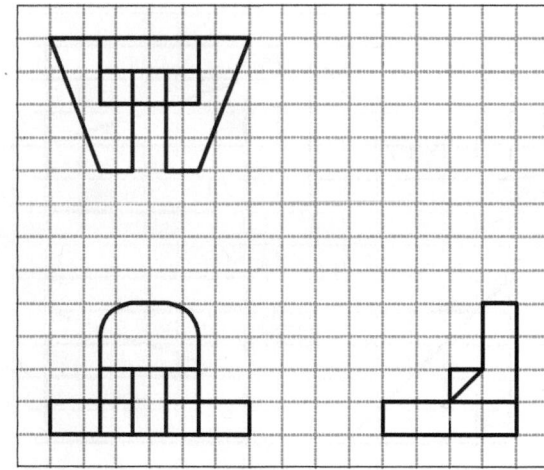

Figure P4-66

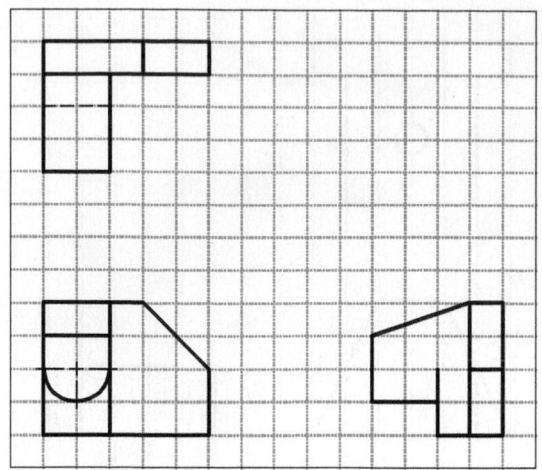

Figure P4-67

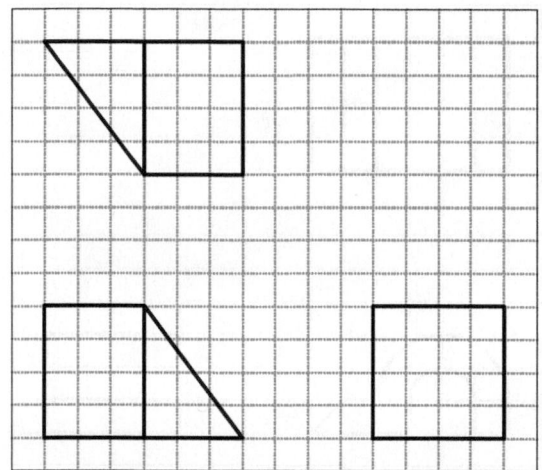

Figure P4-68

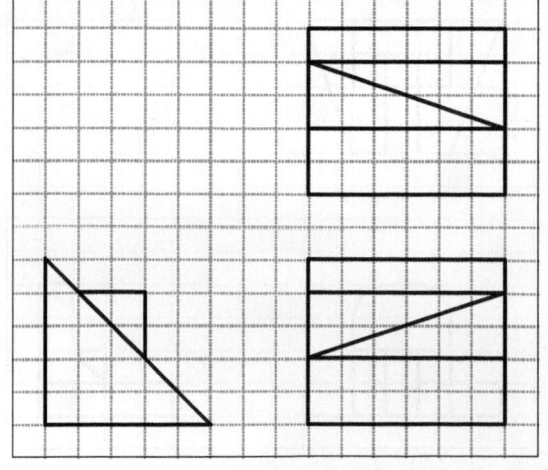

Figure P4-69

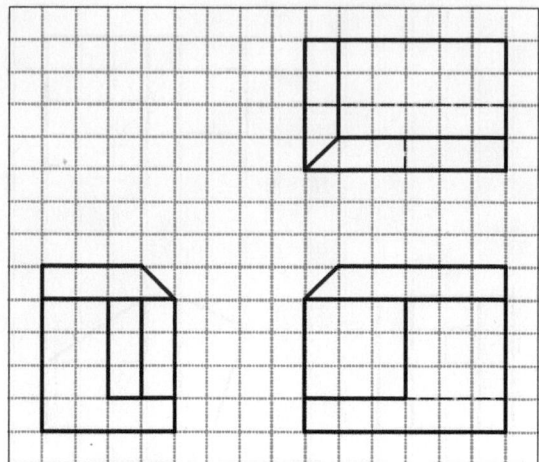

Figure P4-70

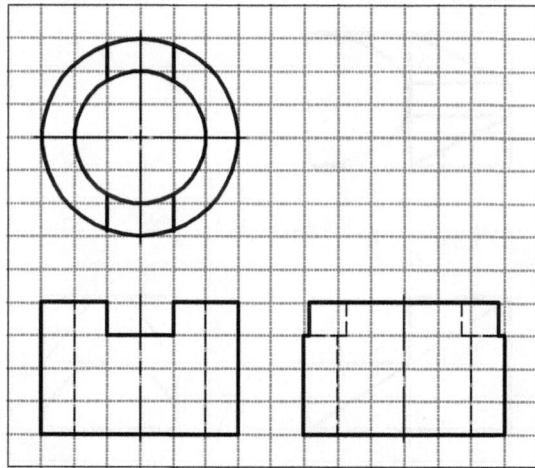

Figure P4-71

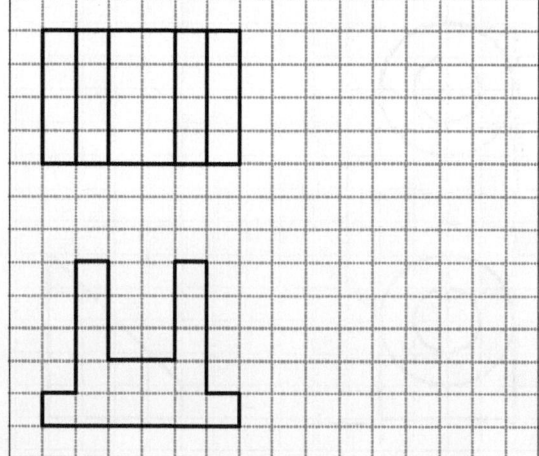

Figure P4-72

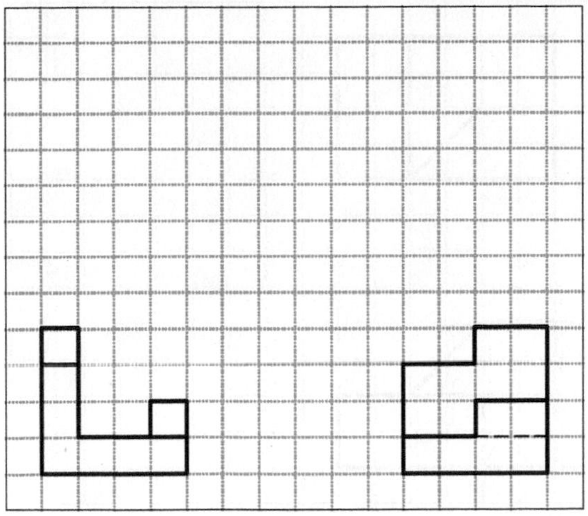

Figure P4-73

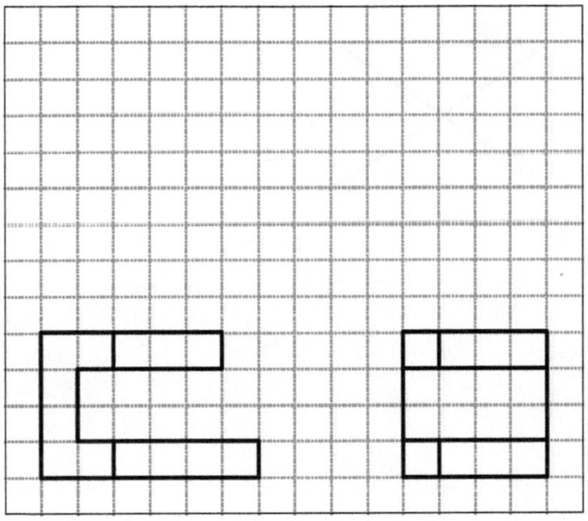

Figure P4-76

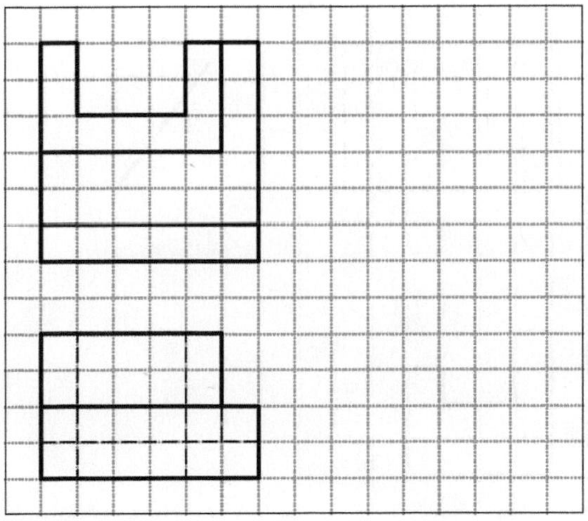

Figure P4-74

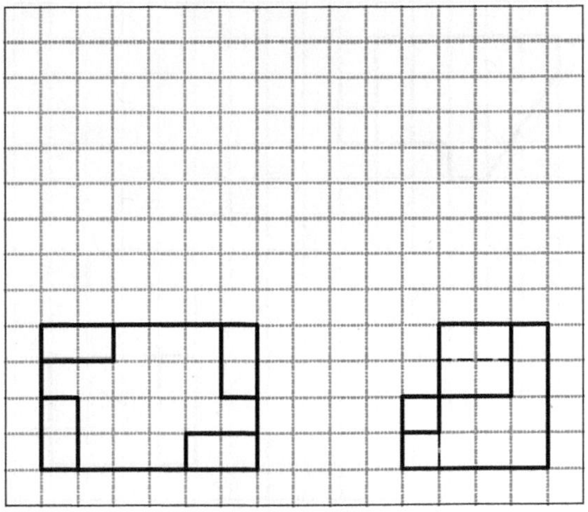

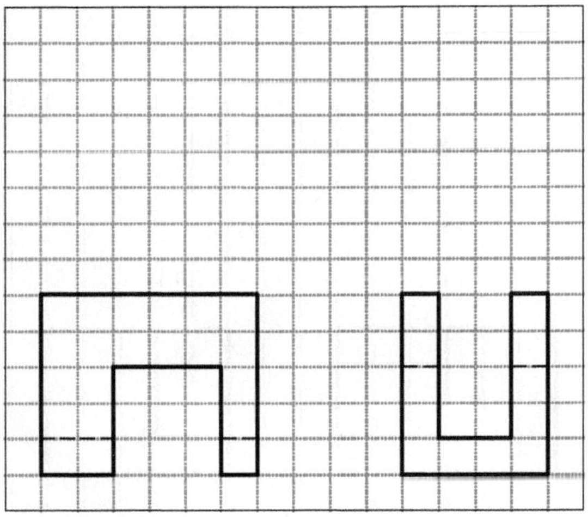

Figure P4-77

Figure P4-75

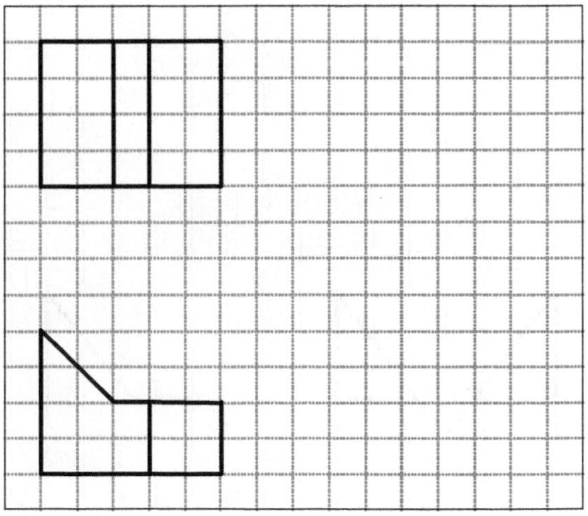

Figure P4-78

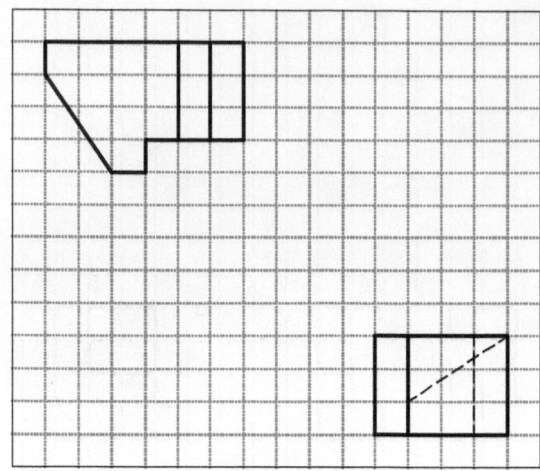

Figure P4-79

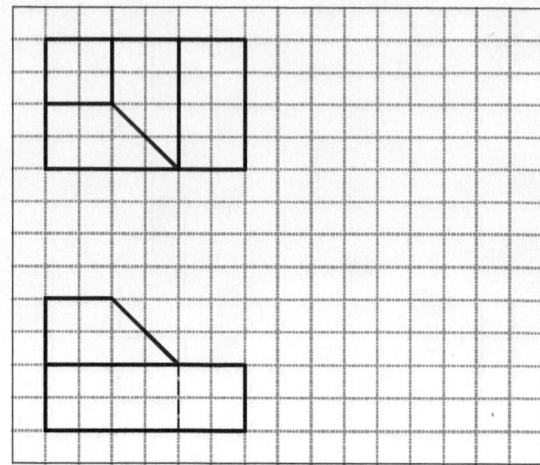

Figure P4-82

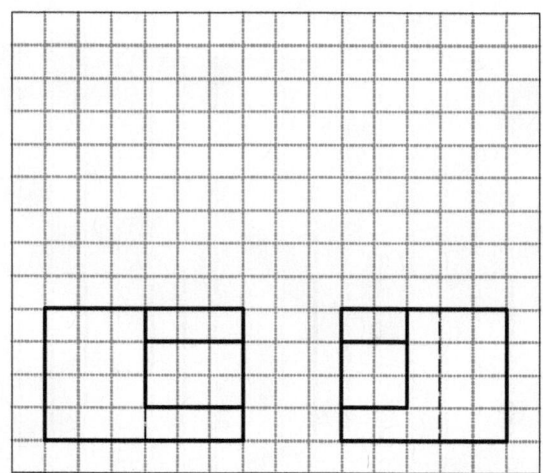

Figure P4-80

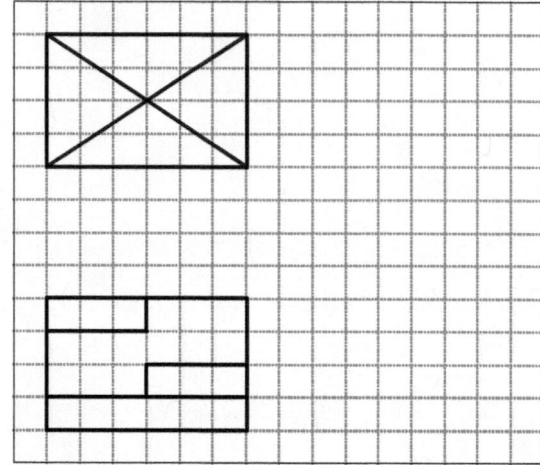

Figure P4-83

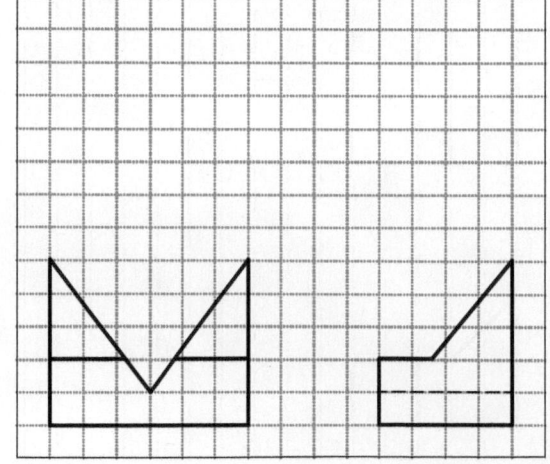

Figure P4-81

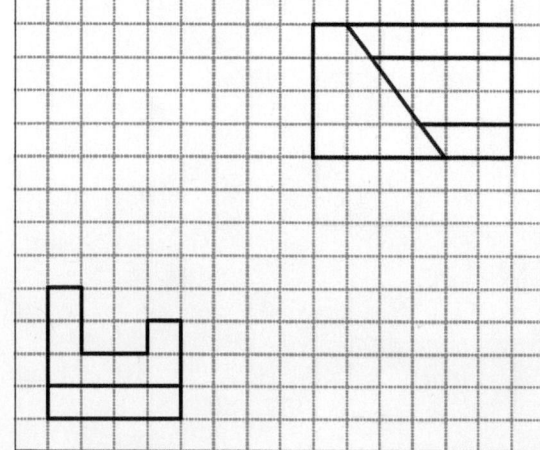

Figure P4-84

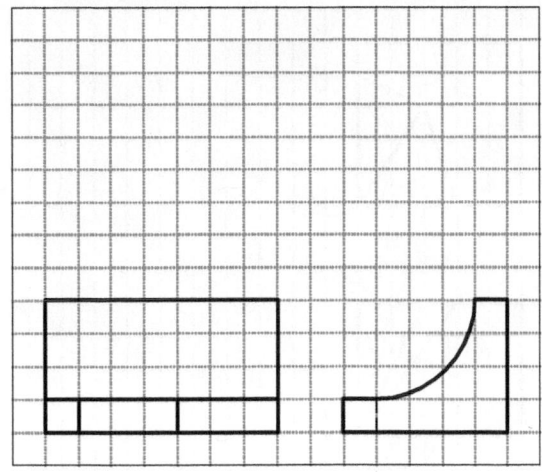

Figure P4-85

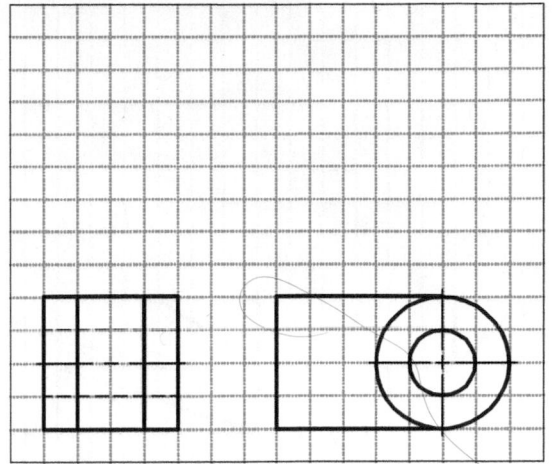

Figure P4-88

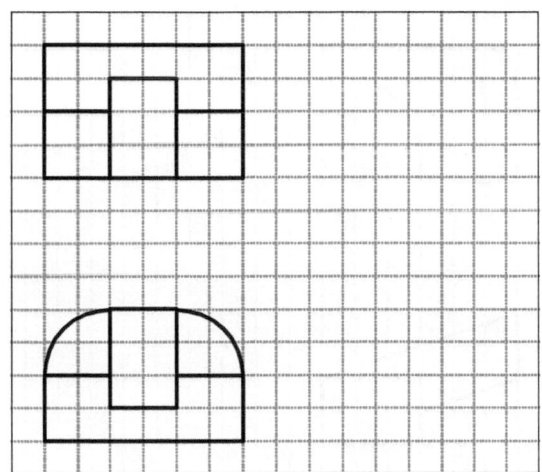

Figure P4-86

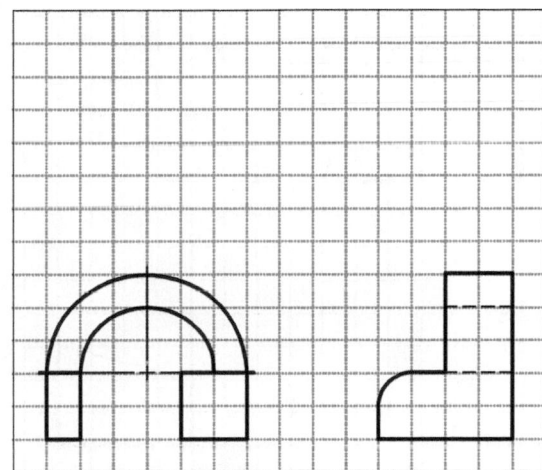

Figure P4-89

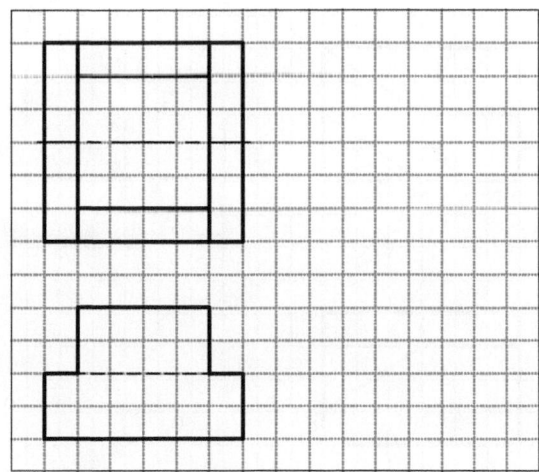

Figure P4-87

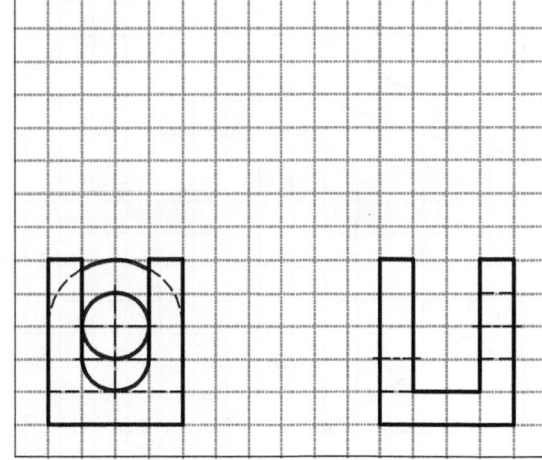

Figure P4-90

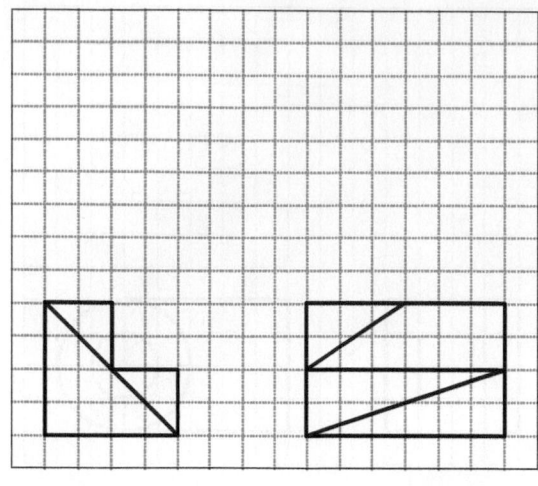

Figure P4-91

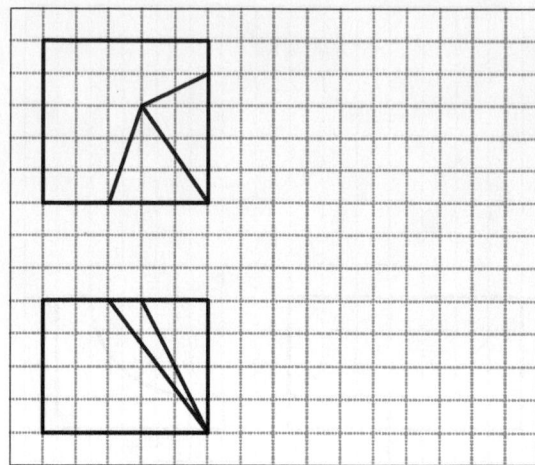

Figure P4-94

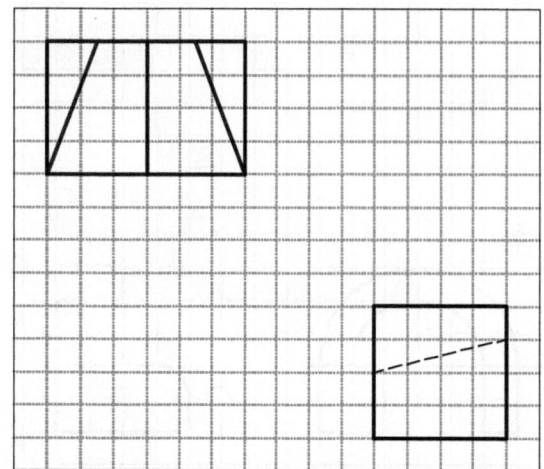

Figure P4-92

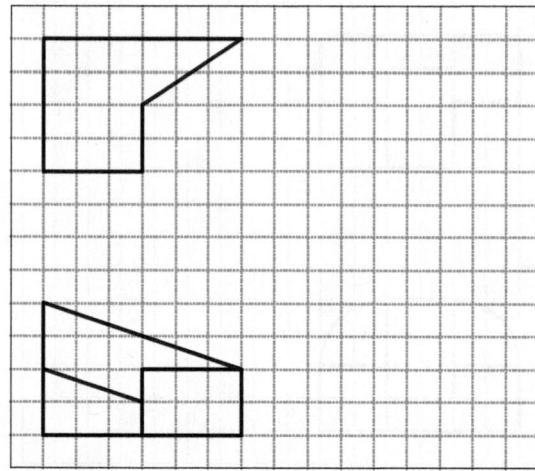

Figure P4-95

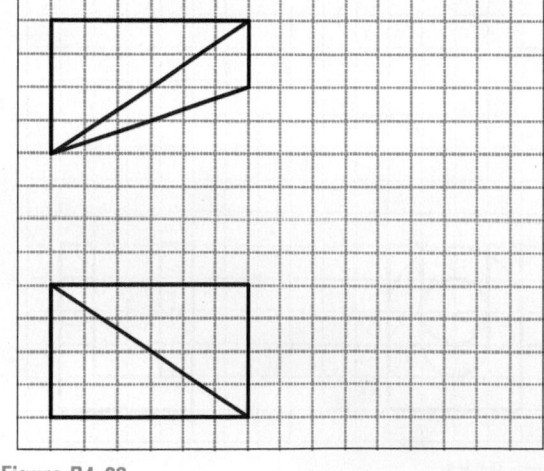

Figure P4-93

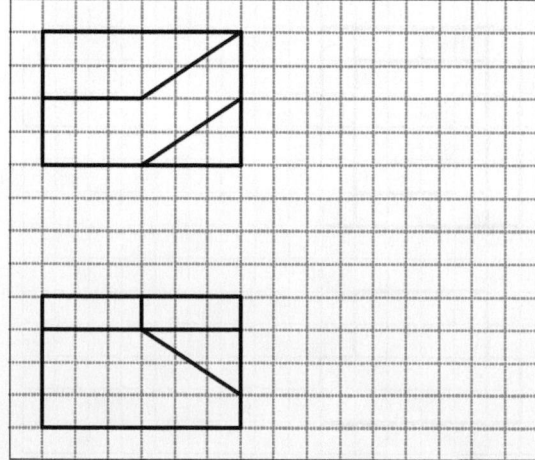

Figure P4-96

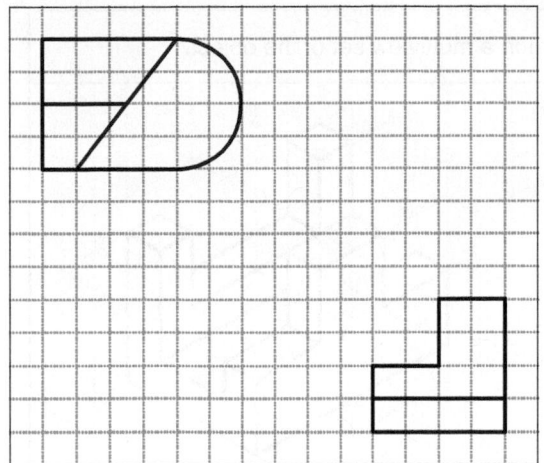

Figure P4-97

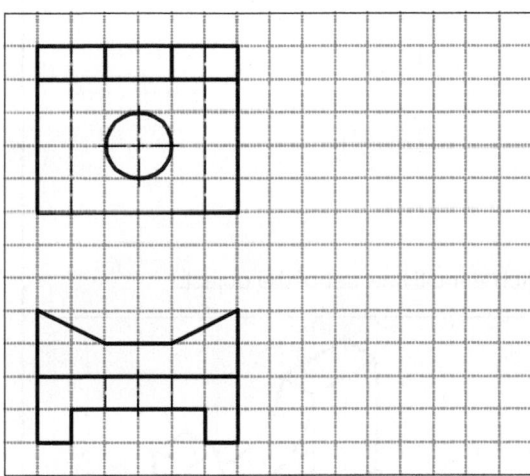

Figure P4-98

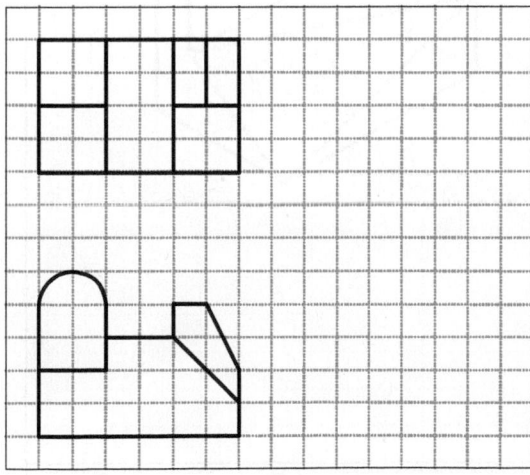

Figure P4-99

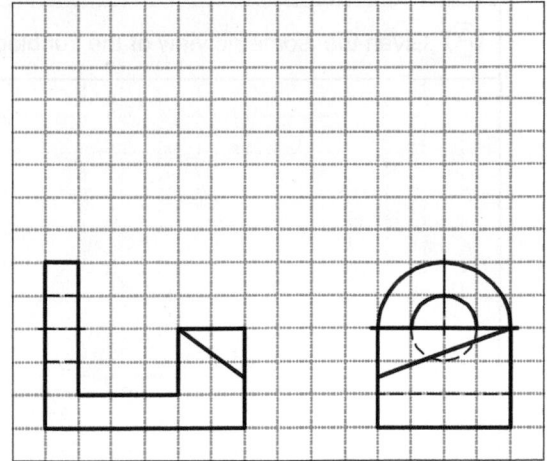

Figure P4-100

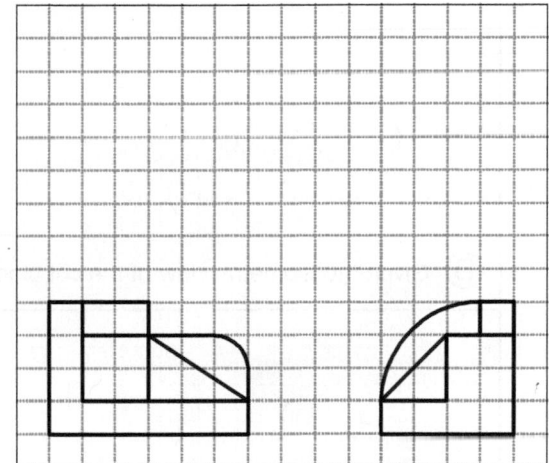

Figure P4-101

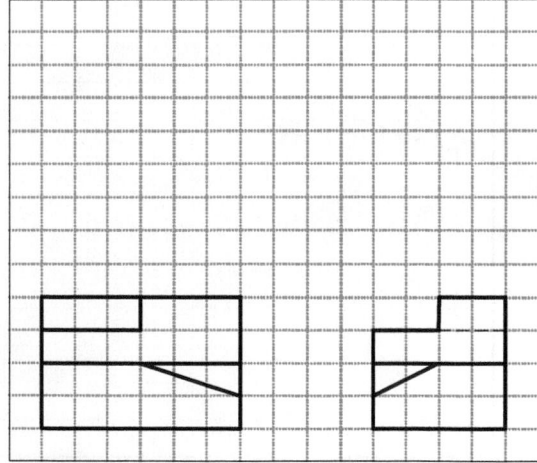

Figure P4-102

Ⓐ Given the isometric view of the cut block, sketch a multiview set of the object.

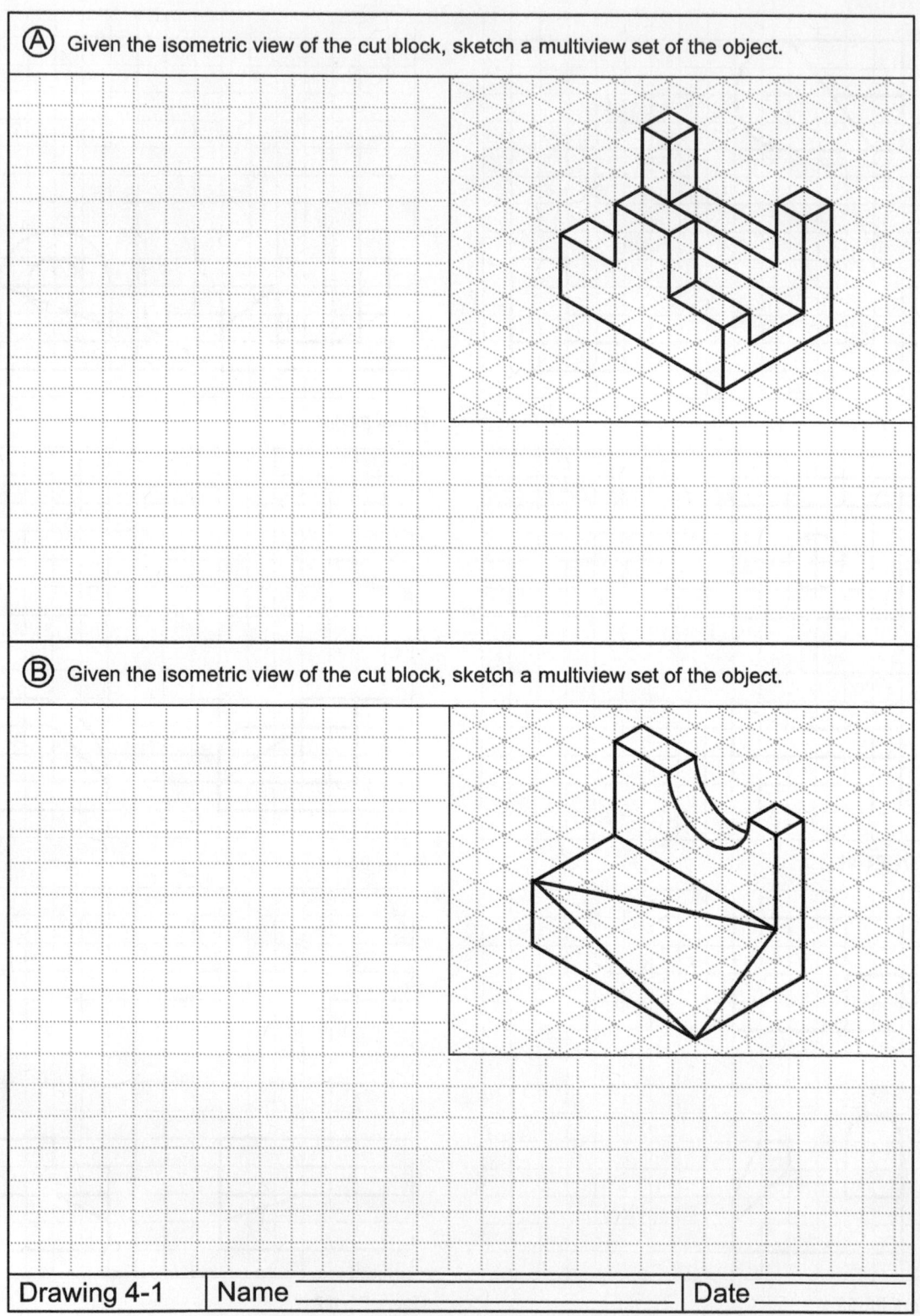

Ⓑ Given the isometric view of the cut block, sketch a multiview set of the object.

Drawing 4-1	Name	Date

Ⓐ Given the isometric view of the cut block, sketch a multiview set of the object.

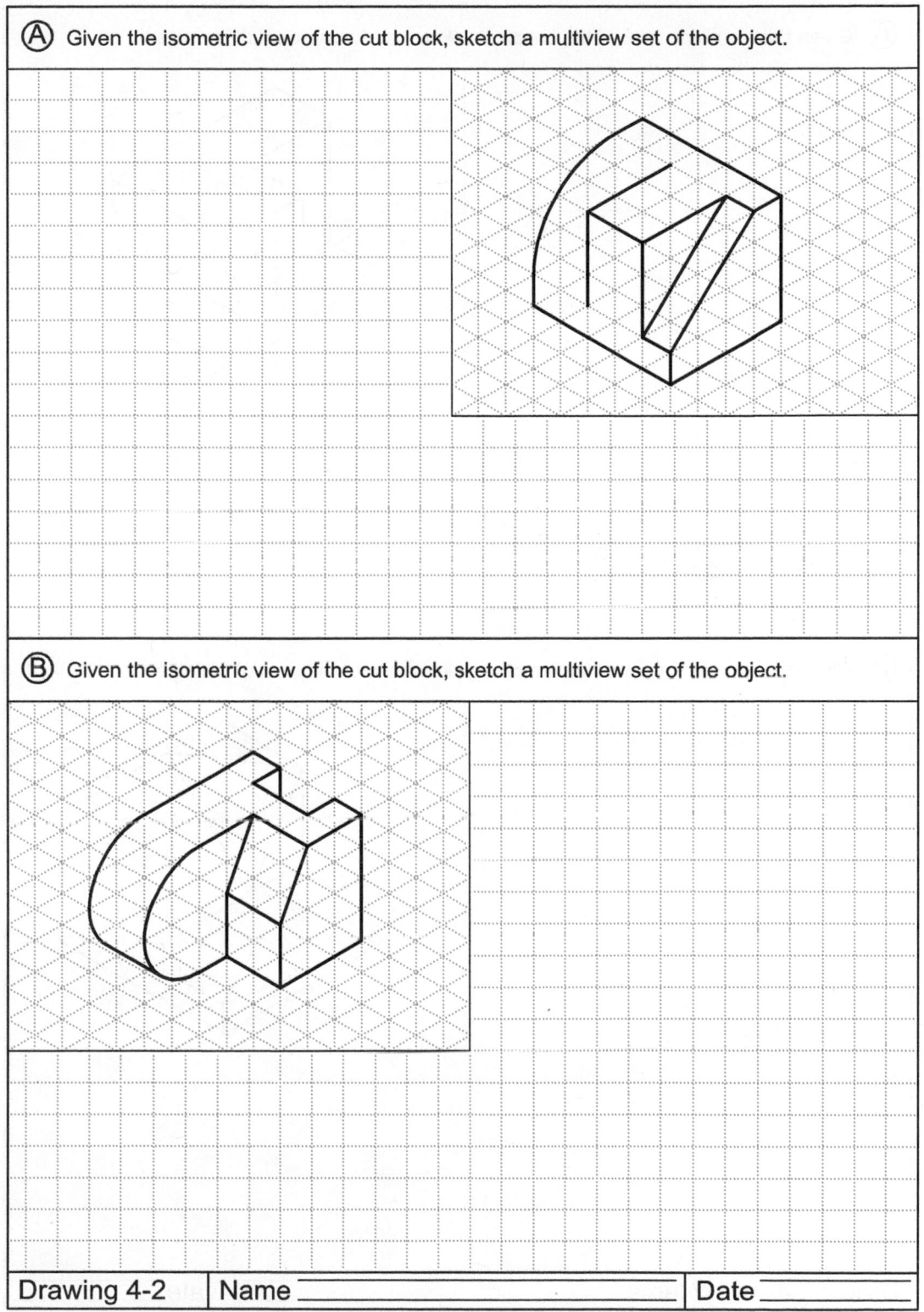

Ⓑ Given the isometric view of the cut block, sketch a multiview set of the object.

Drawing 4-2 | Name _____ | Date _____

Ⓐ Given the isometric view of the cut block, sketch a multiview set of the object.

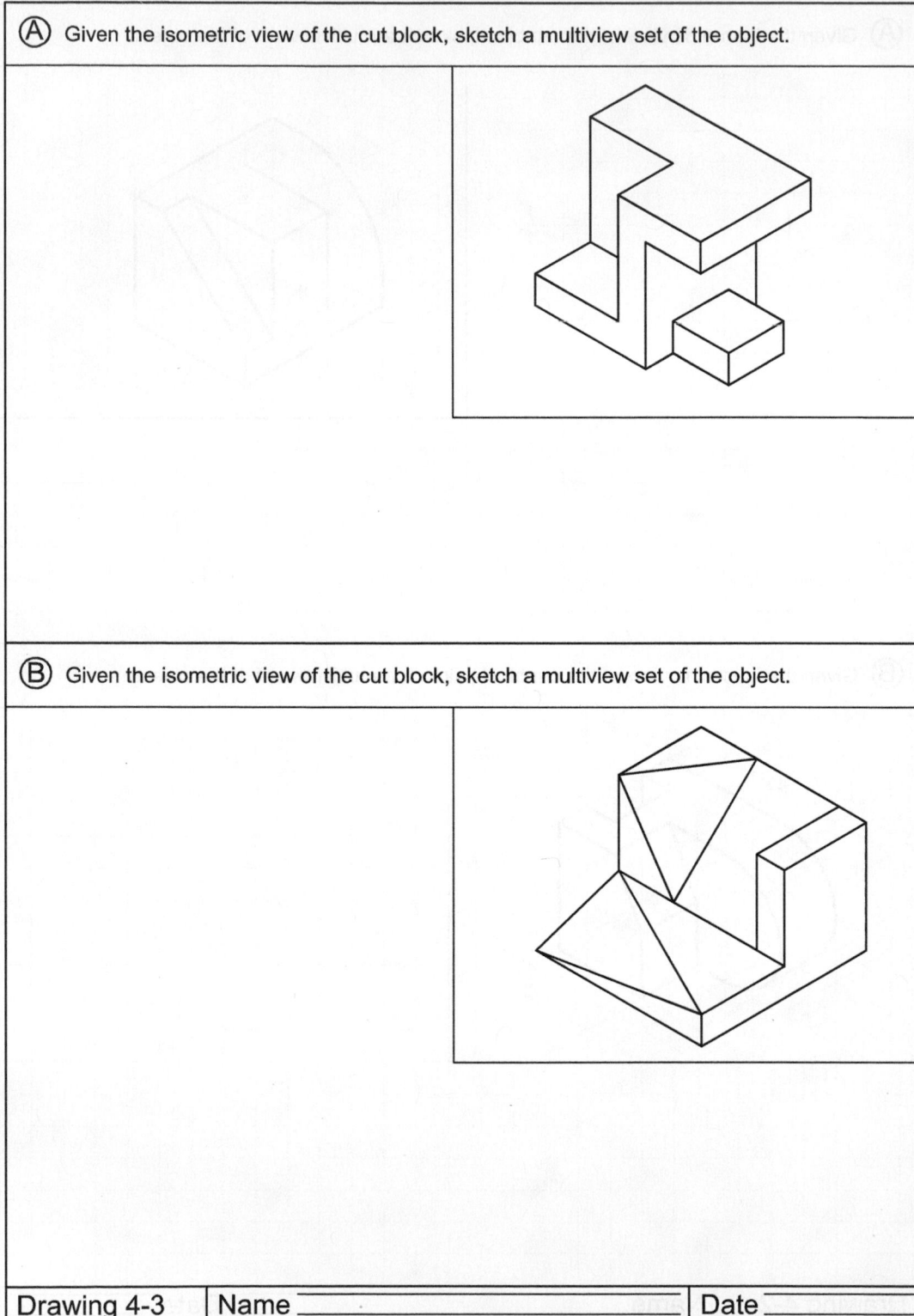

Ⓑ Given the isometric view of the cut block, sketch a multiview set of the object.

Drawing 4-3	Name	Date

Ⓐ Given the isometric view of the cut block, sketch a multiview set of the object.

Ⓑ Given the isometric view of the cut block, sketch a multiview set of the object.

Drawing 4-4 | Name _____ | Date _____

Ⓐ Given the isometric view of the cut block, sketch a multiview set of the object.

Ⓑ Given the isometric view of the cut block, sketch a multiview set of the object.

Drawing 4-5 | Name | Date

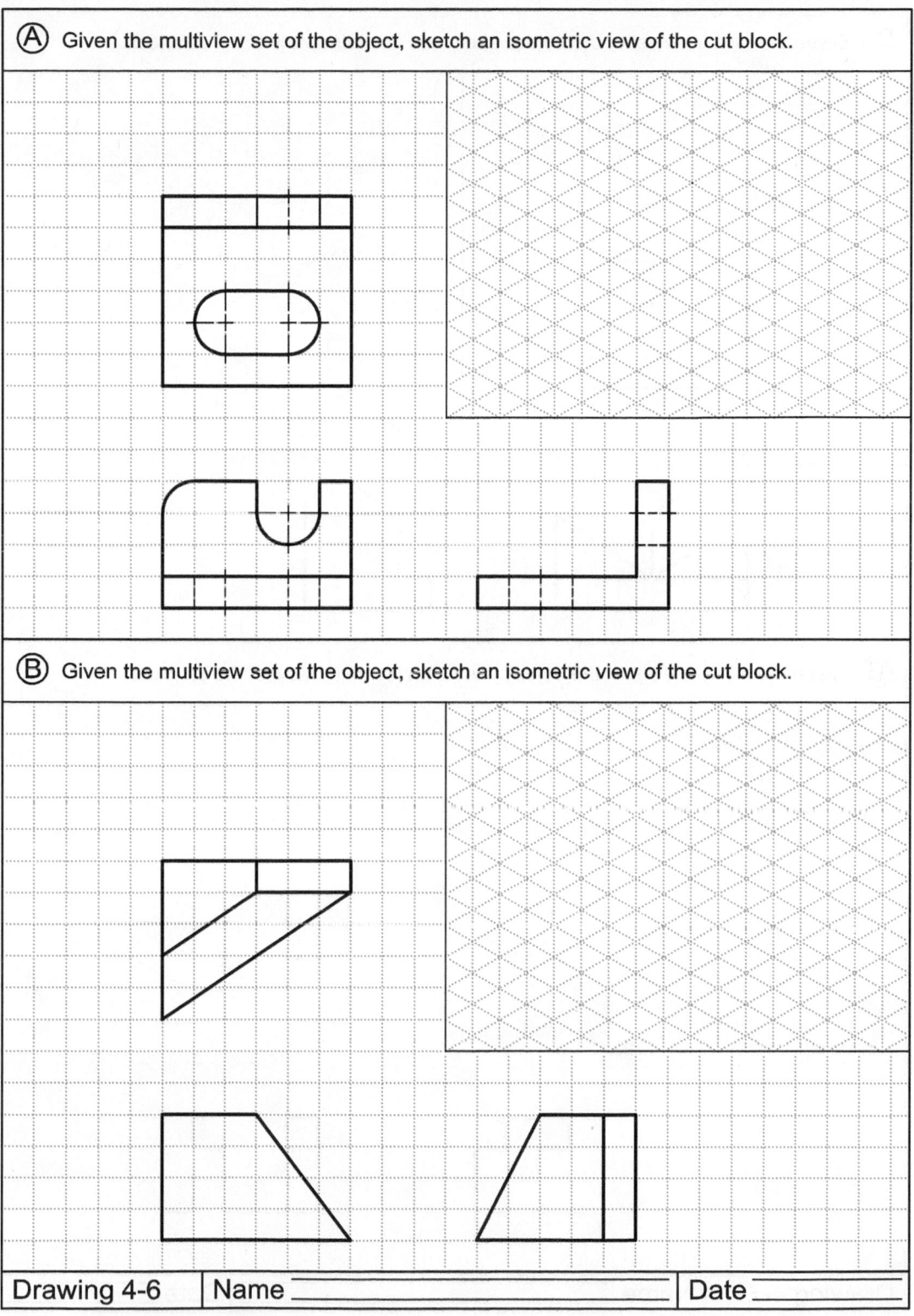

(A) Given the multiview set of the object, sketch an isometric view of the cut block.

(B) Given the multiview set of the object, sketch an isometric view of the cut block.

Drawing 4-6 | Name | Date

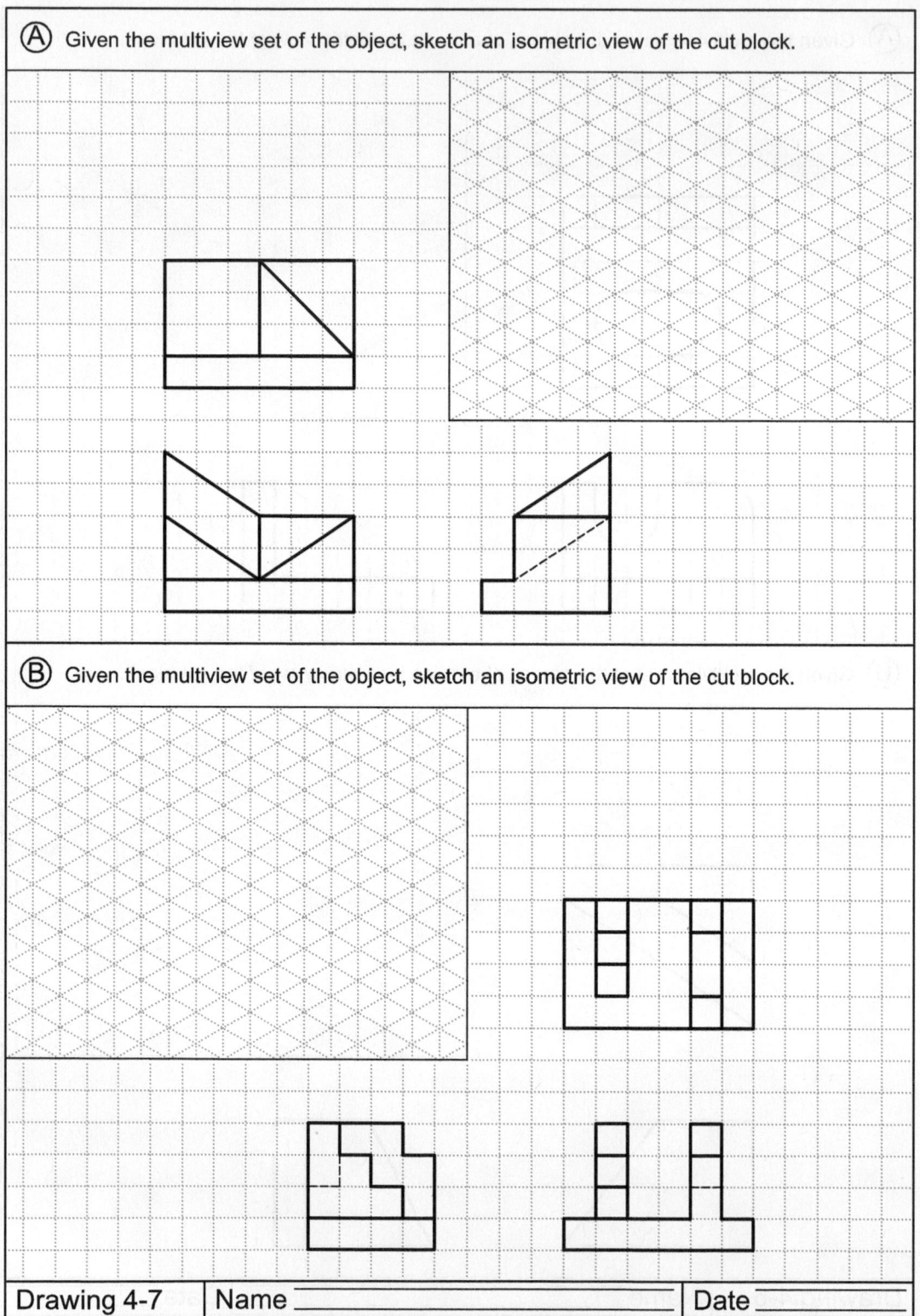

A Given the multiview set of the object, sketch an isometric view of the cut block.

B Given the multiview set of the object, sketch an isometric view of the cut block.

Drawing 4-7 | Name _____ | Date _____

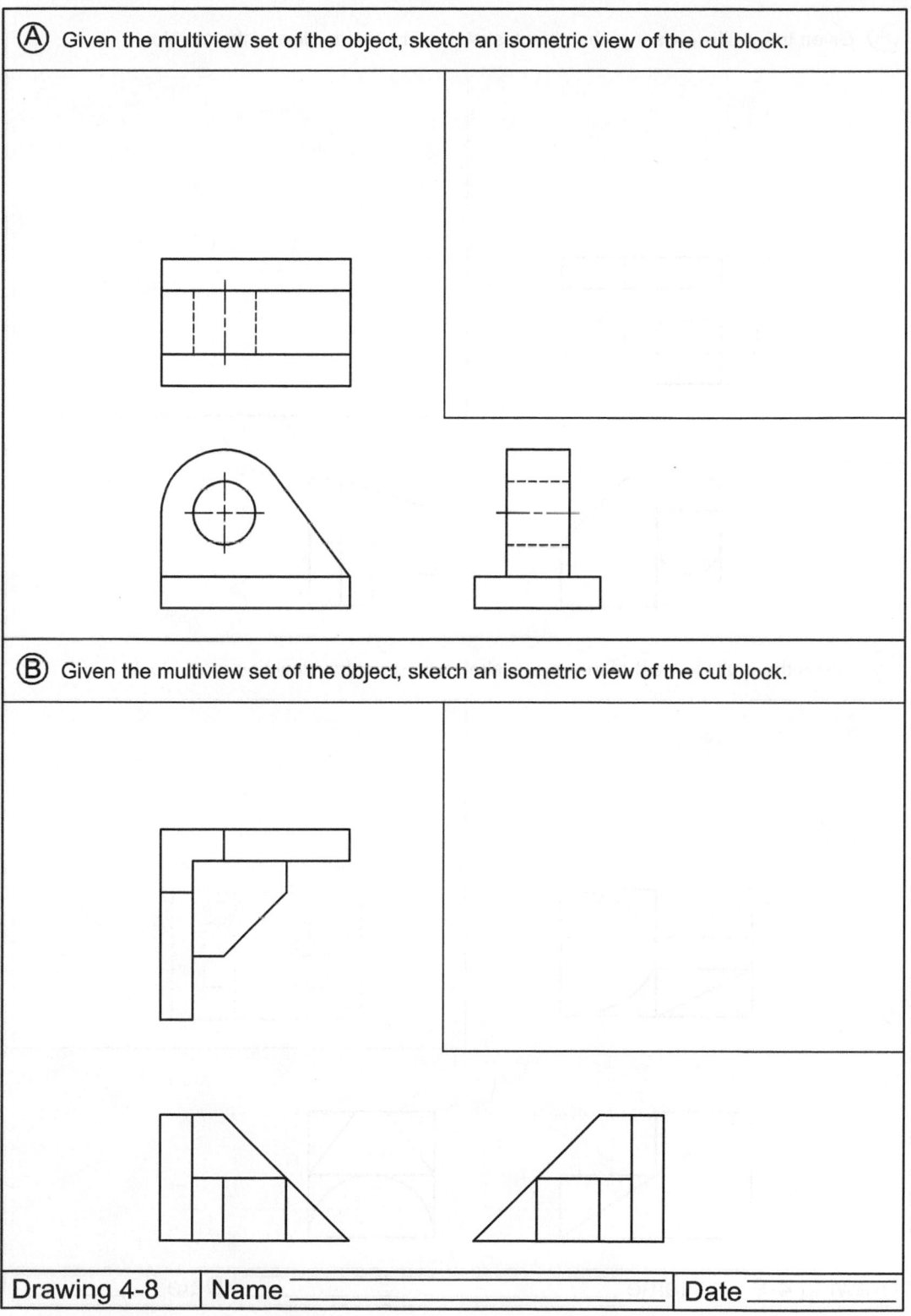

Ⓐ Given the multiview set of the object, sketch an isometric view of the cut block.

Ⓑ Given the multiview set of the object, sketch an isometric view of the cut block.

Drawing 4-8 | Name | Date

Ⓐ Given the multiview set of the object, sketch an isometric view of the cut block.

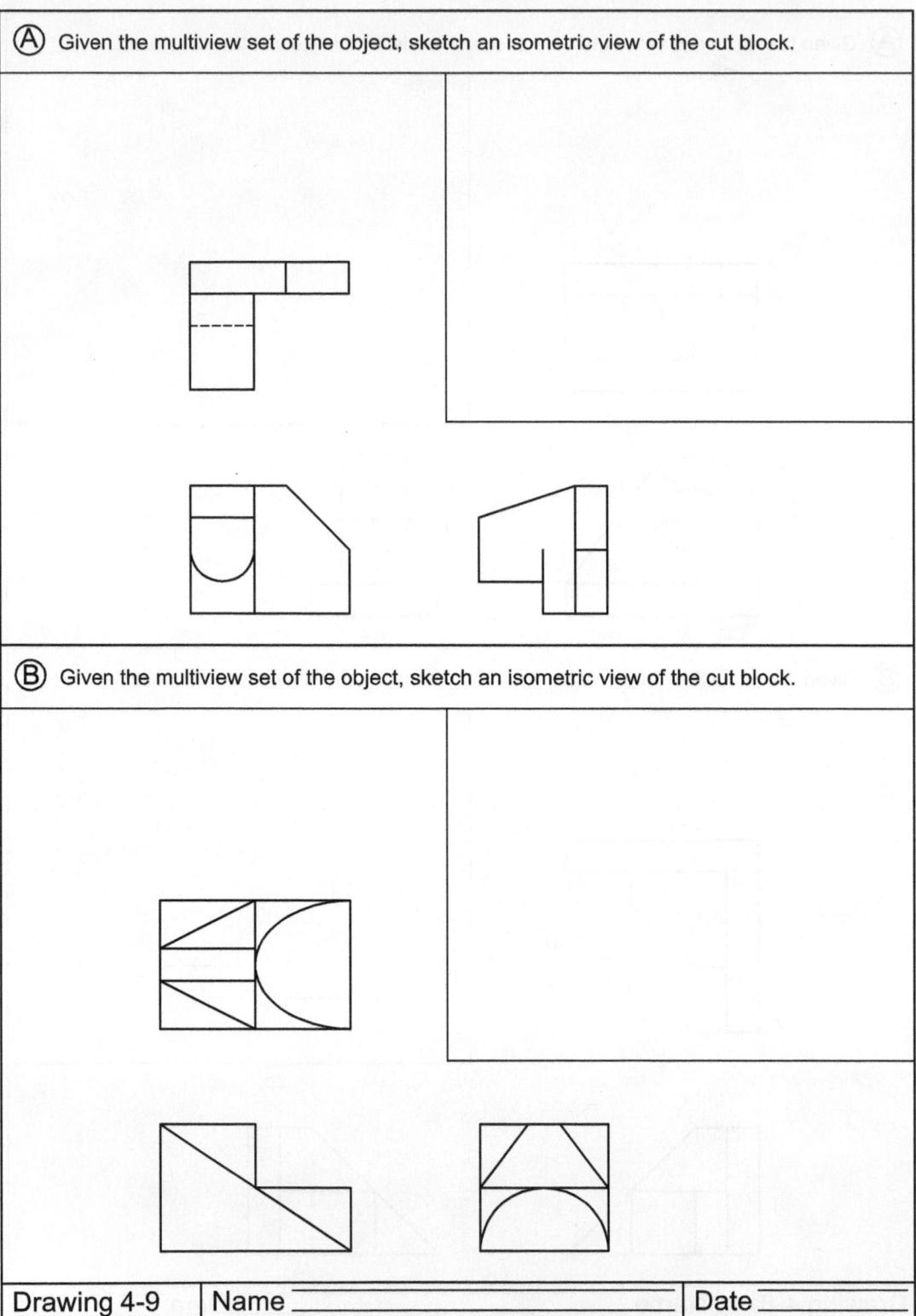

Ⓑ Given the multiview set of the object, sketch an isometric view of the cut block.

| Drawing 4-9 | Name | Date |

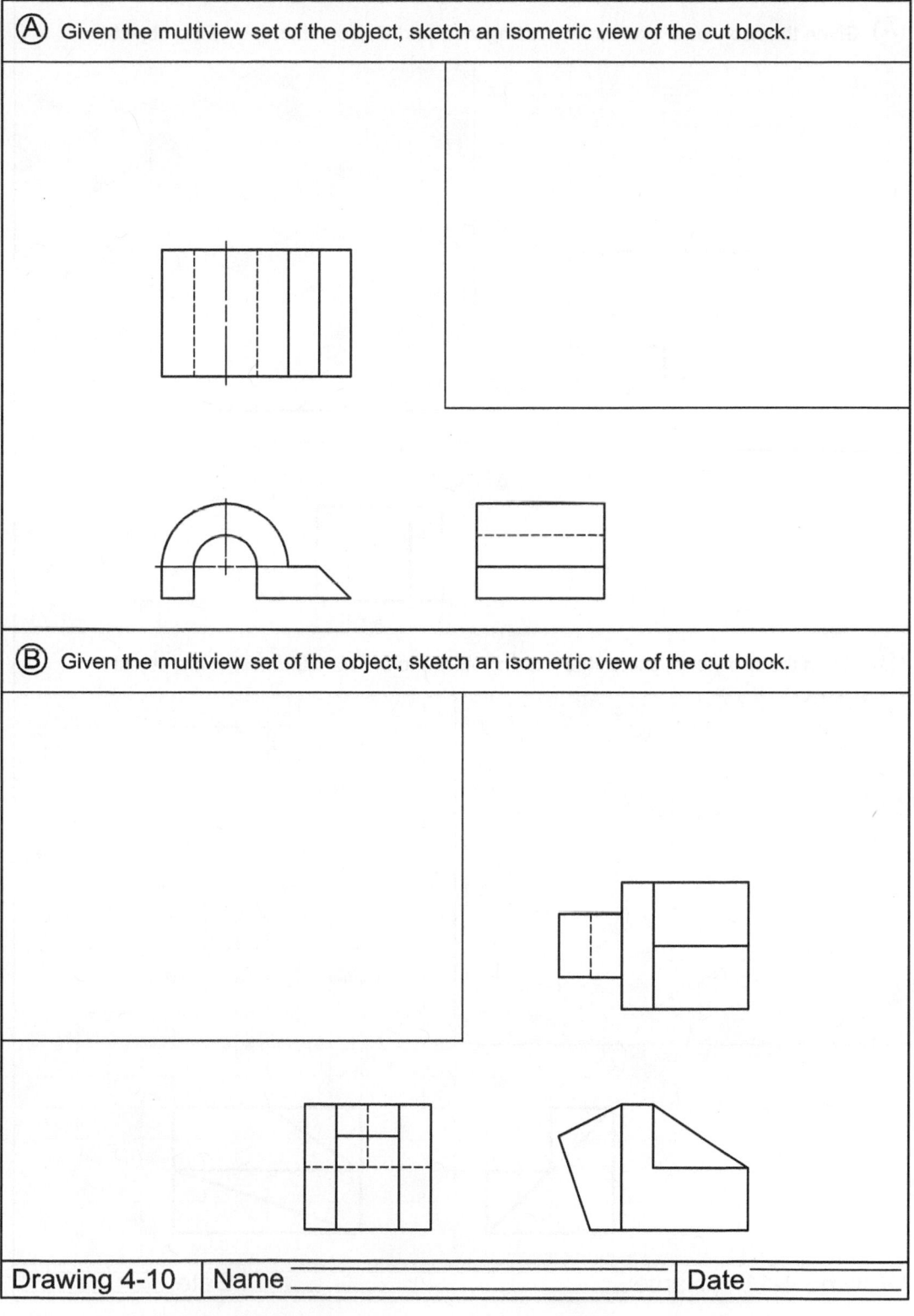

Ⓐ Given the multiview set of the object, sketch an isometric view of the cut block.

Ⓑ Given the multiview set of the object, sketch an isometric view of the cut block.

Drawing 4-10 | Name _____ | Date _____

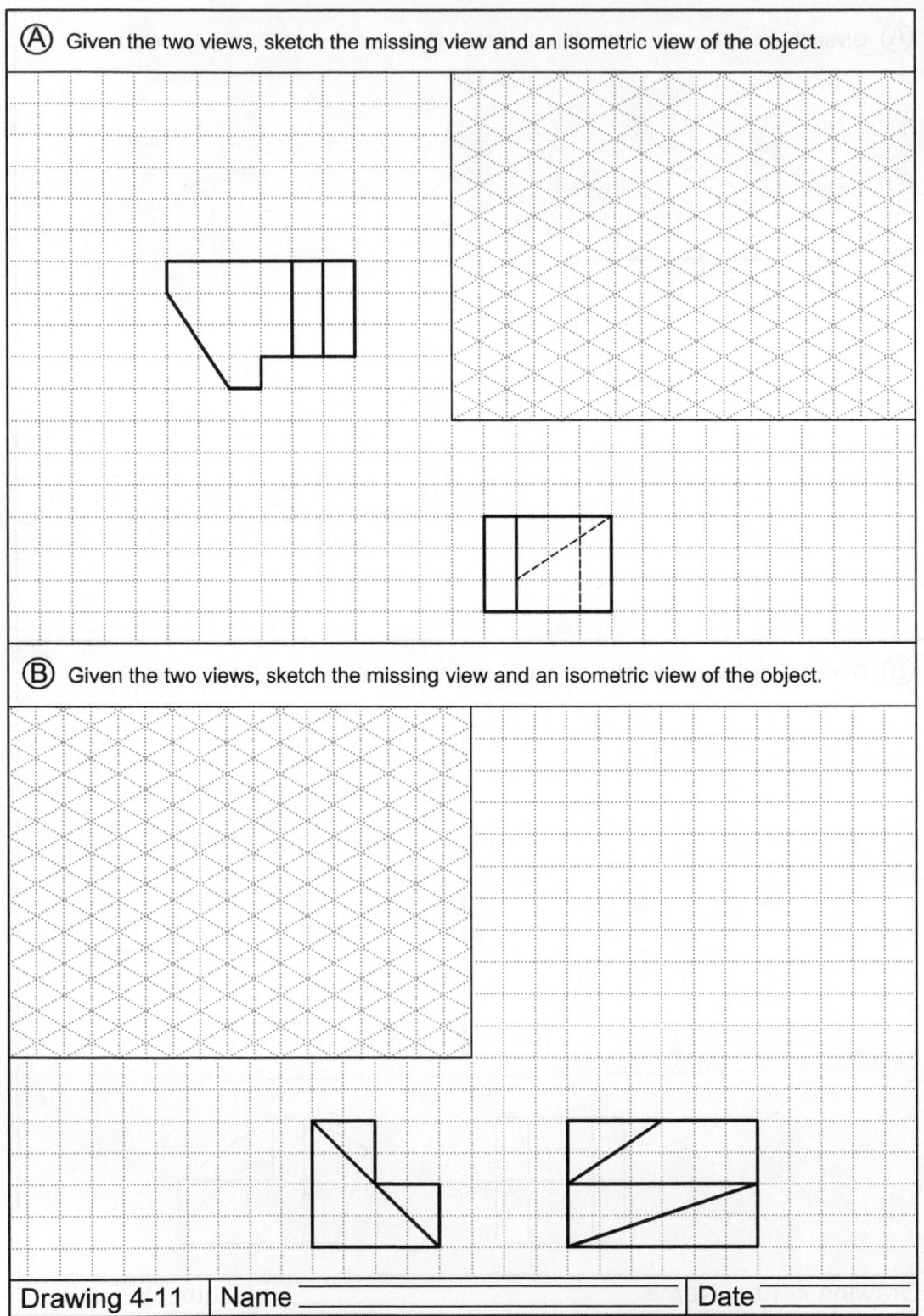

Ⓐ Given the two views, sketch the missing view and an isometric view of the object.

Ⓑ Given the two views, sketch the missing view and an isometric view of the object.

Drawing 4-11 | Name _____ | Date _____

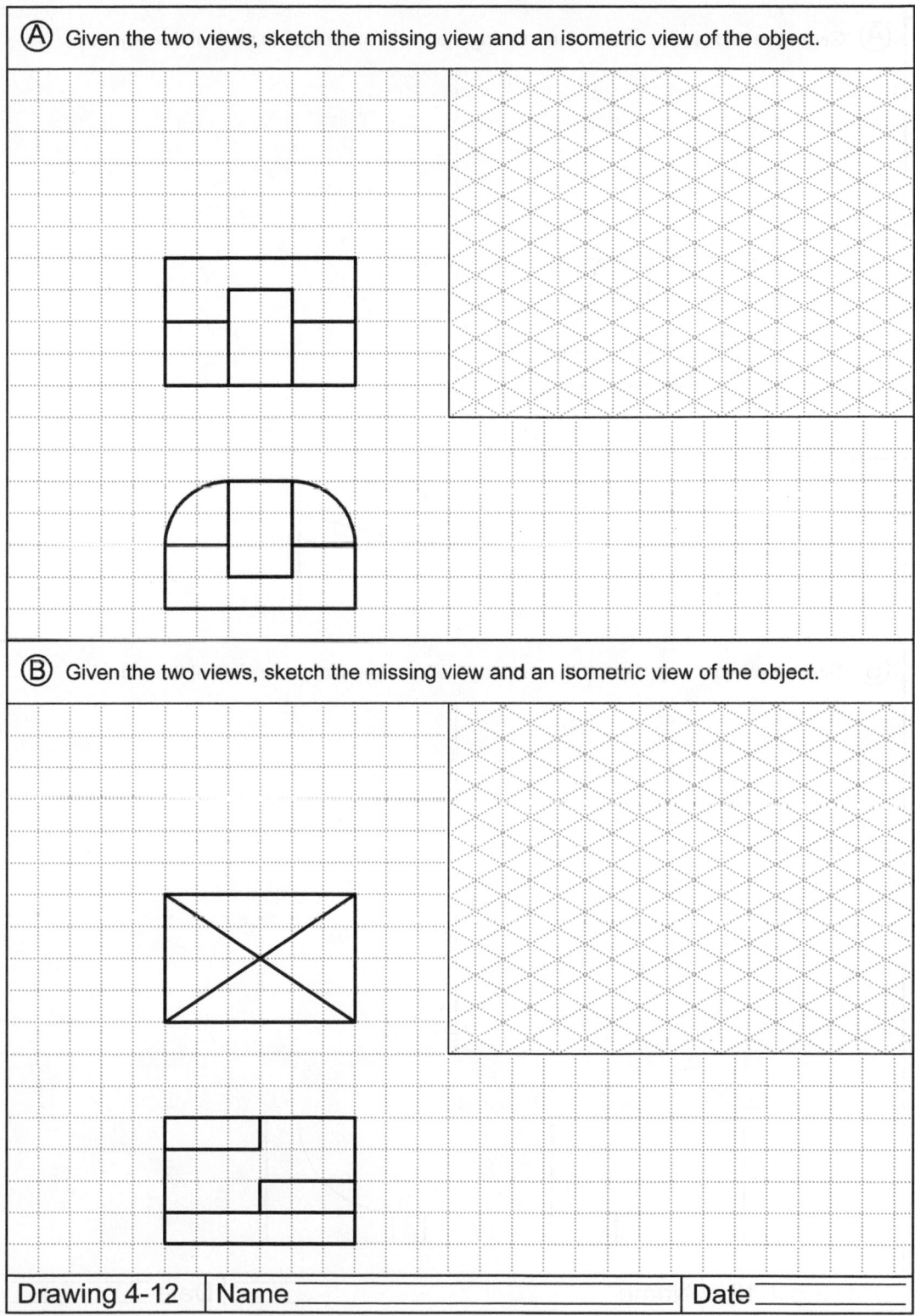

Ⓐ Given the two views, sketch the missing view and an isometric view of the object.

Ⓑ Given the two views, sketch the missing view and an isometric view of the object.

Drawing 4-12 | Name _____ | Date _____

Ⓐ Given the two views, sketch the missing view and an isometric view of the object.

Ⓑ Given the two views, sketch the missing view and an isometric view of the object.

| Drawing 4-13 | Name | Date |

Ⓐ Given the two views, sketch the missing view and an isometric view of the object.

Ⓑ Given the two views, sketch the missing view and an isometric view of the object.

Drawing 4-14 | Name _____ | Date _____

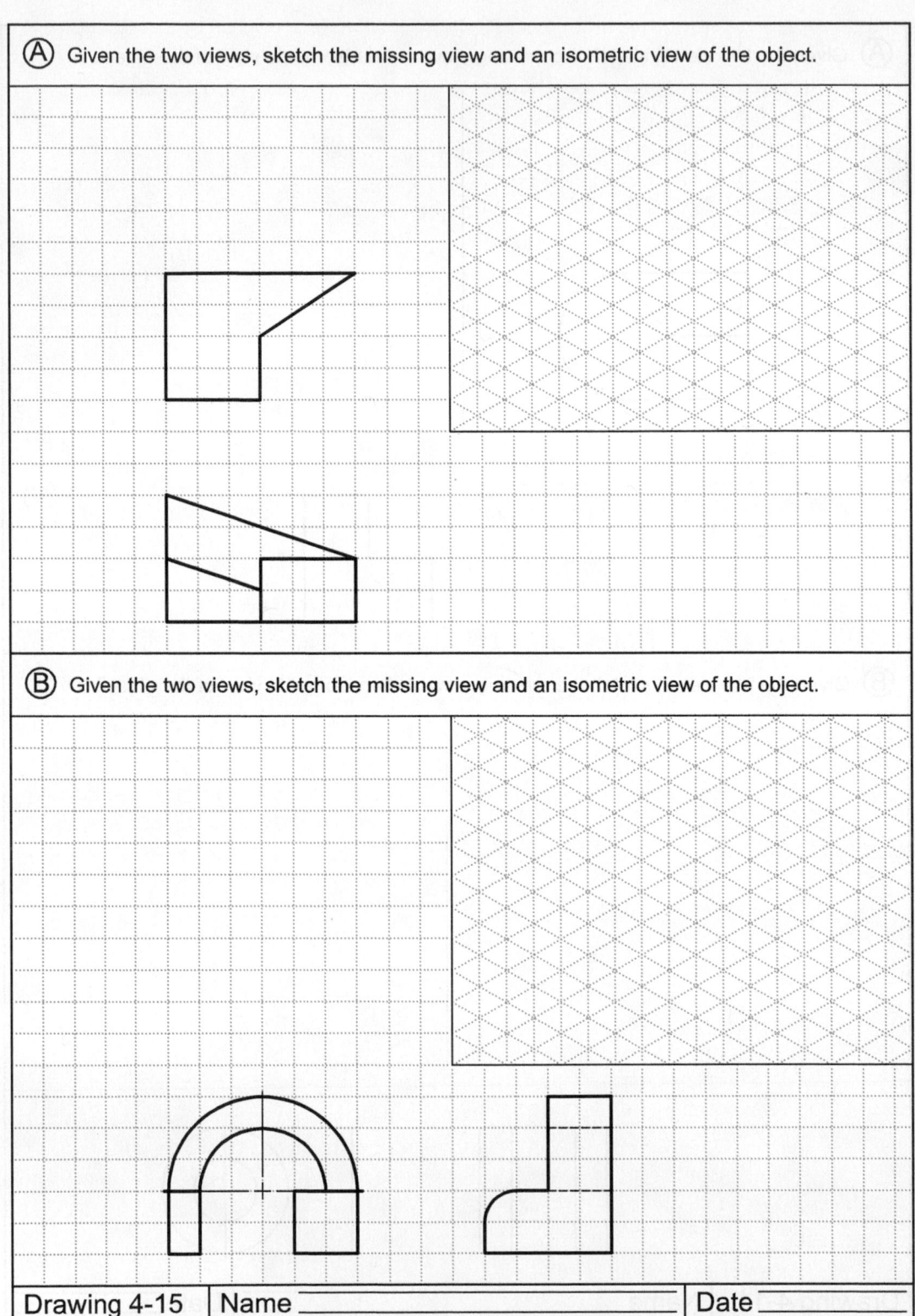

Ⓐ Given the two views, sketch the missing view and an isometric view of the object.

Ⓑ Given the two views, sketch the missing view and an isometric view of the object.

| Drawing 4-15 | Name _____ | Date _____ |

CHAPTER

5

AUXILIARY AND SECTION VIEWS

■ AUXILIARY VIEWS

Introduction

Recall that in a multiview drawing of an object with an inclined surface, in one multiview the inclined surface is seen on edge, whereas in the other two multiviews the inclined surface appears as a foreshortened (i.e., not true-size) surface. In Figure 5-1, for example, the inclined surface labeled A is seen on edge in the top view, and as a foreshortened surface in the front and right views. In some circumstances though, a view showing the true size of an inclined surface is useful.

From **descriptive geometry**[1] it is known that the true size and shape of a planar face (or the true length of a line) can be represented in an orthographic projection only if the line of sight is normal to the planar face, or, equivalently, if the projection plane is parallel to the face. This knowledge will be put to use to find the true size of an inclined surface.

Definitions

In earlier chapters we saw that multiview projection is an orthographic projection technique wherein a three-dimensional object is projected onto one of three mutually perpendicular planes.

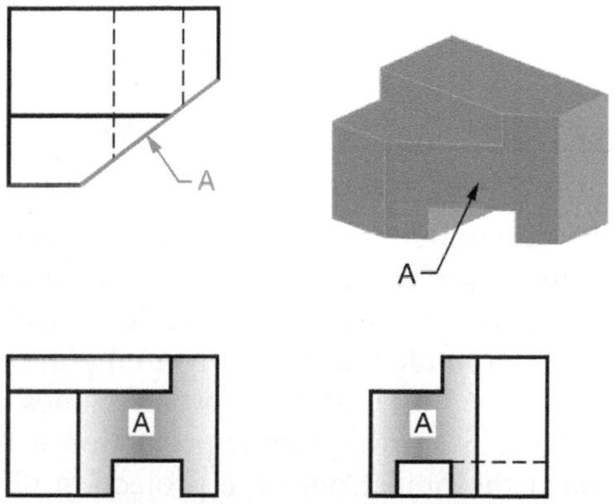

Figure 5-1 Views of cut block with an inclined surface

These are the principal planes: horizontal, frontal, and profile. An **auxiliary view** is an orthographic view that is projected onto any plane other than one of the principal planes. A **primary auxiliary view** is an auxiliary view that is projected onto a plane perpendicular to one of the principal planes and inclined to the other two. A primary auxiliary view can be used to find the true size and shape of an inclined surface. A **secondary auxiliary view** is projected from a primary auxiliary view onto a plane that is inclined to all three principal projection planes. A secondary auxiliary view can be used to find the true size and shape of an oblique surface.

Auxiliary View Projection Theory

Figure 5-2 on page 138 shows an object with an inclined surface that has been placed inside a glass box. Note that the glass box has been modified

---

[1] The term *descriptive geometry* refers to the body of knowledge, developed over the centuries, consisting of mathematical-graphical procedures used to accurately describe 3D geometry within 2D media. The French mathematician Gaspard Monge (1746–1818) is considered the father of descriptive geometry.

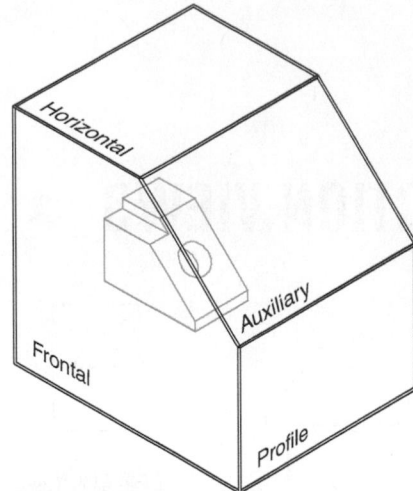

Figure 5-2 Glass box modified for auxiliary view

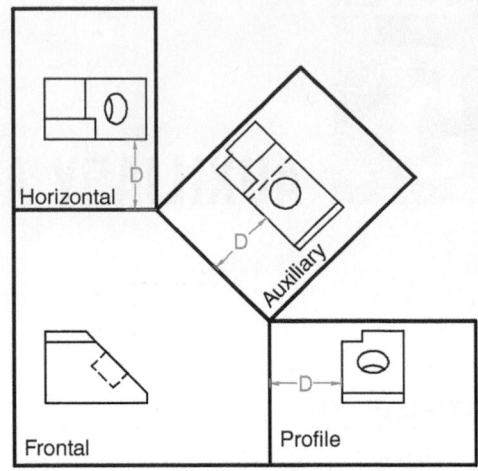

Figure 5-4 Modified glass box sides open and laid flat

by adding an additional plane that is parallel to the inclined surface. For the situation shown in Figure 5-2, the inclined plane (i.e., the auxiliary plane) is perpendicular to the frontal plane and inclined to the horizontal and profile planes.

In Figure 5-3 orthographic projection is used to project the object onto the projection planes (i.e., the sides of the glass box), including the auxiliary plane. Because the auxiliary plane is parallel to the inclined surface, the resulting projection shows the true size and shape of the inclined surface, as well as foreshortened projections of any other visible surfaces. Also notice that the edge

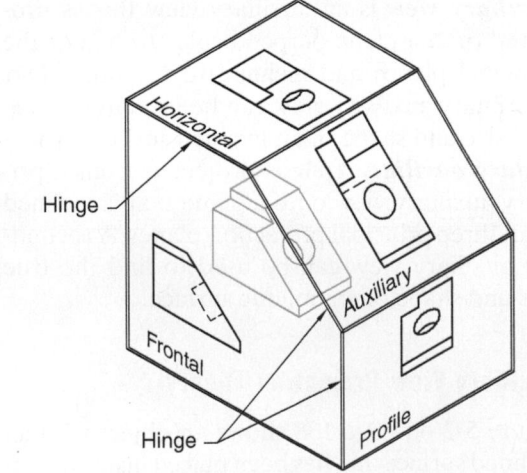

Figure 5-3 Views projected onto sides of glass box

view of the inclined surface appears on the frontal projection plane.

Now imagine that the frontal plane is hinged to the horizontal, profile, and auxiliary planes. If the hinged views are then unfolded so that they lie in the same plane as the frontal view, Figure 5-4 results. Note that the distance D from the hinge line to the near side of the inclined surface is the same for all three projected views (horizontal, profile, and auxiliary). This fact will be used later in the construction of an auxiliary sketch of an inclined surface.

As shown in Figure 5-5, an auxiliary view is in alignment with the principal view that shows the inclined surface on edge. Because these inclined edges are projected true length, the perpendicular distances between the dashed lines shown in Figure 5-5 represent the actual inclined edge lengths on the inclined surface.

Note that the auxiliary view in Figure 5-5 shows only the inclined surface. This is called a ***partial auxiliary view***. Because they are both easier to execute and easier to visualize, partial auxiliary views are frequently employed. In creating an auxiliary projection with a CAD system, however, a full auxiliary view is obtained automatically. The resulting view can always be modified by either hiding or deleting lines in order to obtain a view that shows only the inclined surface of interest.

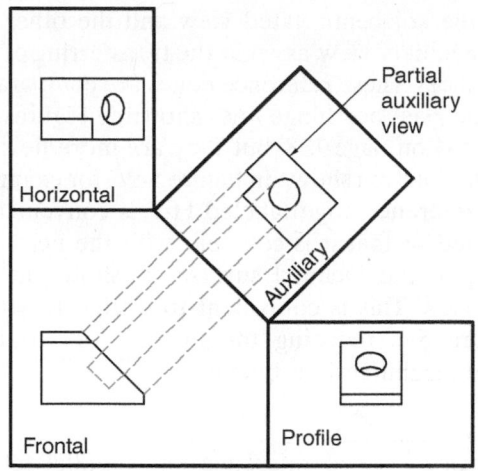

Figure 5-5 Auxiliary view alignment

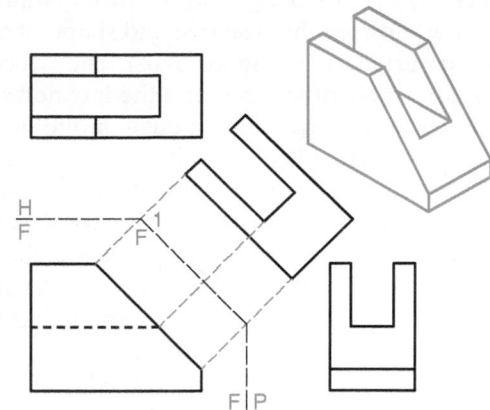

Figure 5-7 Auxiliary view projected from frontal plane

Auxiliary Views: Three Cases

The primary auxiliary view is projected from the principal plane containing the edge view of the inclined surface. If an object with an inclined surface is oriented as shown in Figure 5-6, then the edge view of the inclined surface appears on the horizontal projection plane. In this case, the auxiliary view is projected from the horizontal plane. Note how the hinge lines between the horizontal and frontal (H-F) and the horizontal and auxiliary (H-1) planes are represented.

With the same object oriented as shown in Figure 5-7, the inclined surface appears on edge in the frontal plane. In this case the primary auxiliary view is projected from the frontal plane.

In the third case, shown in Figure 5-8, the edge view of the inclined surface appears in the profile plane. The primary auxiliary view is consequently projected from the profile plane.

General Sketching Procedure for Finding a Primary Auxiliary View

In this section the sketching procedure for obtaining a primary auxiliary view is outlined. The problem can be stated as follows: Given a

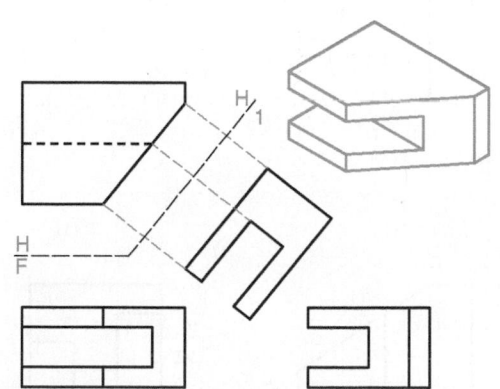

Figure 5-6 Auxiliary view projected from horizontal plane

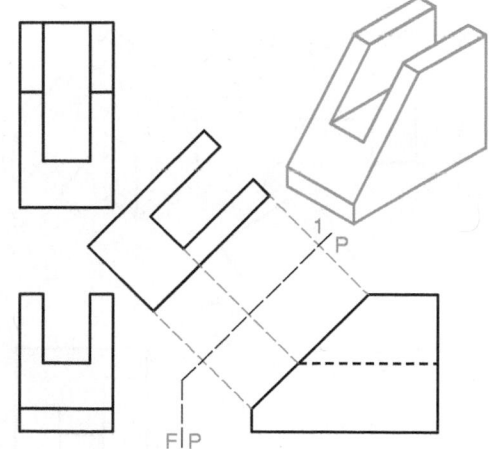

Figure 5-8 Auxiliary view projected from profile plane

multiview drawing of an object with an inclined surface (see Figure 5-9a), find the primary auxiliary view showing the true size and shape of this inclined surface (see Figure 5-9b). The process takes advantage of the fact that the inclined surface is shown as a true-length edge in one of the multiviews. Perpendicular projectors are erected from the edge view to obtain these distances. All that is required to complete a partial auxiliary view are the edge lengths perpendicular to the inclined edge. These distances are available in the views adjacent to the view containing the inclined surface on edge. A trammel can be used to capture and then transfer these edge lengths to the auxiliary view. Two reference edges, one

for the adjacent/related view and the other for the auxiliary view, assist in the transferring of the distances. These reference edges are comparable to the glass box hinge lines shown in Figures 5-3 and 5-4 on page 138, but they are more flexible. In the solution shown in Figure 5-9b, for example, the reference edge labeled H-F is conveniently located so that it is collinear with the near-side edge of the inclined surface, as shown in the top view. This is equivalent to setting D = 0 in Figure 5-4, reducing the number of required trammel dimensions by one.

STEP 1

Sketch a dashed reference line parallel to the edge view of the inclined surface to be projected (see Figure 5-9c).

- The perpendicular distance between the reference line and the inclined edge should be chosen such that the resulting auxiliary view does not interfere with other views.

- This reference line is labeled either H-1, F-1, or P-1, depending on whether the edge view of the inclined surface appears in the horizontal (H), frontal (F), or profile (P) plane. The "1" indicates that a primary auxiliary view is being constructed.

- The reference line represents the hinge that the auxiliary view is rotated 90 degrees about, thus causing it to lie in the same plane as the principal plane it is projected from.

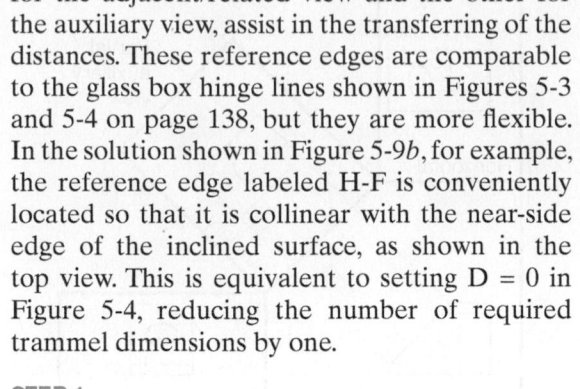

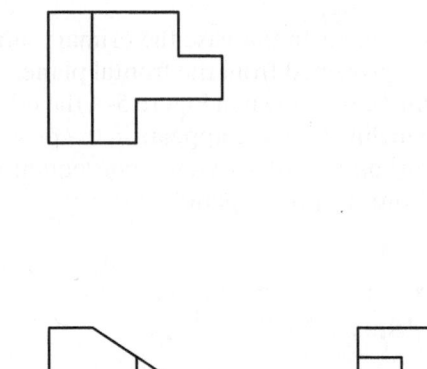

Figure 5-9a Given

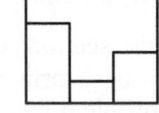

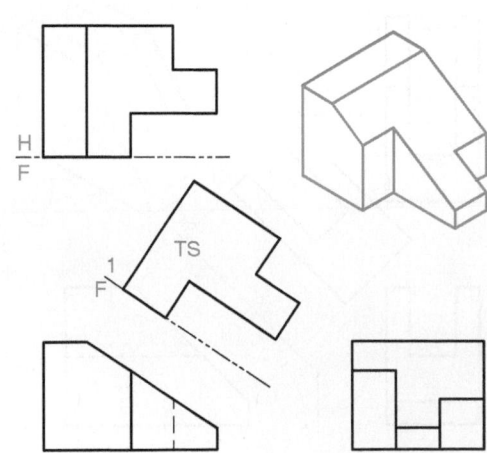

Figure 5-9b Solution

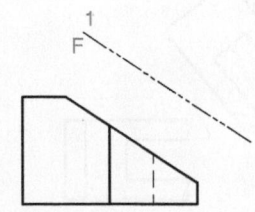

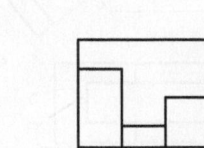

Figure 5-9c Step 1

STEP 2

Sketch perpendicular projectors from the inclined surface (see Figure 5-9d).

- The edge view of the inclined surface shows the true lengths of these edges, because these edges are parallel to the projection plane.

- Steps 1 and 2 may be performed in either order.

STEP 3

Draw a second dashed reference line at a convenient location between the view being projected from and an adjacent or related principal view (see Figure 5-9e).

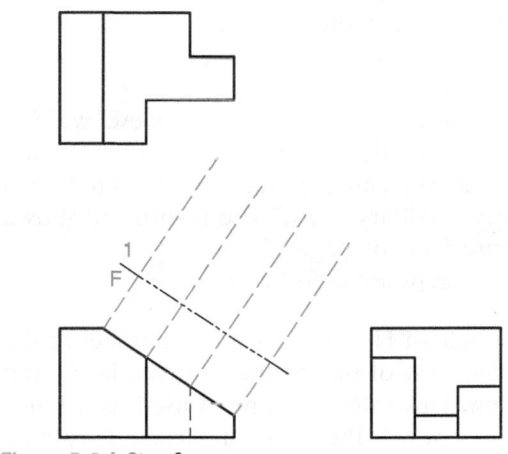

Figure 5-9d Step 2

- Label the second reference line such that the first letter indicates the view being projected from (either H, F, or P), and the second letter (H, F, or P) indicates the adjacent or related view from which the missing dimensional information is to be obtained.

- In Figure 5-9e, reference edge F-H will be used to capture depth information from the top view. Alternatively, a reference edge F-P could have been used to capture this same depth information from the profile view.

- This second reference line represents a hinge, where the adjacent/related view is rotated into the same plane as the edge view.

STEP 4 (OPTIONAL)

In this optional step, label the vertices of the inclined surface in the adjacent/related view, and then transfer the vertex labels to the edge view (see Figure 5-9f).

- This step may be helpful in correctly orienting the auxiliary view.

STEP 5

Using a trammel or dividers, transfer the depth dimensions from the F-H reference line in the top view to the F-1 reference line (see Figure 5-9g on page 142).

- The vertex labels can also be transferred to the auxiliary view.

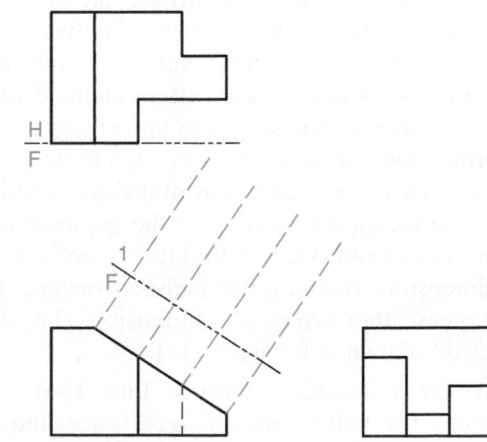

Figure 5-9e Step 3

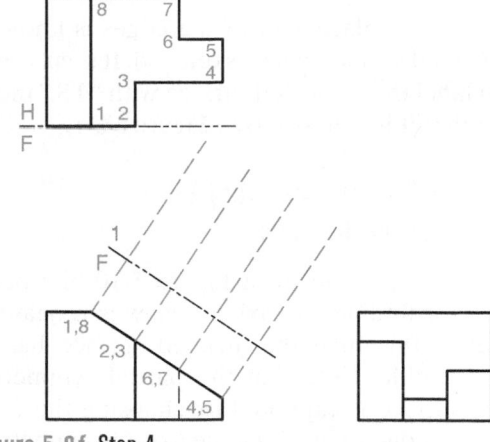

Figure 5-9f Step 4

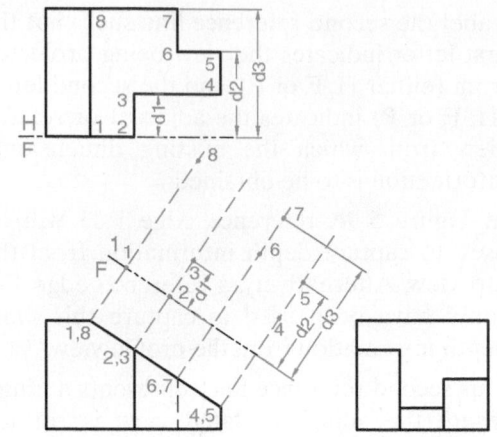

Figure 5-9g Step 5

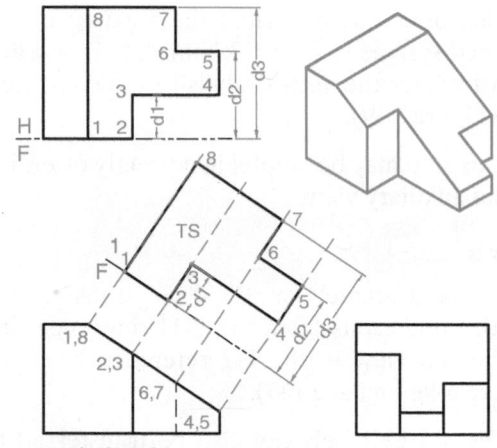

Figure 5-9h Step 6

STEP 6

Now that the placement of the edges is known, the inclined surface can be sketched. It is customary to label the projected surface with "TS," indicating that it is true size (see Figure 5-9h).

Finding a Primary Auxiliary View of a Contoured Surface

In the example shown in Figure 5-10, the procedure for finding an auxiliary view is repeated. However, this time the inclined surface has a curved profile. Note that the partial symmetry of this surface is exploited by choosing the H-1 reference edge to lie along the axis of symmetry.

Finding a Partial Auxiliary View, an Isometric Pictorial, and a Missing View, Given Two Views

For the two-view drawing shown in Figure 5-11, find (1) an auxiliary view of the inclined surface, (2) an isometric pictorial of the object depicted, and (3) the missing right view.

The inclined surface appears on edge as the diagonal line in the top view in Figure 5-11a. Recall that an inclined surface will appear as an edge in one view and as a foreshortened surface in the other two views. Projecting from the inclined edge to the front view as seen in Figure 5-11b, it is clear that the inclined surface has the shape of a backwards "Z." This means that we can expect to see this same Z shape, now true size, in the auxiliary view, as well as a foreshortened Z shape surface in the missing right view.

To find the partial auxiliary view, we follow the steps described in the earlier section "General Sketching Procedure for Finding a Primary Auxiliary View." The results are shown in Figure 5-11c on page 144.

These steps are as follows:

1. Sketch an H-1 reference line parallel to the edge view of the inclined surface in the top view. This reference line is used as a hinge about which the inclined surface is rotated 90 degrees into the plane of the paper.

2. To obtain one set of true length dimensions that define the inclined surface, sketch four perpendicular projectors extending from the edge view of the inclined surface in the top view. Since the edge view of an inclined surface is shown true length in an orthographic projection, the distances between these projectors are the true horizontal edge lengths of the inclined surface. Use the adjacent (in this case, front) view to find the second set of dimensions defining the inclined surface. In this case, they are height dimensions (i.e., d1, d2, d3, shown in Figure 5-11c).

3. Sketch a second reference line, H-F, between the two views. This reference line is used when trammeling the missing height

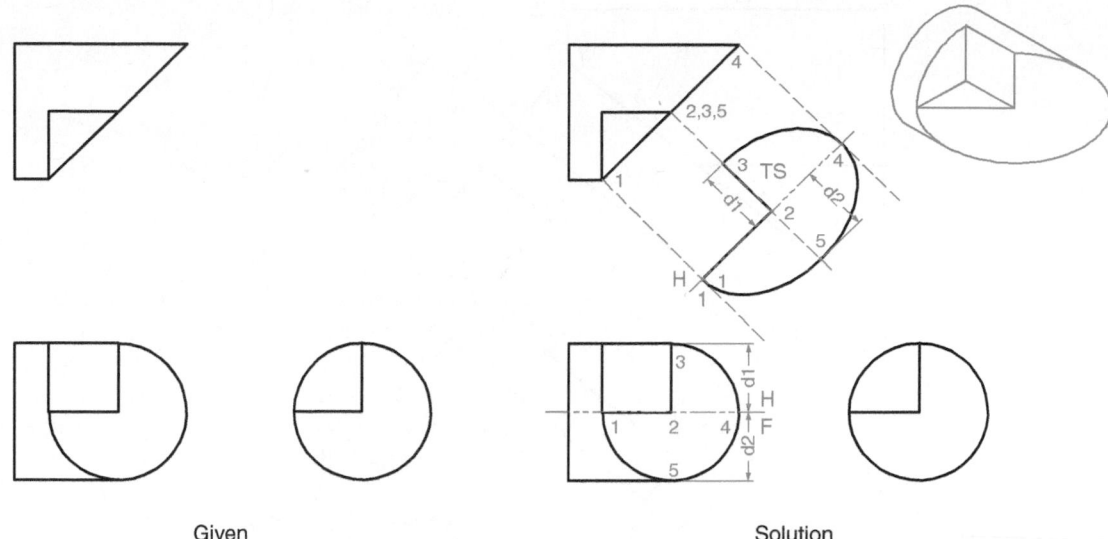

Given   Solution

Figure 5-10 Auxiliary view process applied to a contoured, symmetrical surface

dimensions. It is convenient to locate this horizontal reference line so that it passes through the upper edge of the inclined surface. In this way, the vertices defining the upper edge of the inclined surface will not need to be trammeled, because they will lie on the hinge line.

4. Label the vertices of the inclined surface in the front view, and then project them to the top view.

5. Use a trammel to transfer the height dimensions from H-F in the front view to H-1 in the auxiliary view. Vertex labels can also be transferred to the auxiliary view.

6. Sketch the true size inclined surface by connecting the vertices.

Having completed the partial auxiliary view, the next step is to sketch an isometric pictorial of the object. Because the inclined surface has the shape of a backwards "Z," it makes sense to start

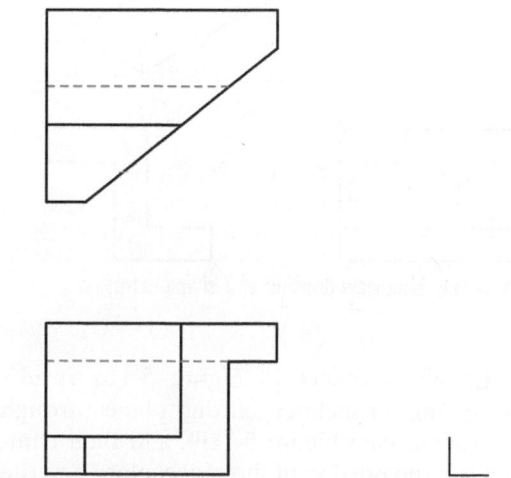

Figure 5-11a Given

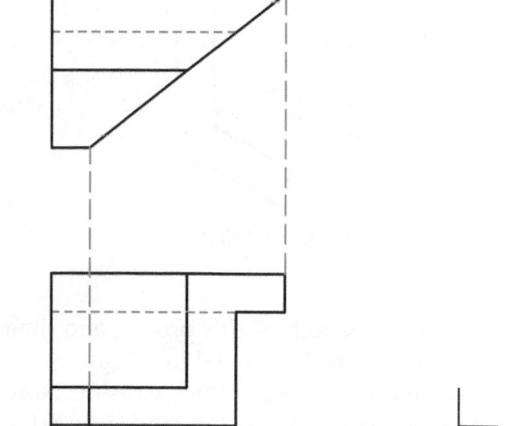

Figure 5-11b Inclined surface in top and front views

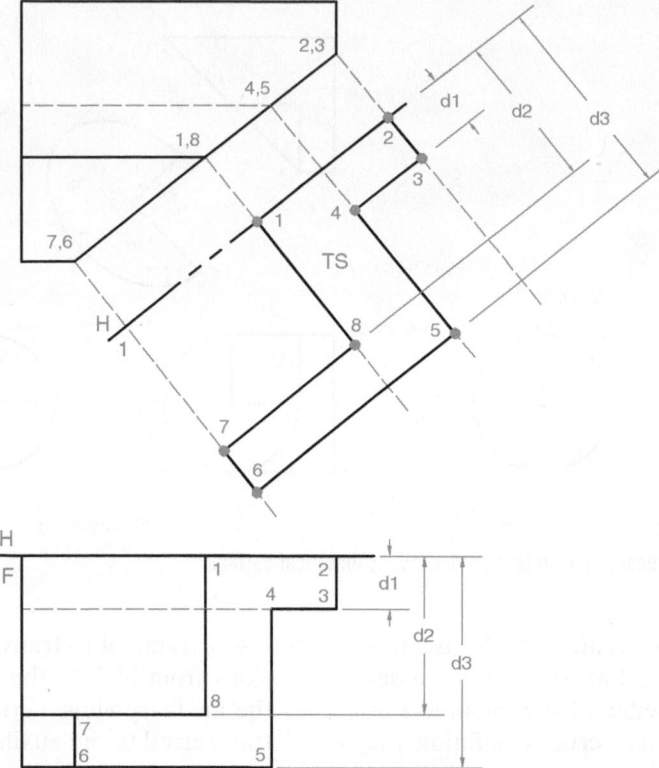

Figure 5-11c Finding the partial auxiliary view

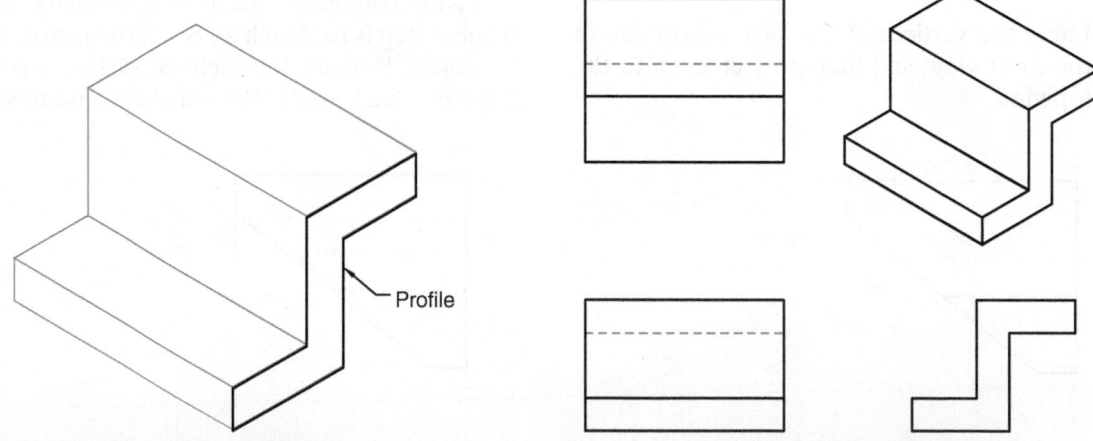

Figure 5-11d Z shape profile and extrusion

Figure 5-11e Multiview drawing of Z shape extrusion

with an isometric sketch of this profile, and then extrude it, as seen in Figure 5-11d. Figure 5-11e shows a multiview drawing of this extrusion. Note the similarity between the top view in Figure 5-11e and the top view in Figure 5-11a. This suggests

that the given object in Figure 5-11a results from passing an inclined cutting plane through the extrusion (see Figure 5-11f), and then trimming away the portion of the object closest to the viewer.

To sketch this, draw lines on the isometric sketch where the cutting plane passes through the extrusion, as seen in Figure 5-11g. Whenever there is a change in planes, the cut line will change direction. Once the closed profile formed by the intersection of the plane with the solid is drawn, erase everything in front of the profile to reveal the object, as seen in Figure 5-11h.

Having sketched a pictorial of the given object, constructing the missing right view is straightforward. From the isometric sketch, it is clear that there will be two visible surfaces in the right view, one being the inclined "Z" shaped surface, and the other, a square surface in the upper right corner of the view. The dimensions of these surfaces in the right view can be found by projecting from the adjacent front and related top views, as seen in Figure 5-11i. Figure 5-11j shows the completed solution.

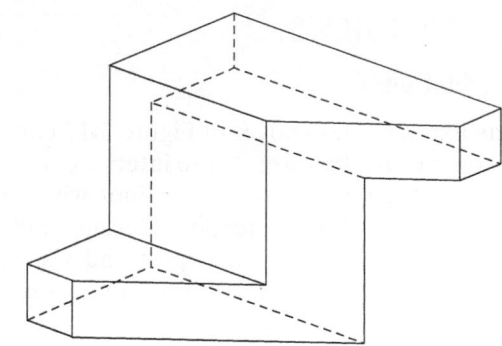

Figure 5-11*h* Pictorial view of object

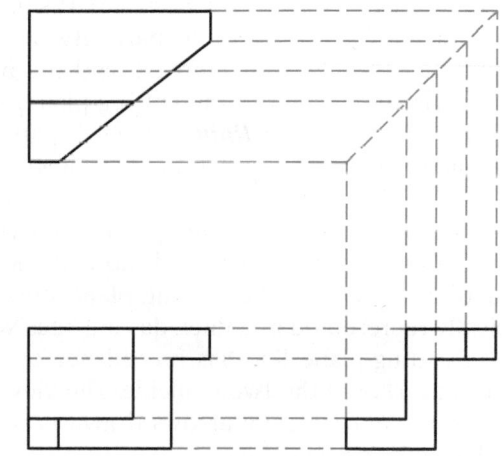

Figure 5-11*i* The missing view

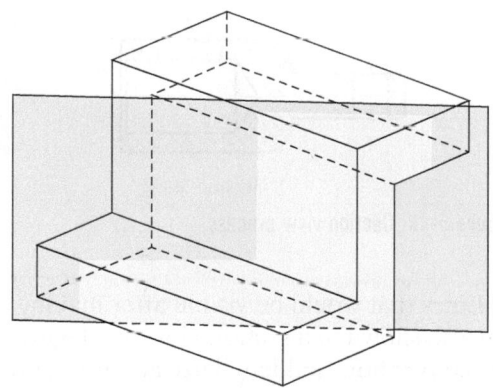

Figure 5-11*f* Cutting plane passed through Z shape extrusion

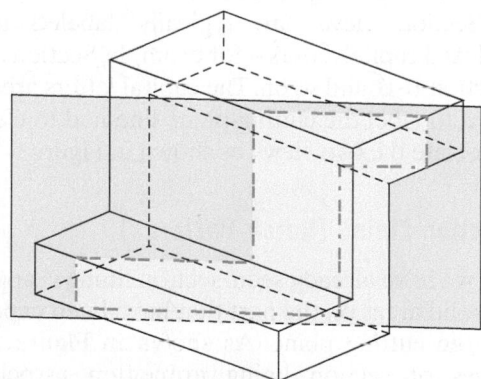

Figure 5-11*g* Profile of cut revealed

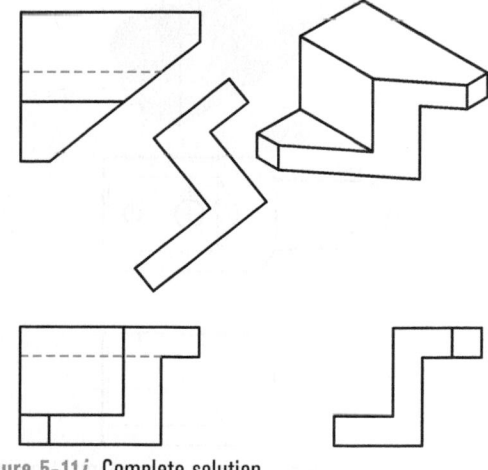

Figure 5-11*j* Complete solution

▍SECTION VIEWS

Introduction

Parts like the ones shown in Figure 5-12 contain several internal features. These interior construction details show up as hidden lines when in a standard multiview projection. Because hidden lines can be difficult to interpret and visualize, *section views* are frequently used to expose the internal features of a part.

Section View Process

In a section view, an imaginary *cutting plane* is passed through a part, often along a part's plane of symmetry. The portion of the part between the viewer and the plane is removed, and the part is then viewed normal to the cutting plane (see Figure 5-13). A *section lining*, or hatch pattern, is applied to the surfaces that make contact with the cutting plane.

The cutting plane on edge appears in a view adjacent to the section view, and shows the location of the section. The cutting-plane edge is typically represented as a thick dashed line. Note that a cutting-plane line has precedence over a centerline, should the two coincide. The viewing direction is indicated by arrows drawn perpendicular to the cutting plane.

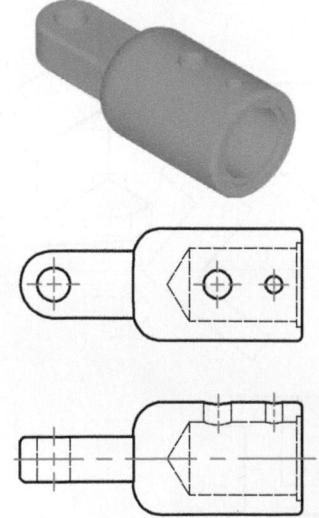

Figure 5-12 Views of a part with multiple internal features

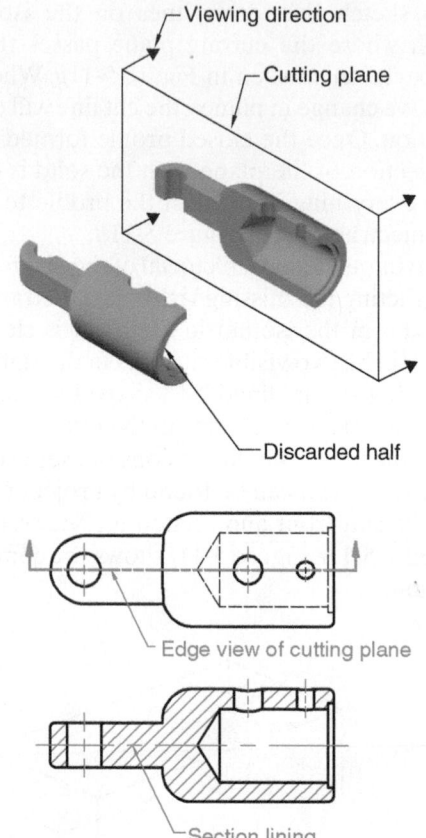

Figure 5-13 Section view process

Lines that would be visible after making a cut are also shown in a section view; see Figure 5-14. By convention, hidden lines are normally not shown in a section view. Exceptions are occasionally made, however, if it is felt that the clarity of the drawing is improved.

Section views are typically labeled using indexed capital letters—for example, Section A-A, Section B-B, and so on. The capital letters are also used to label the cutting-plane line and to clearly associate the two views, as shown in Figure 5-15.

Section Lining (Hatch Patterns)

As we have already seen, section lining is applied to solid areas on the part that have been exposed by the cutting plane. As shown in Figure 5-16, types of section lining are often associated with different materials. The most commonly

CHAPTER 5 AUXILIARY AND SECTION VIEWS

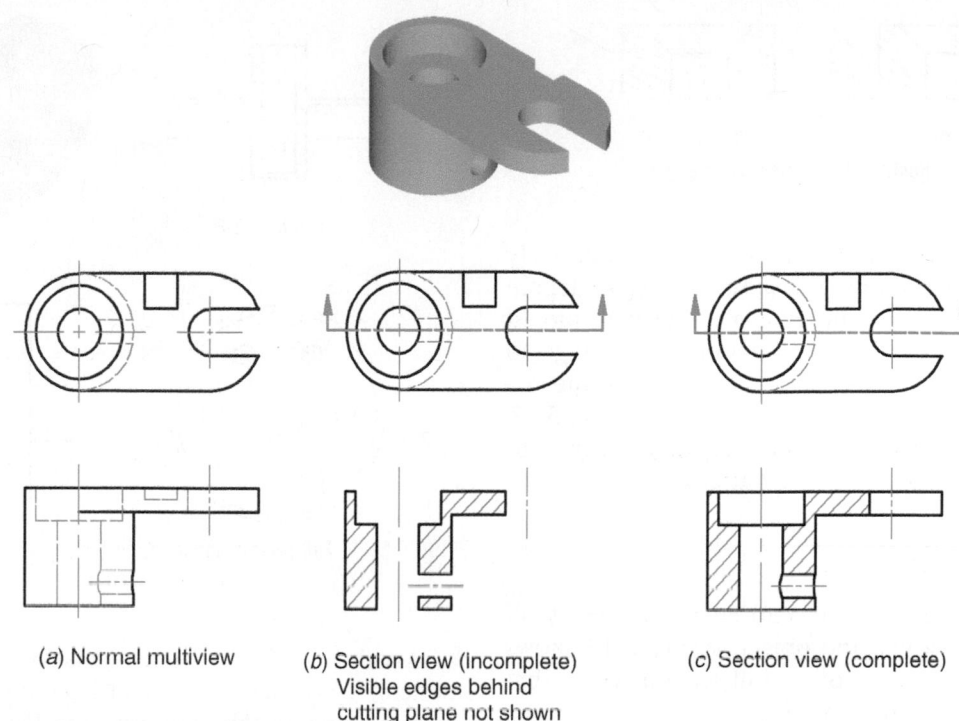

(a) Normal multiview

(b) Section view (Incomplete)
Visible edges behind
cutting plane not shown

(c) Section view (complete)

Figure 5-14 Treatment of visible edges behind the cutting plane in a section view

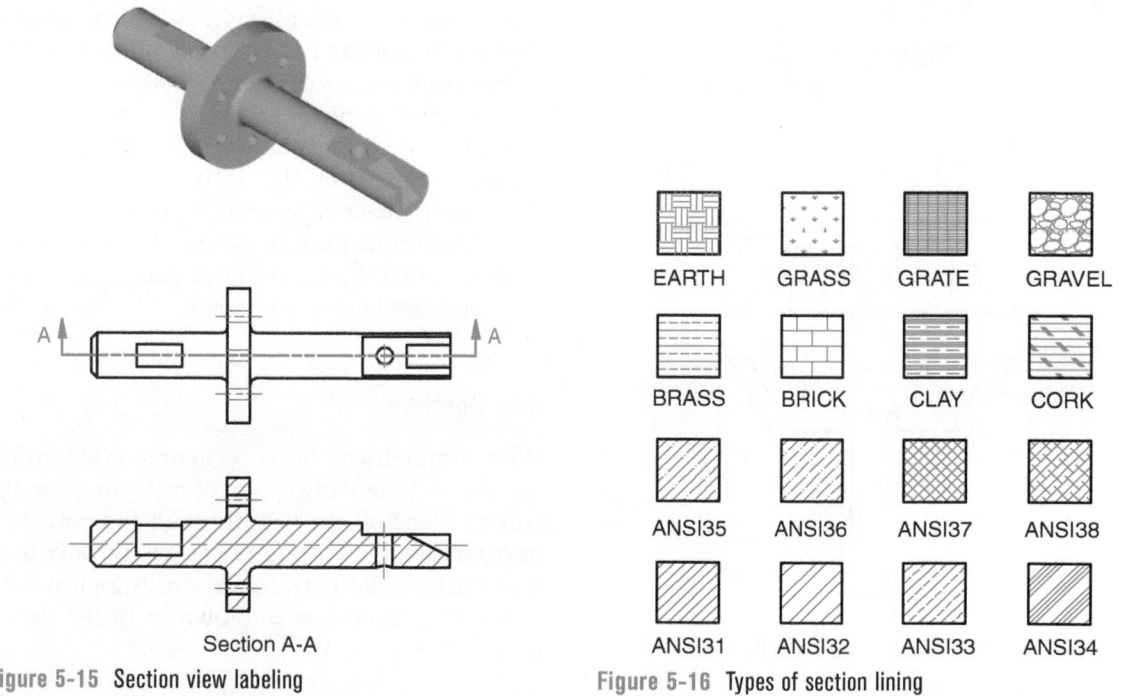

Section A-A

Figure 5-15 Section view labeling

EARTH	GRASS	GRATE	GRAVEL
BRASS	BRICK	CLAY	CORK
ANSI35	ANSI36	ANSI37	ANSI38
ANSI31	ANSI32	ANSI33	ANSI34

Figure 5-16 Types of section lining

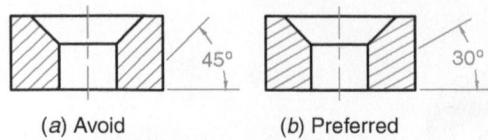

(a) Avoid   (b) Preferred

Figure 5-17 Adjusting the section lining angle

employed section lining consists of uniformly spaced continuous lines, set at a 45-degree angle. This section-lining angle, however, should be adjusted to avoid the section lines being too close to parallel with or perpendicular to the visible lines that bound them, as seen in Figure 5-17. In CAD software programs, section lines are typically referred to as hatch patterns.

Full Sections

In a **full section**, the cutting plane passes all the way through the object. Figure 5-18 shows another example of a full-section view that appears in the front view.

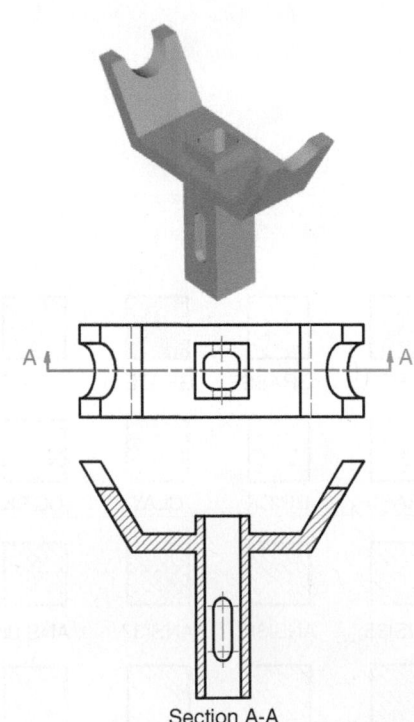

Figure 5-18 Full section appearing in the front view

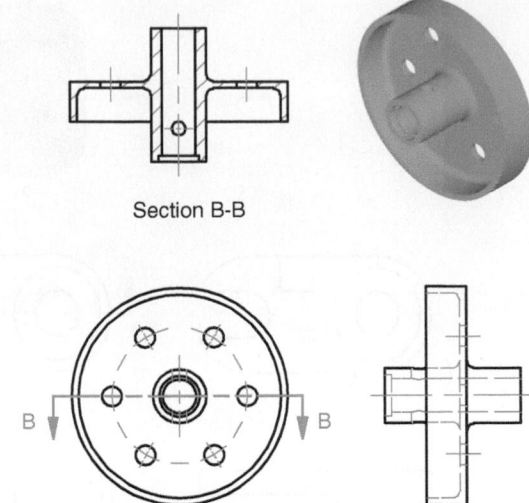

Section B-B

Figure 5-19 Full section appearing in the top view

Another possibility—one that is typically more difficult to visualize—uses a horizontally oriented cutting plane. In this situation (see Figure 5-19), the cutting plane appears on edge in the front view, while the section appears in the top view. Note the direction of the cutting-plane arrows in the front view. The arrow direction represents the viewing direction; the observer is looking down at the bottom portion of the object, the top portion having been removed.

Yet another possible orientation of a full-section view appears in Figure 5-20. Here the section appears in the (left) side view, and the cutting-plane edge appears in the top view. Once again, note the direction of the sight arrows. Also note that the cutting edge could just as well have appeared in the front view.

Half Sections

With symmetrical or very nearly symmetrical objects, it is not always necessary to pass the cutting plane all the way through the part. In a **half section**, the cutting plane passes only halfway through the part, as shown in Figure 5-21. In a half-section view, one-quarter of the part is removed.

Half sections possess the advantage of showing both the interior and the exterior of the part in the same view. External features are included on the unsectioned half. A centerline is used to separate the two halves. Hidden lines are normally omitted in both halves, but they may be shown in the unsectioned half.

Offset Sections

An **offset section** is a modified full section that is used when important internal features do not lie in the same plane. In an offset section the cutting plane is stepped, or offset, in order to pass through these features. Figure 5-22 shows an example of an offset section. Note that any steps (90-degree bends) in the cutting-plane line are not shown in the section view. The section is drawn as if these offsets all lie in the same plane. Also note that the offsets should be located in regions where there are no features.

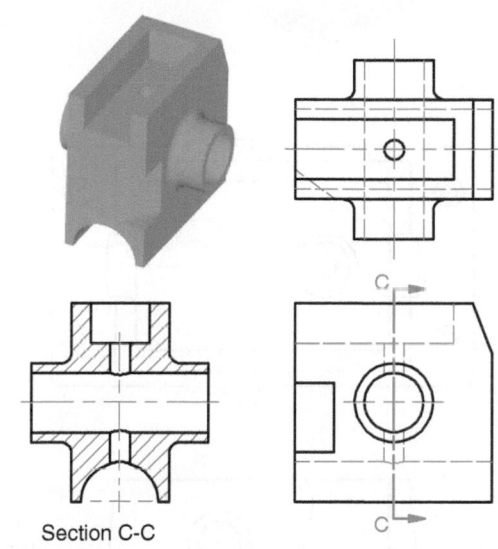

Section C-C

Figure 5-20 Full section appearing in the side view

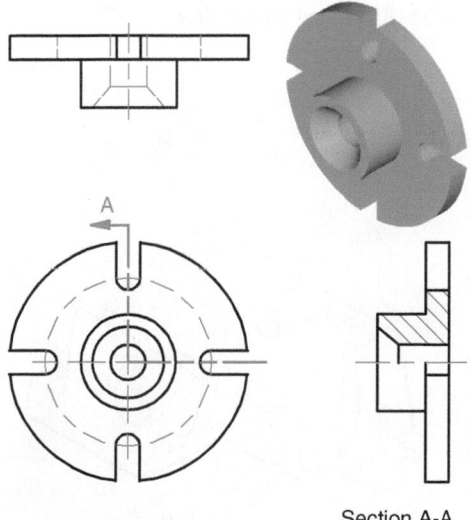

Section A-A

Figure 5-21 Half-section view

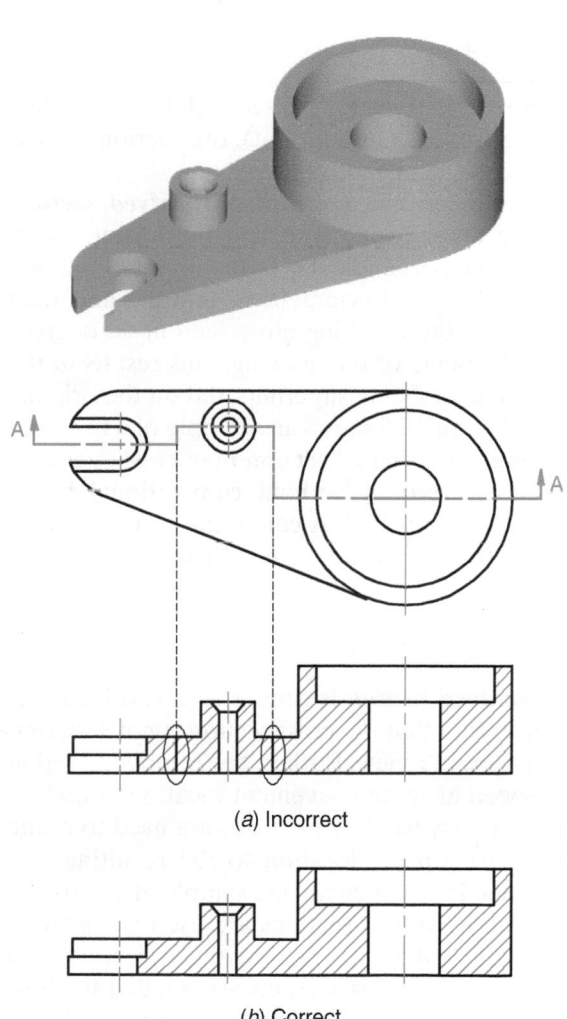

(a) Incorrect

(b) Correct

Figure 5-22 Offset section view

Broken-Out Sections

A ***broken-out section*** is used when only a portion of the part needs to be sectioned. See Figure 5-23 for an example of a broken-out section. A jagged, freehand break line is used to separate the sectioned from the unsectioned portion of the drawing. Like half sections, broken-out sections have the advantage of showing internal and external features in the same view. In addition to being used on multiview drawings, broken-out sections are also used on pictorial views, particularly when they are executed in CAD (see Figure 5-24).

Revolved Sections

In all of the sections discussed thus far (full, half, offset, and broken-out), the section view is projected from the adjacent view in which the cutting-plane line appears. A ***revolved section***, on the other hand, is created by passing a cutting plane perpendicular to the longitudinal axis of an elongated symmetrical feature, and then revolving the resulting cross section 90 degrees into the plane of the drawing. This results in the cross section being superimposed on the original view. Figure 5-25 shows an example of a revolved section. The original section may be shown with (Figure 5-25*a*) or without conventional break lines (Figure 5-25*b*). A centerline is used to represent the axis of the revolved section.

Removed Sections

A ***removed section*** is similar to a revolved section, except that the cross section is not superimposed on the view. Rather, the removed section is placed at some convenient location. Standard section view labeling practices are used to relate the cutting-plane location to the resulting section. See Figure 5-26 for an example of a removed section. Removed sections are used when there is insufficient room for a revolved section, and when several cross sections are needed to show the transition of an elongated feature from one shape to another (see Figure 5-27 on page 152).

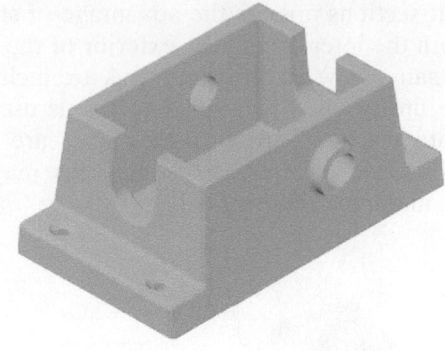

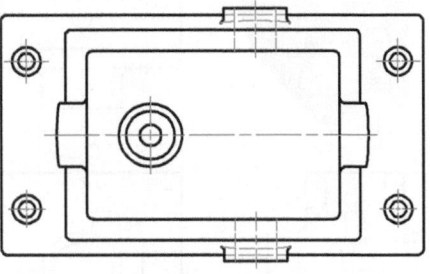

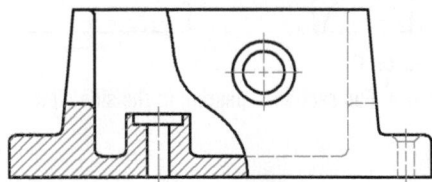

Figure 5-23 Broken-out section view

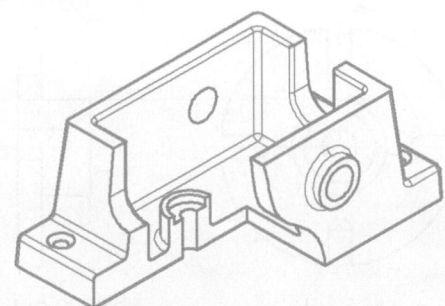

Figure 5-24 Broken-out section view–pictorial

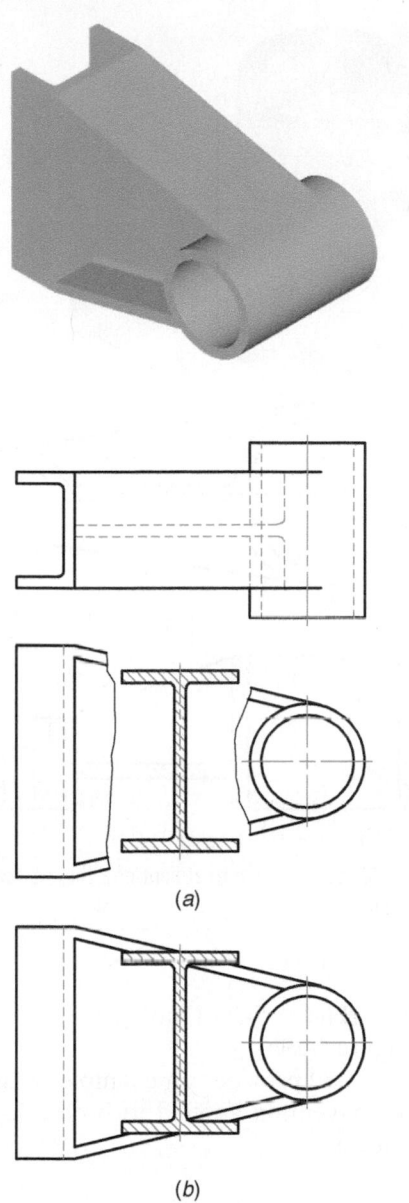

Figure 5-25 Revolved section

(a)

(b)

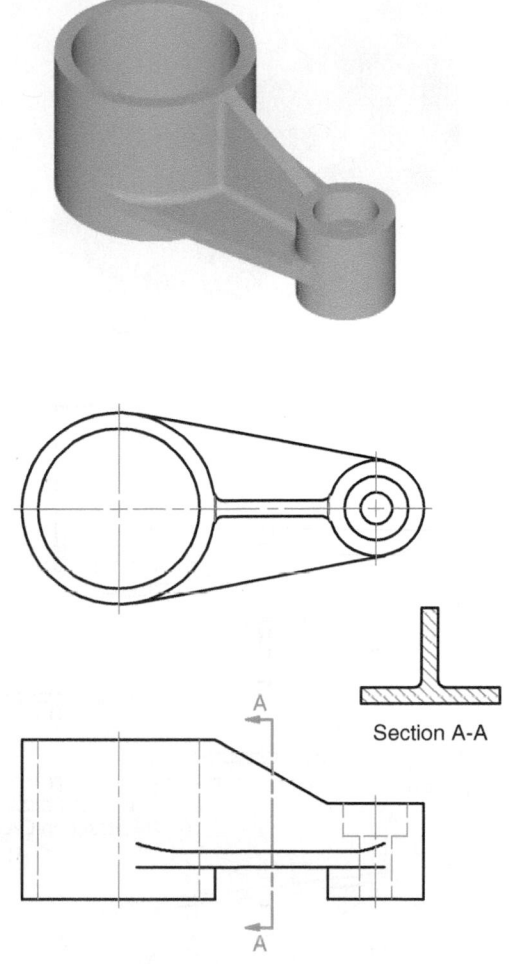

Section A-A

Figure 5-26 Removed section

Conventional Representations: Section Views

In order to simplify the construction and improve the clarity of section views, conventional representations are sometimes employed in place of true orthographic projections. These simplified representations are widely recognized and accepted as being a part of standard drawing practice. Conventional representations associated with section views include the treatment of thin features, radially distributed and off-angle features, and section lining in assembly sections. Note that when using CAD, it may actually be easier to obtain a true projection, rather than a simplified representation. Although this has to some extent reduced the usage of some conventional representations, it is still necessary for engineers to be familiar with this aspect of the language of engineering graphics.

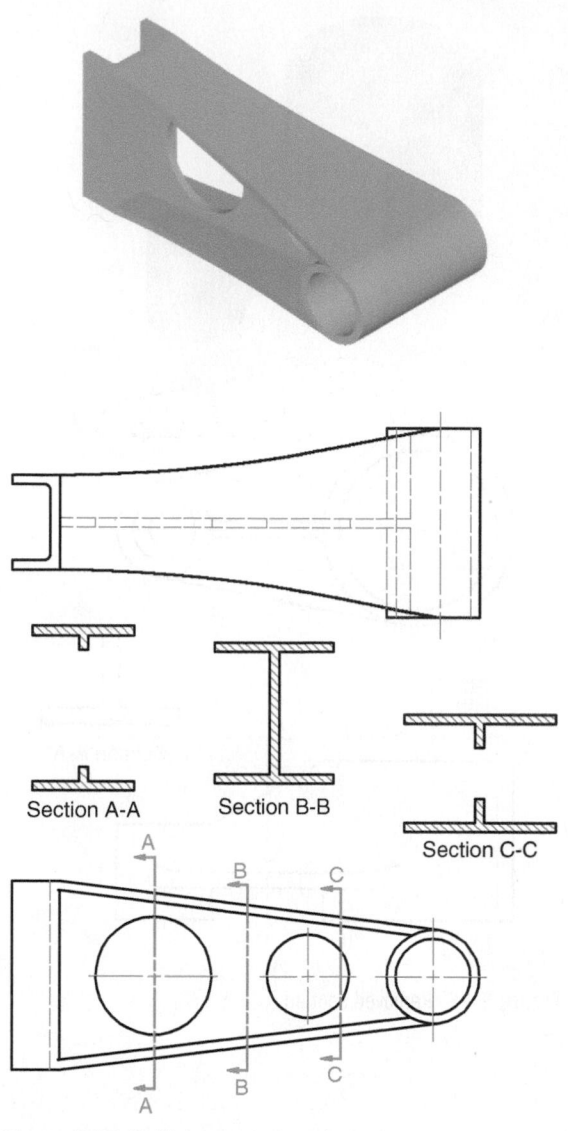

Section A-A   Section B-B

Section C-C

Figure 5-27 Multiple removed sections

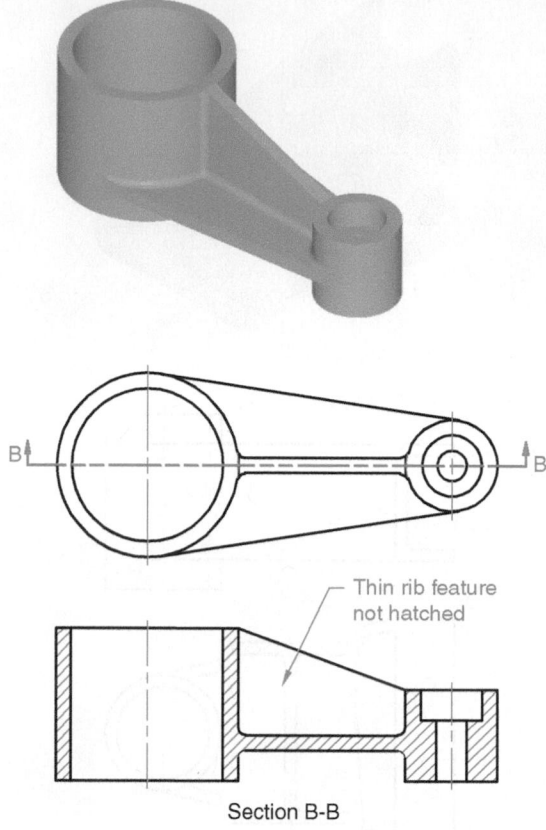

Thin rib feature
not hatched

Section B-B

Figure 5-28 Conventional treatment of a thin rib feature in a section view

Conventional Representations: Thin Features

In an effort to make some section views more readable, section lining is not applied to the outline of thin features like ribs, webs, and lugs when the cutting plane passes along the length of the feature. Figure 5-28 provides an example of this convention applied to a rib feature.

Note that without this convention, the section shown in Figure 5-29 could be incorrectly interpreted as depicting a part with uniform thickness (Figure 5-29a), rather than as a ribbed part (Figure 5-29b).

Figure 5-30 provides an example of this thin-feature convention applied to both a lug and a web feature.

Section View Construction Process—Example 1

Figure 5-31 on page 154 illustrates the process of constructing a full-section view of a part. In this particular problem, only the top and right-side principal views are given, and we are asked to draw the cutting plane on edge and to find the front-section view. In a less difficult variant of this problem, a pictorial view of the object is given, along with the cutting-plane location.

In Step 1, the cutting plane is drawn in the top view along the object's axis of symmetry. The

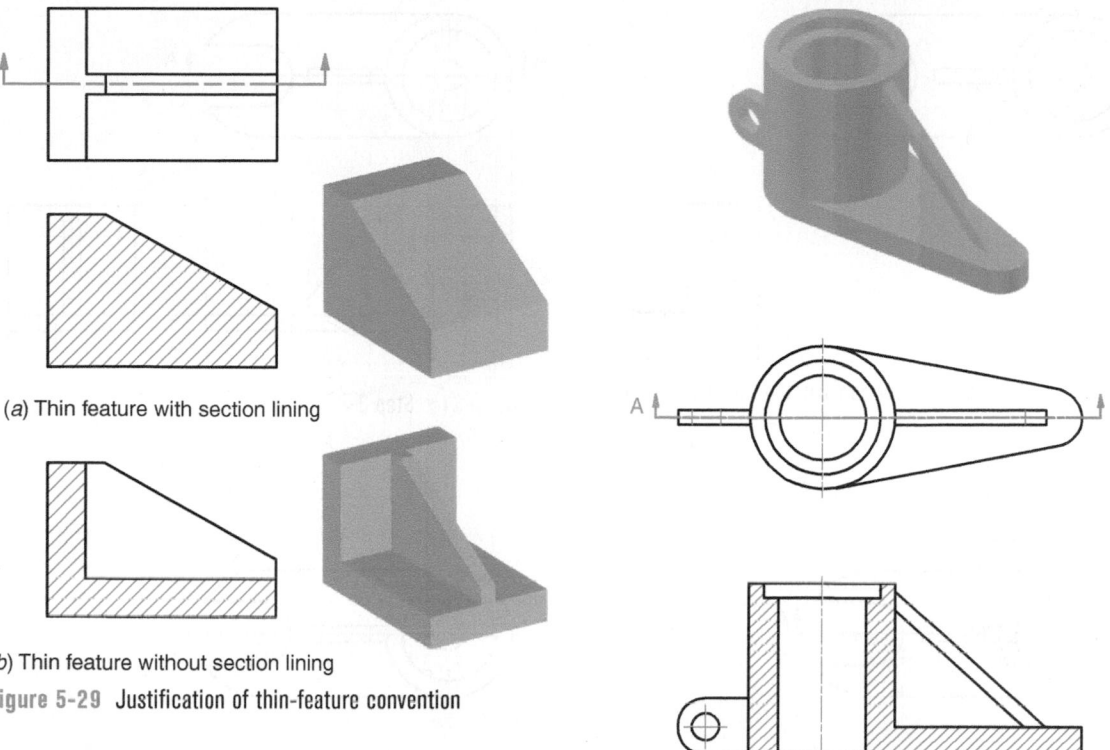

(a) Thin feature with section lining

(b) Thin feature without section lining

Figure 5-29 Justification of thin-feature convention

Section A-A

Figure 5-30 Thin-feature convention applied to a part with a lug and web

arrows should point as shown, to indicate that the section will appear in the front view. Parallel construction lines, or projectors, are drawn from the given views to help locate the extents of the object features in the front view.

Without the benefit of a pictorial view, one must now employ visual reasoning, reading between the given top and side views, to help piece together a mental image of the object. In the top view on the left there are three concentric circles. A raised cylinder with either a counter-drilled, counterbore, or countersunk hole feature (see Figure 4-31 in Chapter 4, for example) will appear like this in the circular view. Looking now at the right view, it appears that this is a counterdrilled hole feature. Reading between the two given views, it also appears that a horizontally oriented hole is drilled through the raised cylinder, intersecting with the vertically oriented counterdrill feature. Note that these two holes account for all of the hidden lines appearing in the two given views.

Moving now to the right side of the object, as shown in the top view on the right, we see a feature tangent to the raised cylinder on the left, with a fillet and a semicircular cut on the right. But what are the vertical extents of this feature? Looking at the right view, we see three horizontally aligned rectangles a little way up from the bottom. Together these views suggest a plate, rounded at one end with a semicircular cutout, attached to the raised cylinder.

At this point (Step 2), these features can be laid out. The only remaining feature, represented by the thin rectangles centered on the centerline in the given views, top and right, suggests a support rib. Note that, based on the information provided in the given views, the rib profile could have been either straight-sided (as shown in Figure 5-31c) or contoured (curved).

In Step 3, section lining is applied to the solid areas through which the imaginary cutting plane

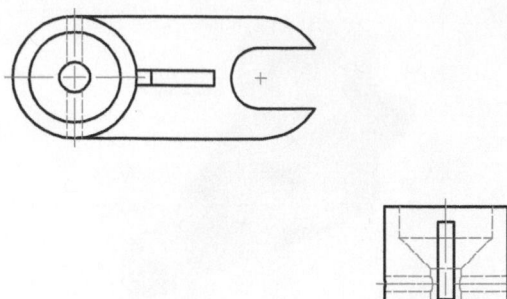

Figure 5-31*a* Full-section view construction process: given

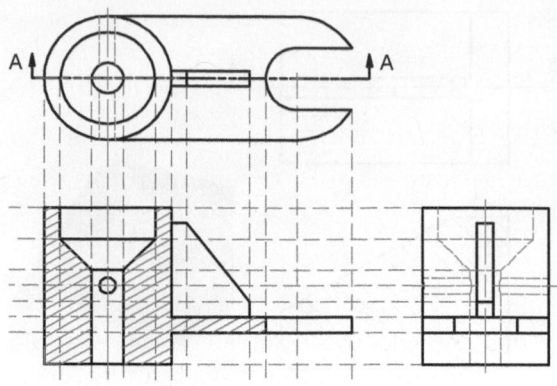

Figure 5-31*d* Step 3

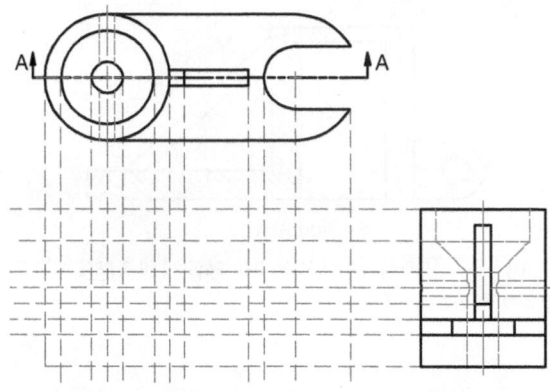

Figure 5-31*b* Step 1

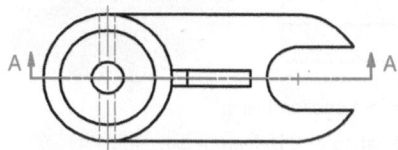
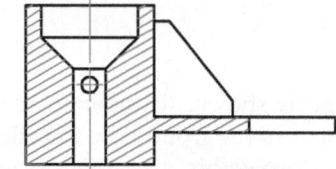
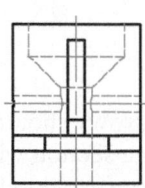

Figure 5-31*e* Completed sketch

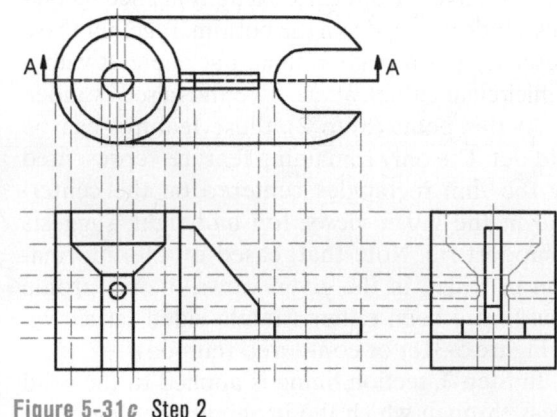

Figure 5-31*c* Step 2

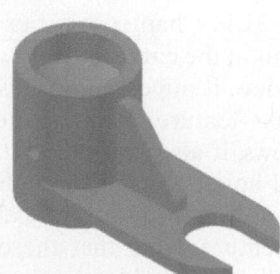

Figure 5-31*f* Shaded view

passes, except for the rib feature, which is left without hatching by convention. Note that the section-lining angle has been adjusted from the default 45 to 30 degrees, as shown in Figure 5-17, because of the inclined lines describing the counterdrill feature. Centerlines have also been added in Step 3.

Section View Construction Process—Example 2

In this example, two views are given, as shown in Figure 5-32a. The problem calls for finding an offset section in the top view.

Start with the visualization of the object. Cut features include, from top to bottom, a countersink hole, a counterbore hole, and an open slot, as seen in Figure 5-32b.

In Figure 5-32c, three (normal) surfaces are labeled in the front view. The corresponding edge views of these surfaces are also labeled in the left view.

From this, the object can now be visualized, as seen in Figure 5-32d.

Next, draw an offset cutting plane in the front view, as shown in Figure 5-32e. Note that in moving from left to right, the offsets (90 degree bends) are chosen so that they occur after one internal feature ends and before the next feature begins. Note

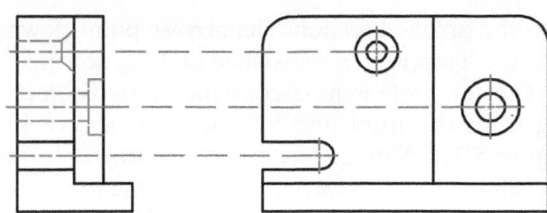

Figure 5-32b Projection of cut features

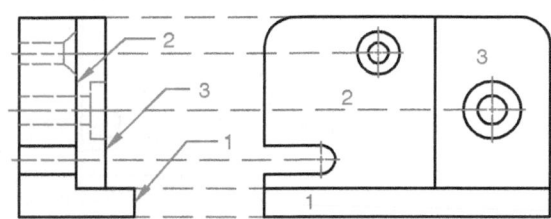

Figure 5-32c Surface labeling

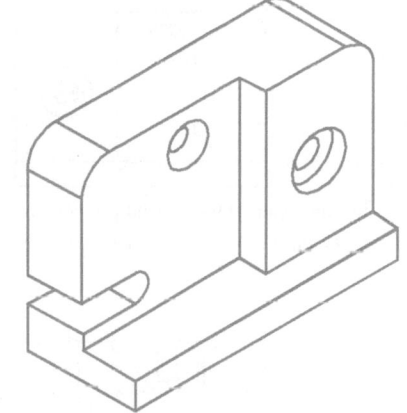

Figure 5-32d Isometric sketch of object

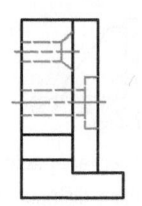

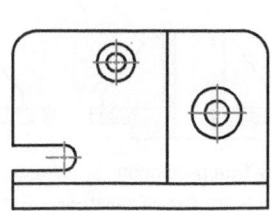

Figure 5-32a Given

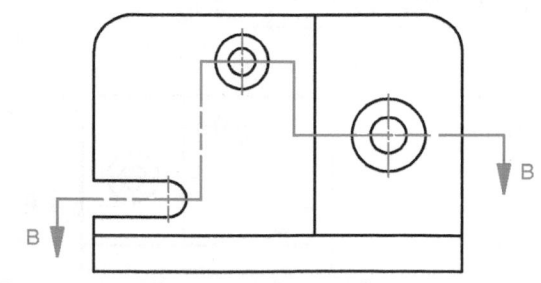

Figure 5-32e Offset cutting plane on edge

also the arrow direction. The arrows point down because a top view is seen when looking down.

The top view is next constructed by projecting from the front and left views, as shown in Figure 5-32f. A miter line is used to project from the left to the top view. All features in the top view are visible because it is an offset section view.

Figure 5-32g shows the solution. Section lining is applied to those areas where the cutting plane passes through solid material.

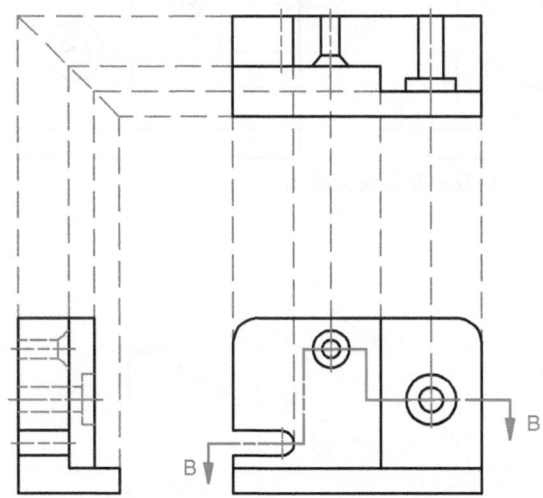

Figure 5-32f Top view construction using projection from adjacent and related views

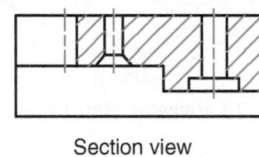

Section view

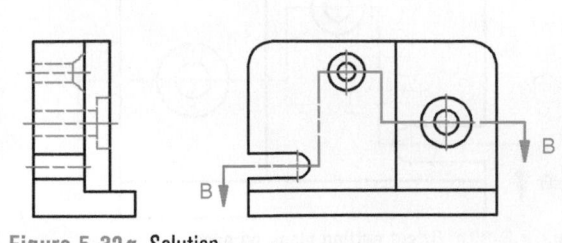

Figure 5-32g Solution

Conventional Representations: Aligned Sections

This convention is used to simplify the construction of section views containing radially distributed and off-angle features like holes, ribs, and lugs. In the object shown in Figure 5-33a, for example, a true section view based on orthographic projection results in a difficult-to-interpret, foreshortened view of the rib on the left. To eliminate these problems, by convention this feature is rotated into the cutting plane, or, alternatively, the cutting plane is bent to pass through the feature. Similarly, the lug and hole features shown on the right side in Figure 5-33a are rotated into the cutting plane. The result is a clearer representation of the geometry, as shown in Figure 5-33b.

Another example of features being rotated or aligned to simplify the representation is shown in Figure 5-34.

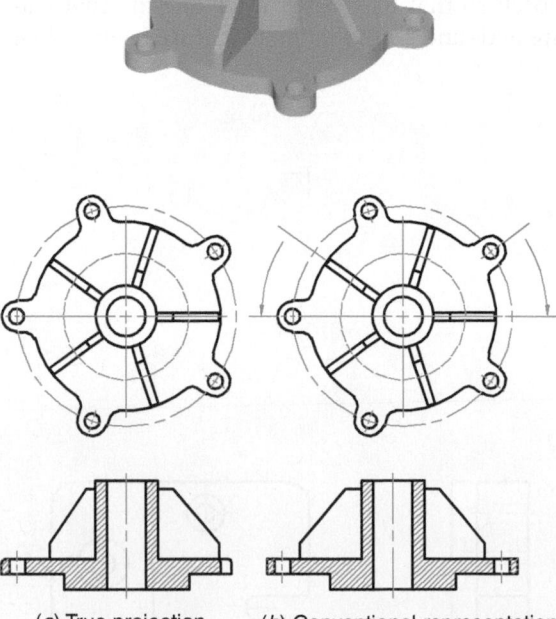

(a) True projection   (b) Conventional representation

Figure 5-33 Conventional representation of radially distributed features

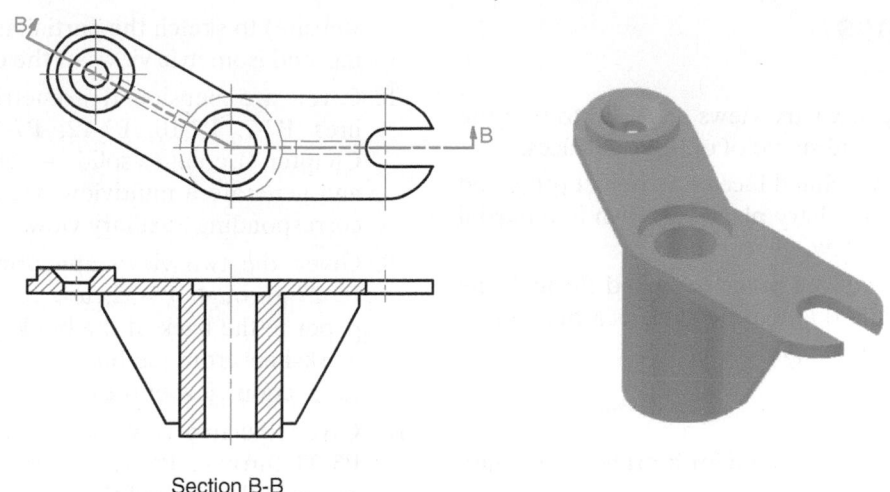

Section B-B

Figure 5-34 Conventional representation of an off-angle feature

Assembly Section Views

In a section view of an assembly, different hatch patterns are applied to different parts. See, for example, Figure 5-35, where several different section linings are used to represent different parts. Note that thin-walled parts (such as shafts and nuts) are not sectioned. In addition, parts like washers, bushings, gaskets, bolts, screws, keys, rivets, pins, bearings, spokes, and gear teeth are not sectioned where the cutting plane lies along the longitudinal axis of the part.

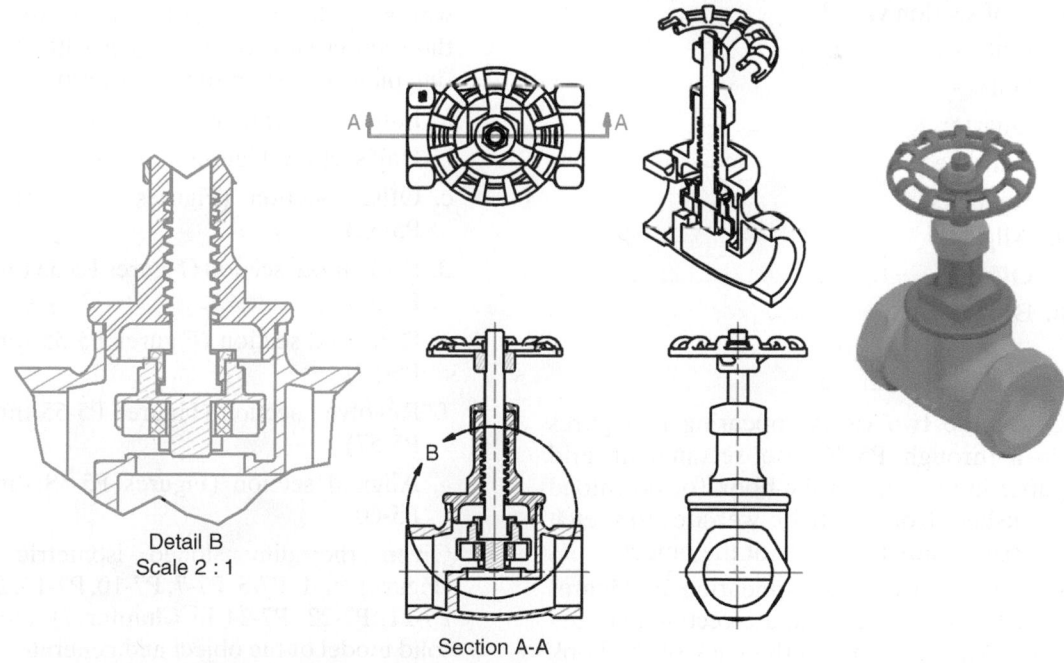

Detail B
Scale 2 : 1

Section A-A

Figure 5-35 Assembly section view (Courtesy of Alexander H. Hays)

▌QUESTIONS

TRUE OR FALSE

1. Primary auxiliary views are used to find the true size and shape of oblique surfaces.

2. Only the inclined face of an object projected onto an auxiliary plane is shown in a partial auxiliary view.

3. When a cutting plane is passed through the length of a thin feature such as a rib, section lining is not applied.

MULTIPLE CHOICE

4. The projection plane for a primary auxiliary view is:
   a. Parallel to one of the principal projection planes and perpendicular to the other two
   b. Perpendicular to one of the principal projection planes and inclined to the other two
   c. Perpendicular to two of the principal projection planes and inclined to the other one
   d. Inclined to all three of the principal projection planes

5. Which of the following is not a legitimate type of section view?
   a. Full
   b. Half
   c. Quarter
   d. Removed
   e. Revolved
   f. Aligned
   g. Offset
   h. Broken-out

SKETCHING AND MODELING

6. Given the two views appearing in Figures P5-1 through P5-26, use rectangular grid paper in the back of the book (or download worksheet from the book website) to sketch the partial auxiliary view of the object.

7. Given the two views appearing in Figures P5-1 through P5-26, use use rectangular and isometric grid paper in the back of the book (or download worksheets from the book website) to sketch the partial auxiliary, missing, and isometric views of the object.

8. Given a dimensioned isometric view (Figures P7-7, P7-10, P7-12, P7-15, P7-23 in Chapter 7), create a solid model of the object and generate a multiview drawing with the corresponding auxiliary view.

9. Given the two views appearing in Figures P5-27 through P5-32, use rectangular grid paper in the back of the book (or download worksheet from the book website) to sketch the section view of the object.

10. Given the two views appearing in Figures P5-33 through P5-45, use rectangular grid paper in the back of the book (or download worksheet from the book website) to sketch the section view of the object with the cutting plane shown in the appropriate view.
    a. Full section (Figures P5-33 through P5-37)
    b. Half section (Figures P5-38 through P5-41)
    c. Offset section (Figures P5-42 through P5-45)

11. Given the two views appearing in Figures P5-46 through P5-60, use rectangular grid paper in the back of the book (or download worksheet from the book website) to sketch the section view of the object with the cutting plane shown in the appropriate view.
    a. Full section (Figures P5-46 through P5-48)
    b. Half section (Figures P5-49 through P5-50)
    c. Offset section (Figures P5-51 through P5-52)
    d. Broken-out section (Figures P5-53 through P5-54)
    e. Removed section (Figures P5-55 through P5-57)
    f. Revolved section (Figures P5-55 through P5-57)
    g. Aligned section (Figures P5-58 through P5-60)

12. Given the dimensioned isometric view (Figures P7-1, P7-3, P7-7, P7-10, P7-13, P7-20, P7-21, P7-22, P7-24 in Chapter 7), create a solid model of the object and generate a multiview drawing that includes a section view.

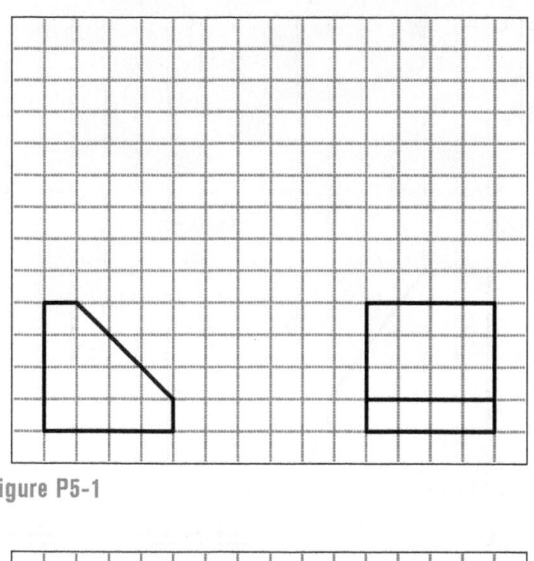

Figure P5-1

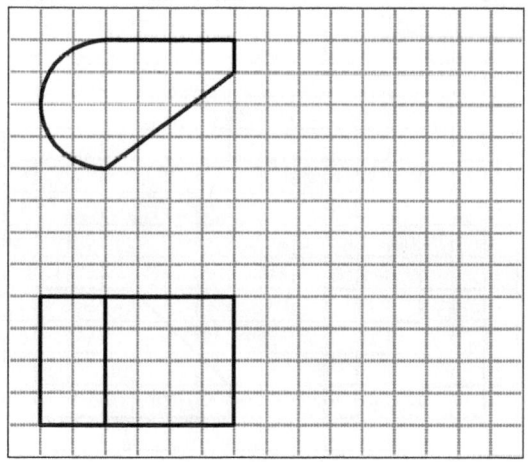

Figure P5-2

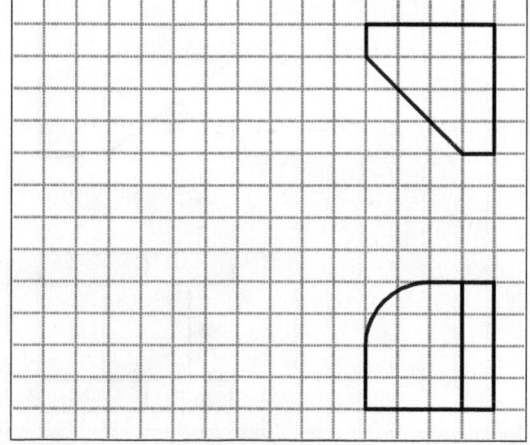

Figure P5-3

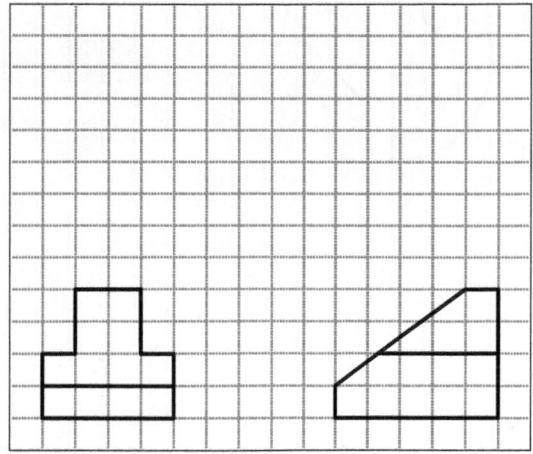

Figure P5-4

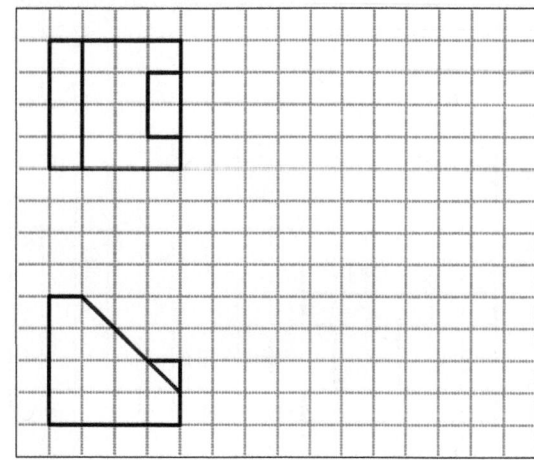

Figure P5-5

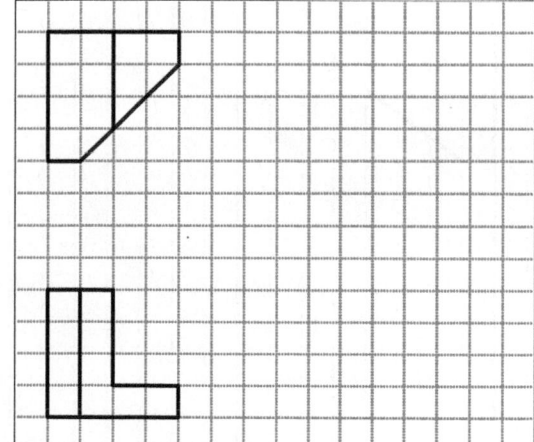

Figure P5-6

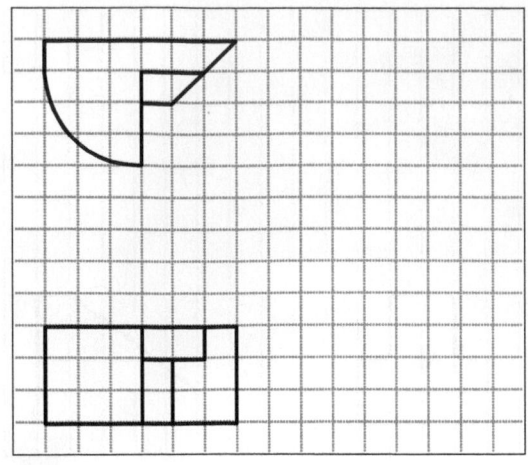

Figure P5-7

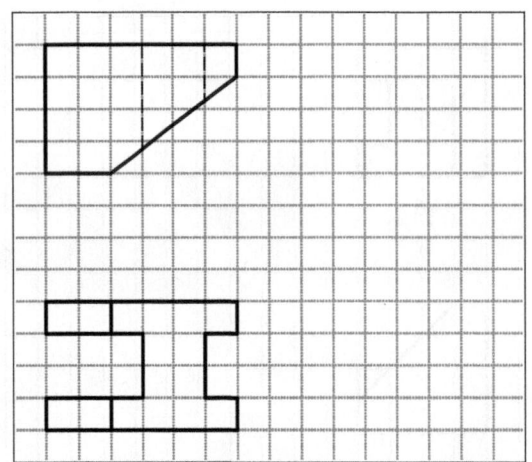

Figure P5-8

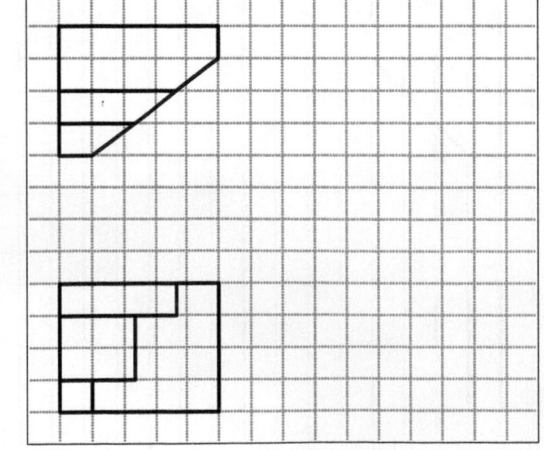

Figure P5-9

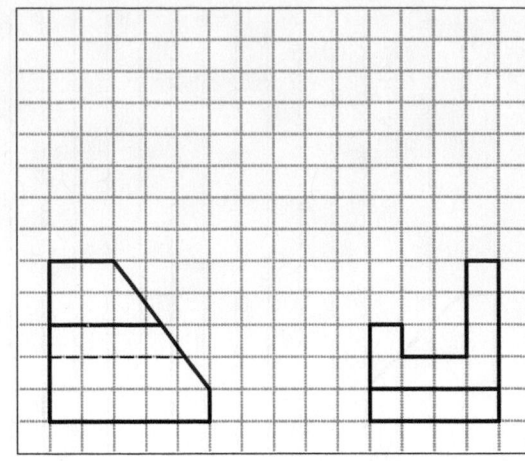

Figure P5-10

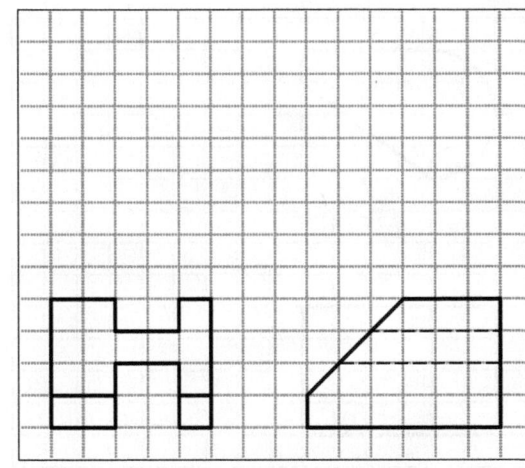

Figure P5-11

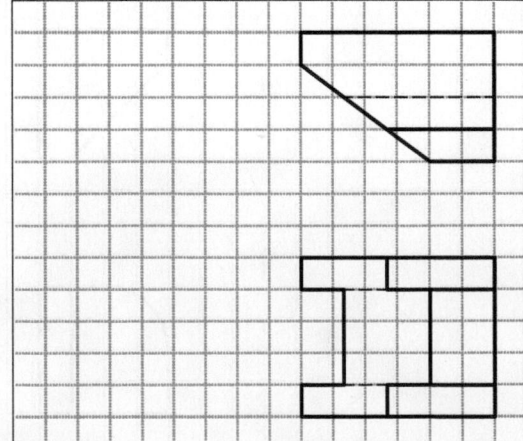

Figure P5-12

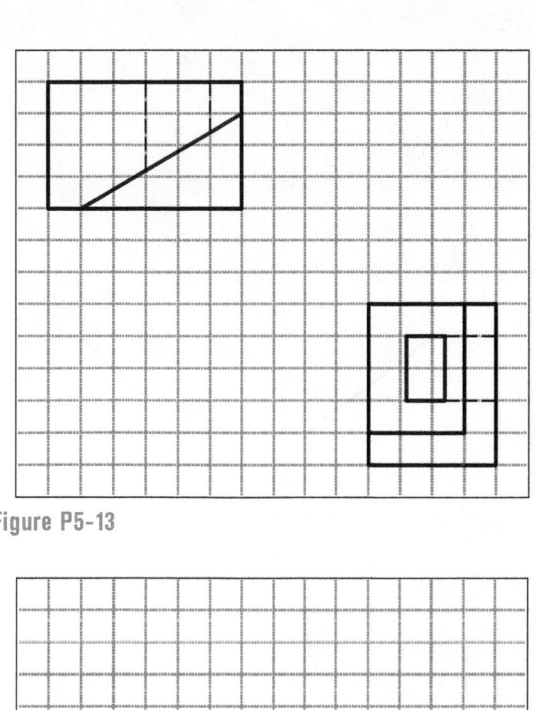

Figure P5-13

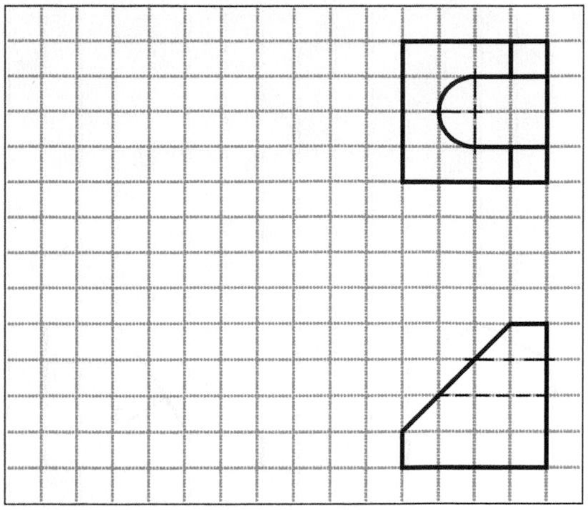

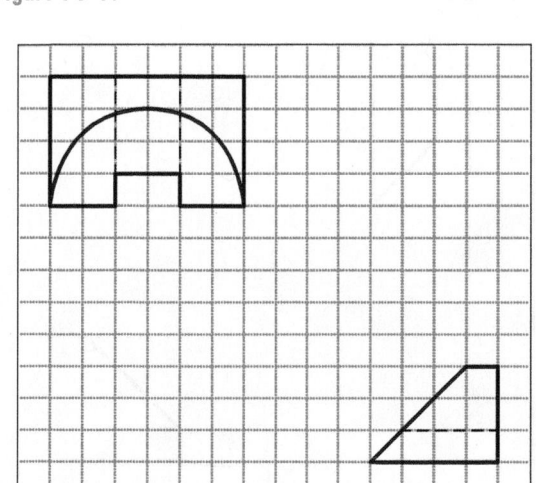

Figure P5-14

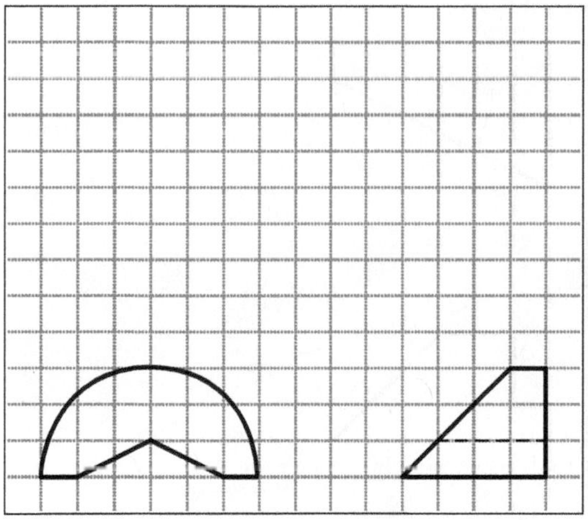

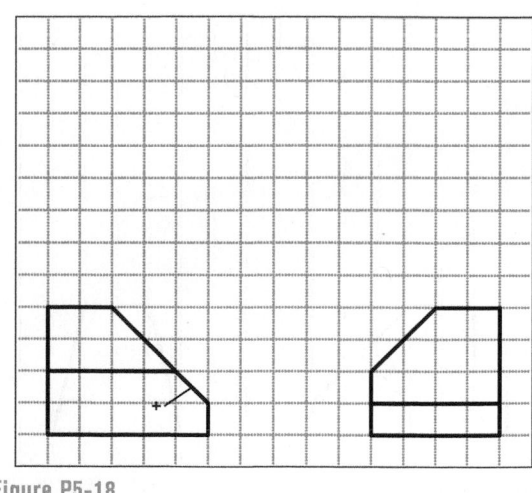

Figure P5-15

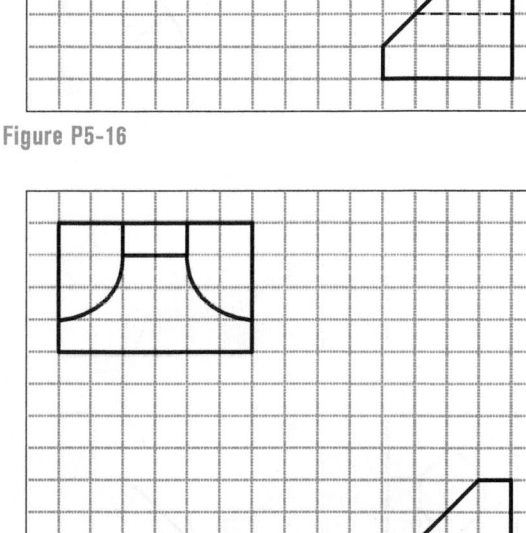

Figure P5-16

Figure P5-17

Figure P5-18

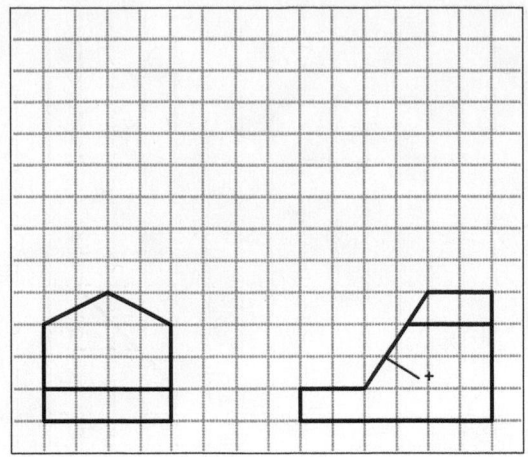

Figure P5-19

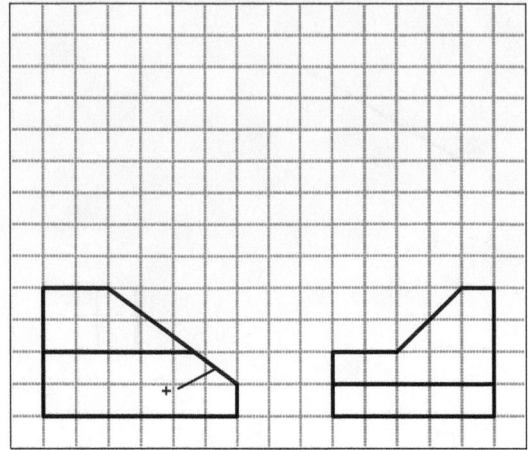

Figure P5-22

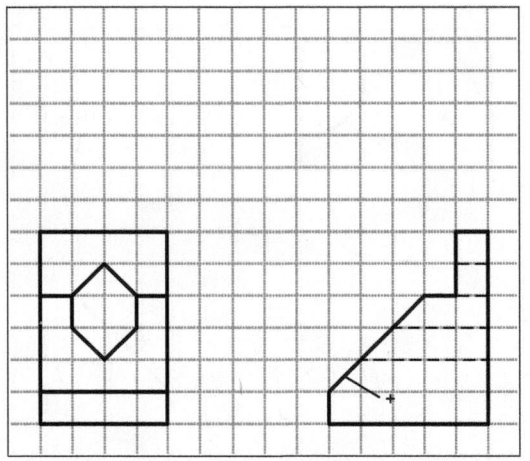

Figure P5-20

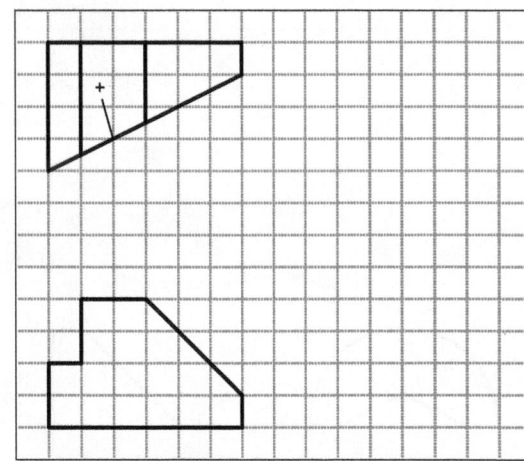

Figure P5-23

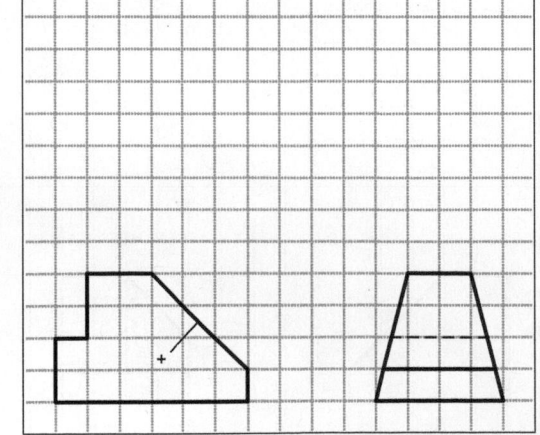

Figure P5-21

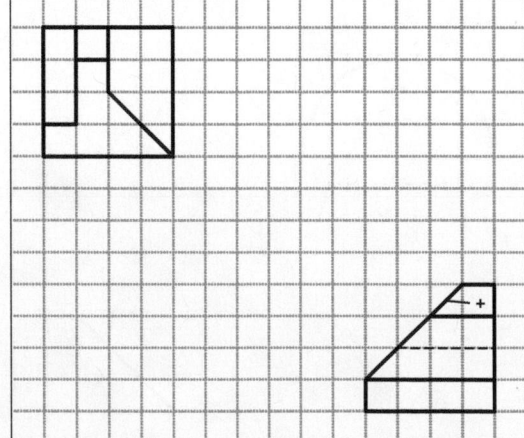

Figure P5-24

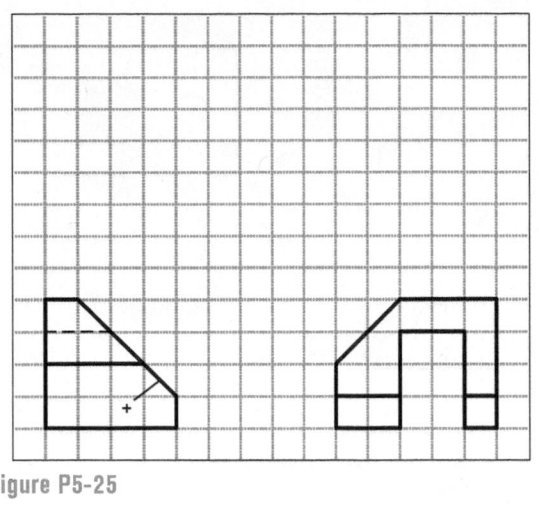

Figure P5-25

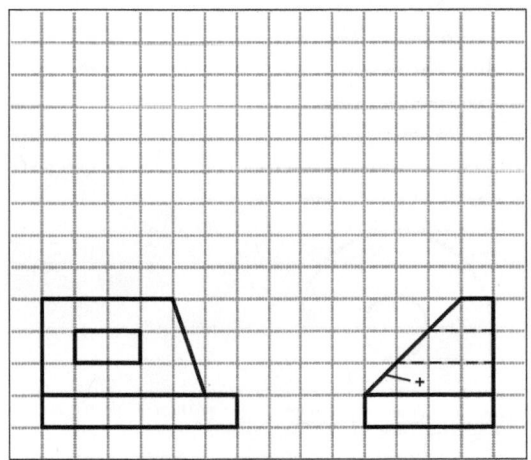

Figure P5-26

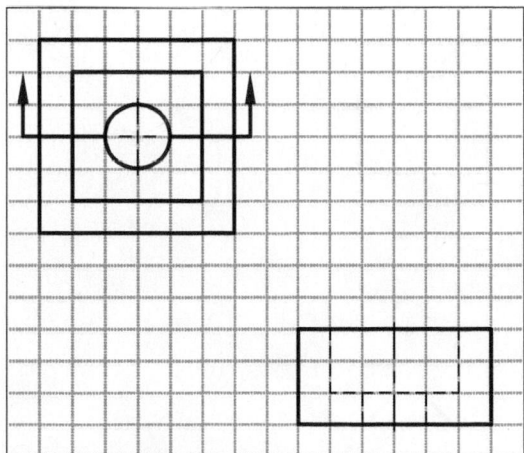

Figure P5-27

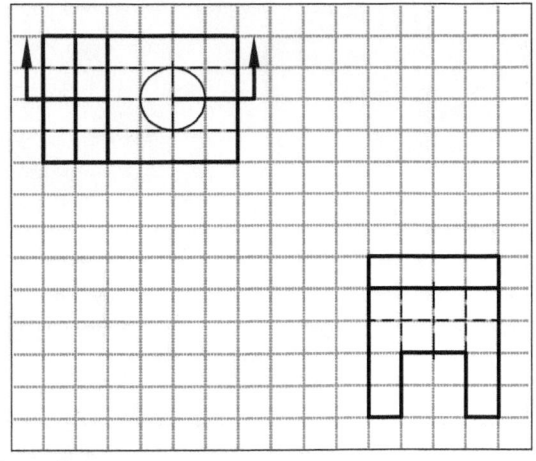

Figure P5-28

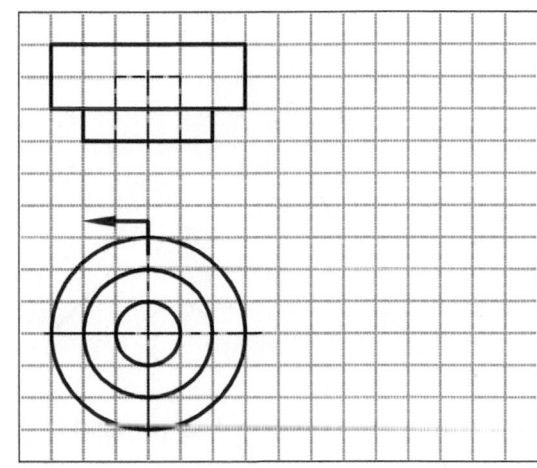

Figure P5-29

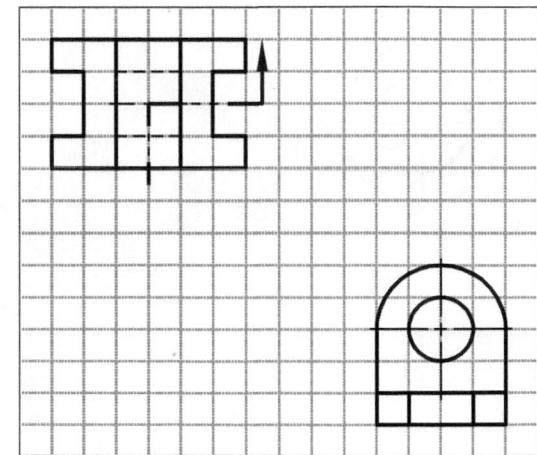

Figure P5-30

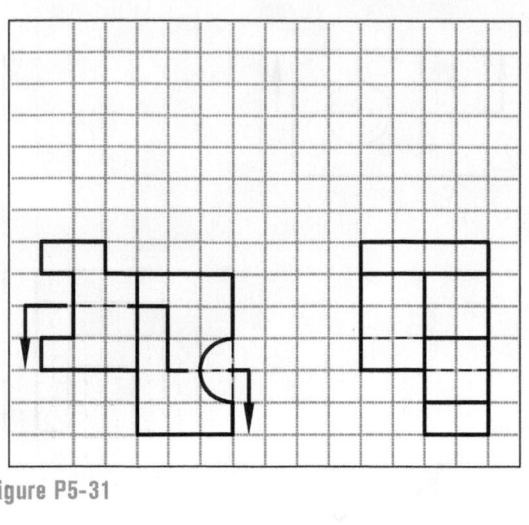

Figure P5-31

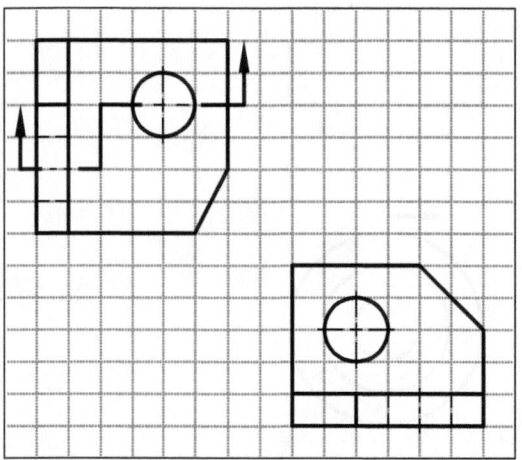

Figure P5-32

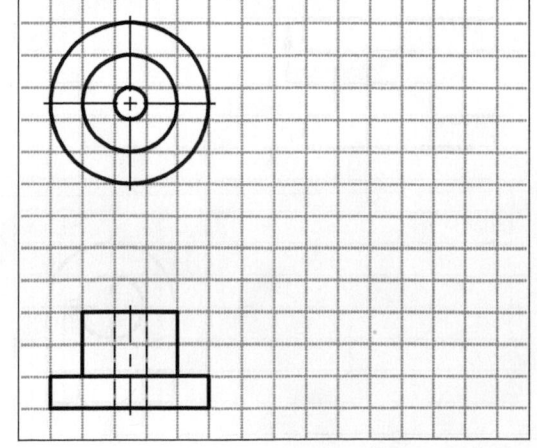

Figure P5-33

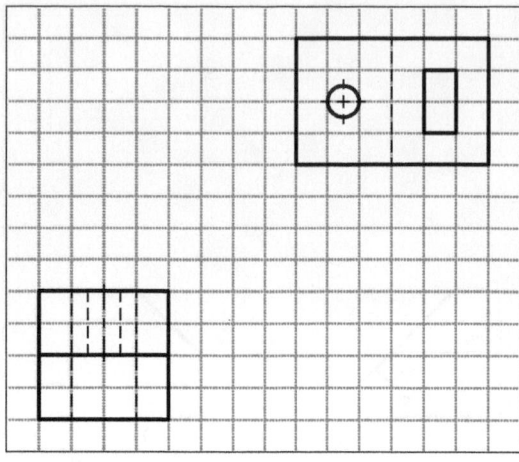

Figure P5-34

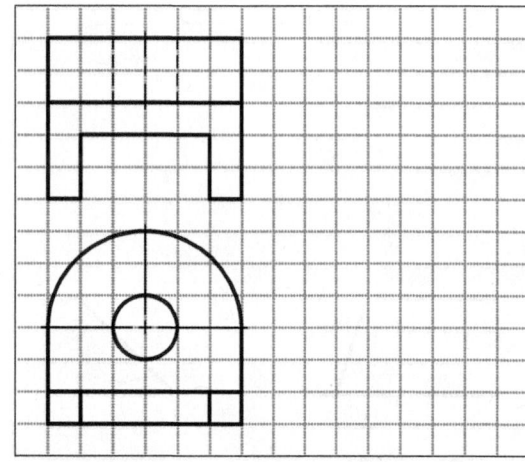

Figure P5-35

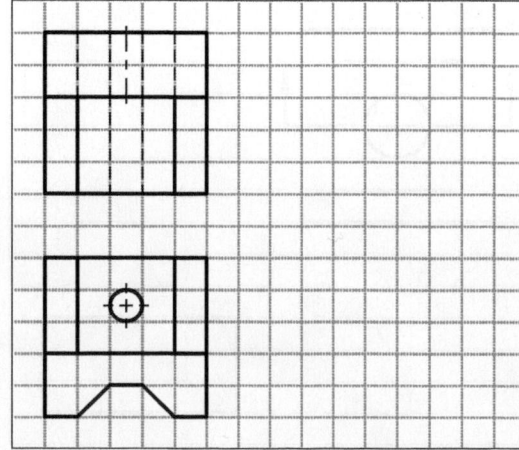

Figure P5-36

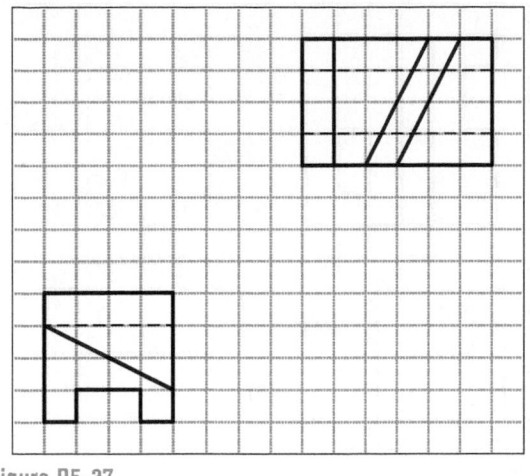

Figure P5-37

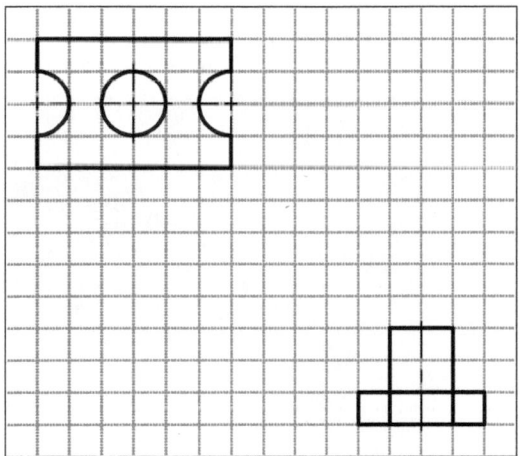

Figure P5-38

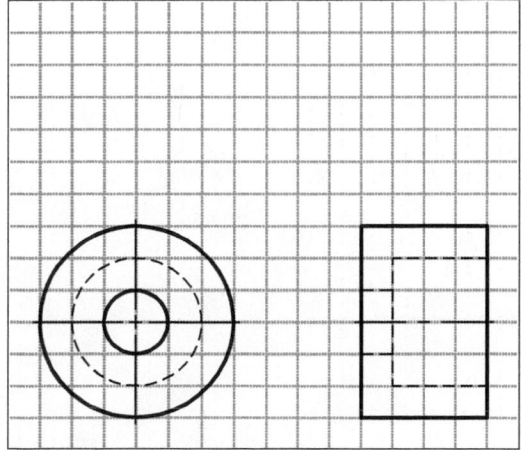

Figure P5-39

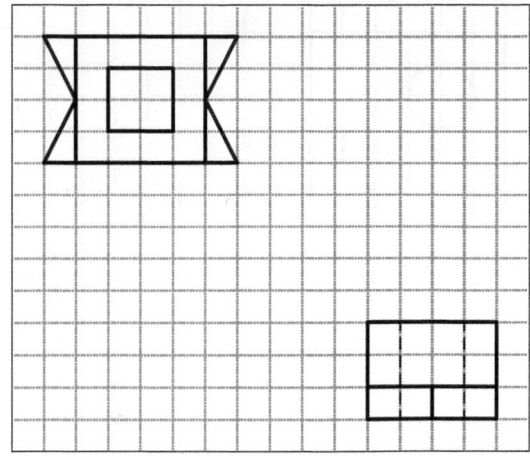

Figure P5-40

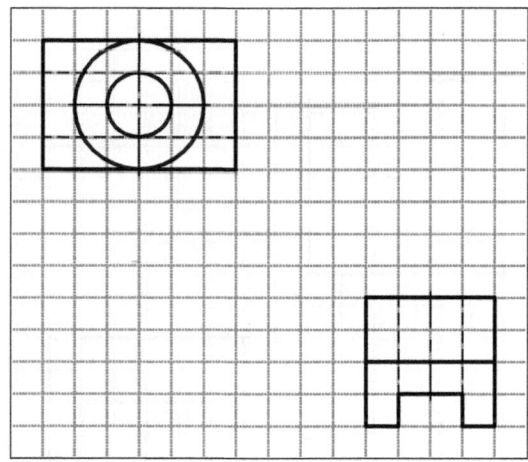

Figure P5-41

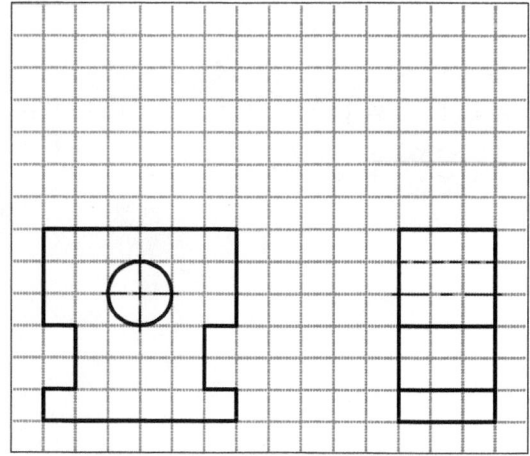

Figure P5-42

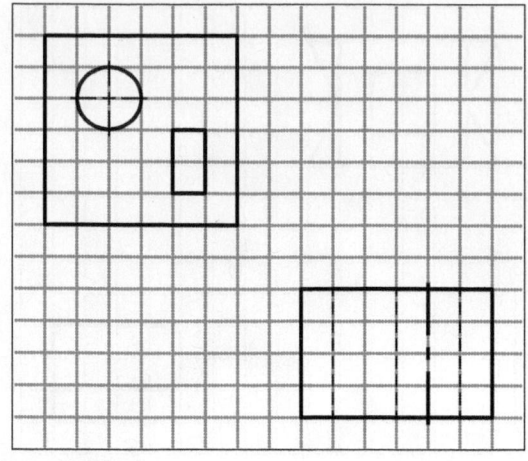

Figure P5-43

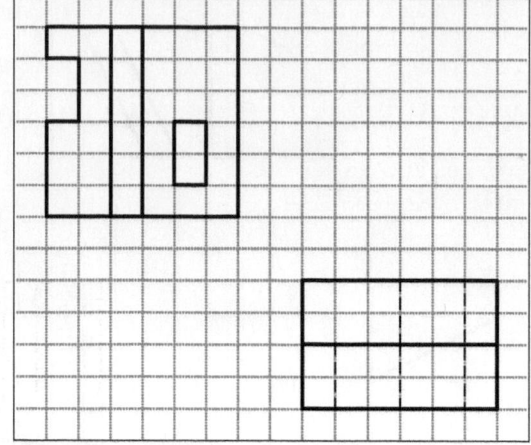

Figure P5-45

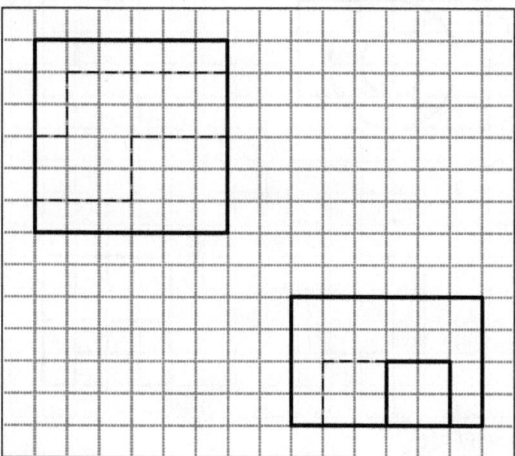

Figure P5-44

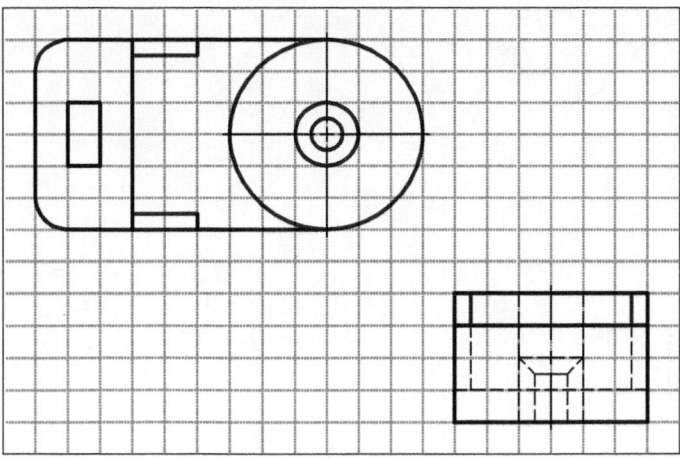

Figure P5-46

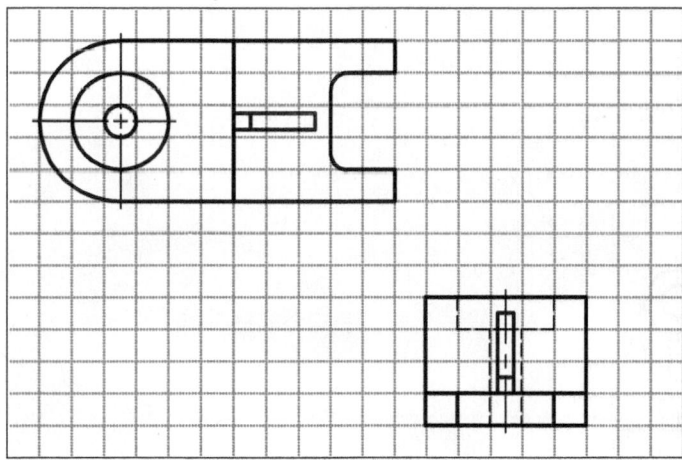

Figure P5-47

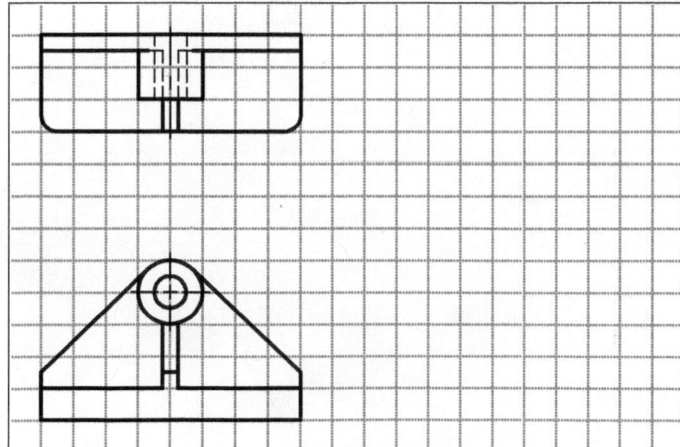

Figure P5-48

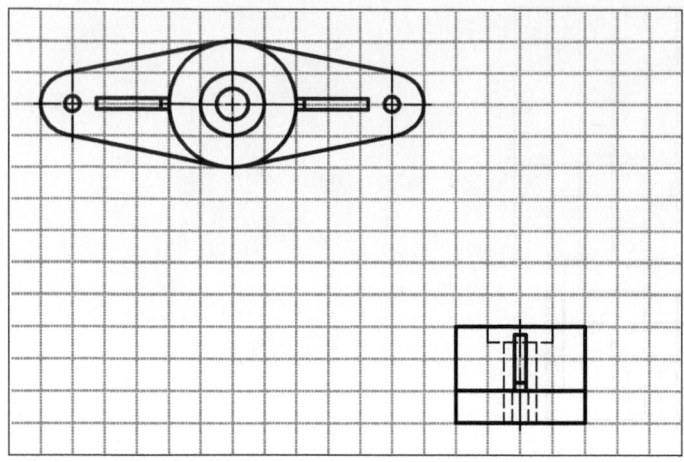

Figure P5-49

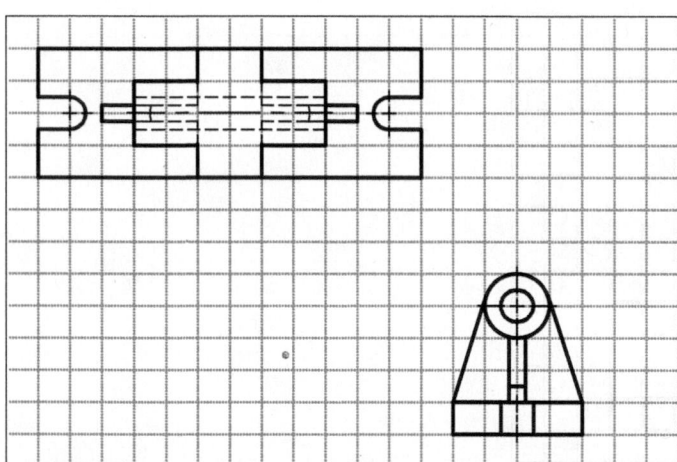

Figure P5-50

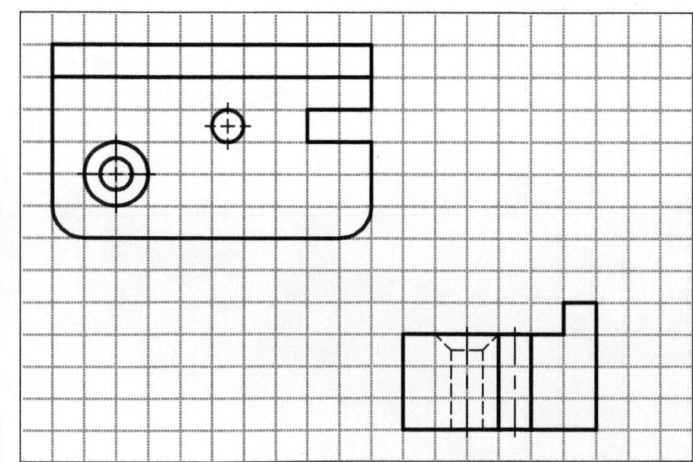

Figure P5-51

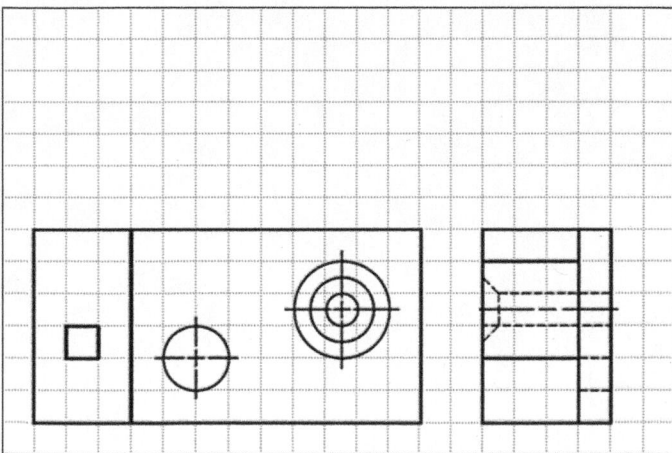

Figure P5-52

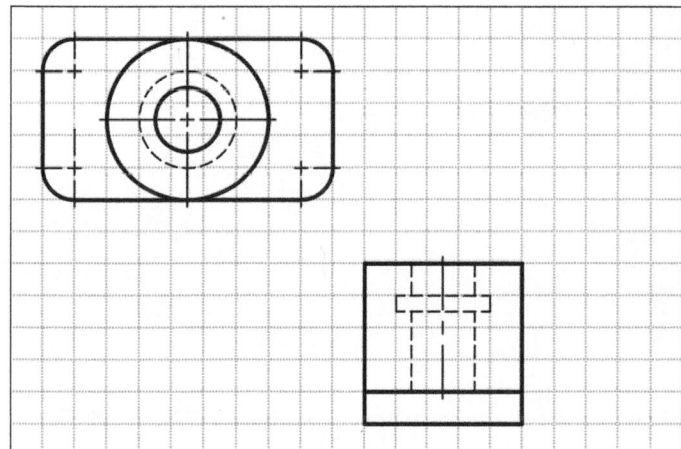

Figure P5-53

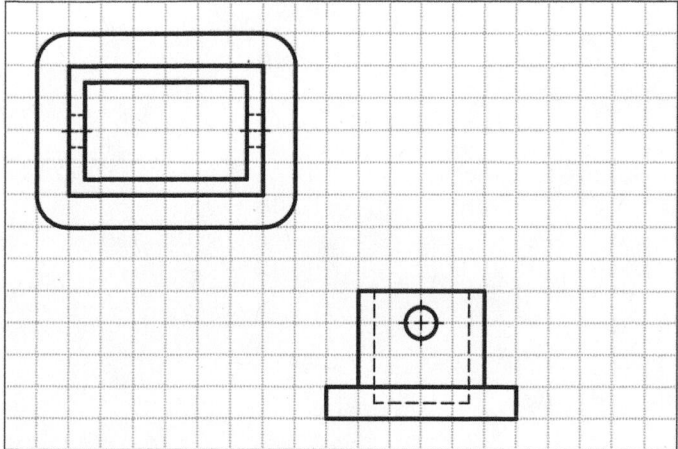

Figure P5-54

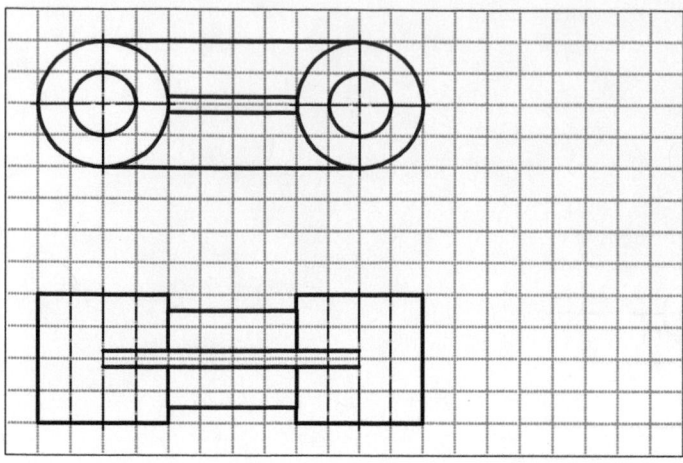

Figure P5-55

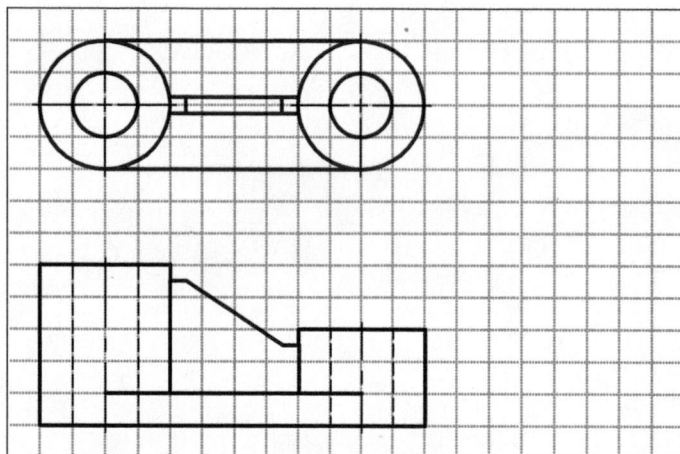

Figure P5-56

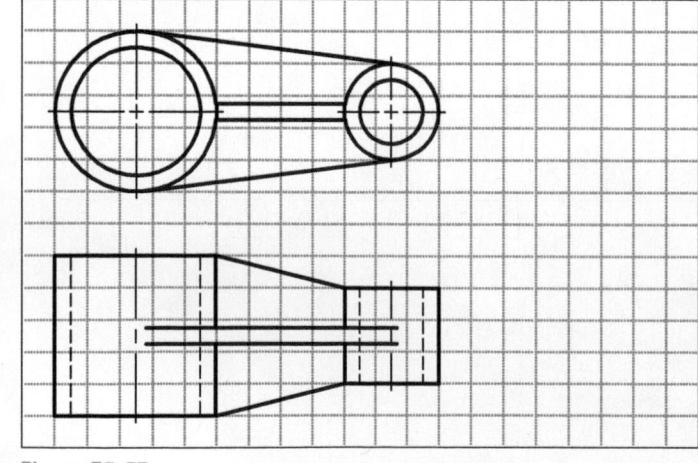

Figure P5-57

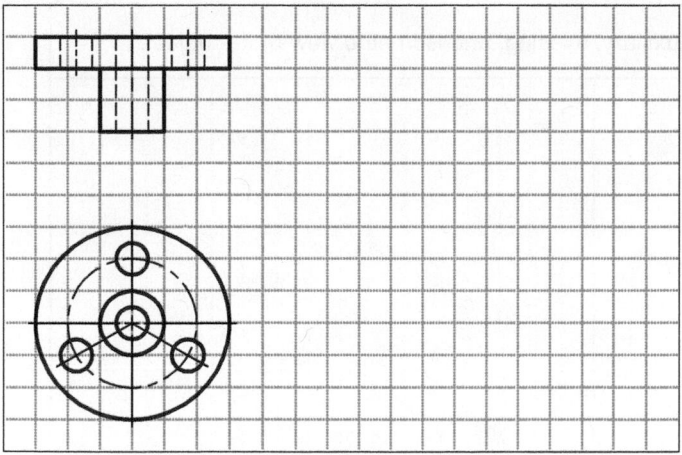

Figure P5-58

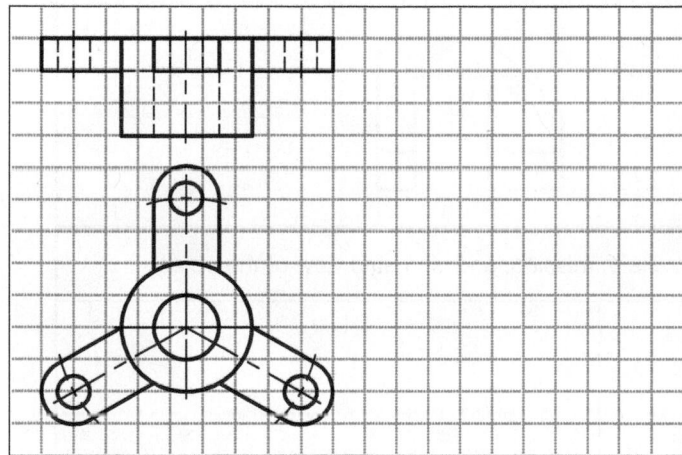

Figure P5-59

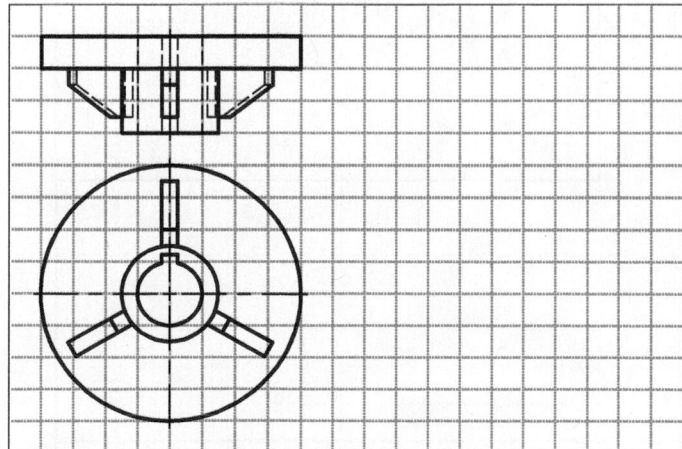

Figure P5-60

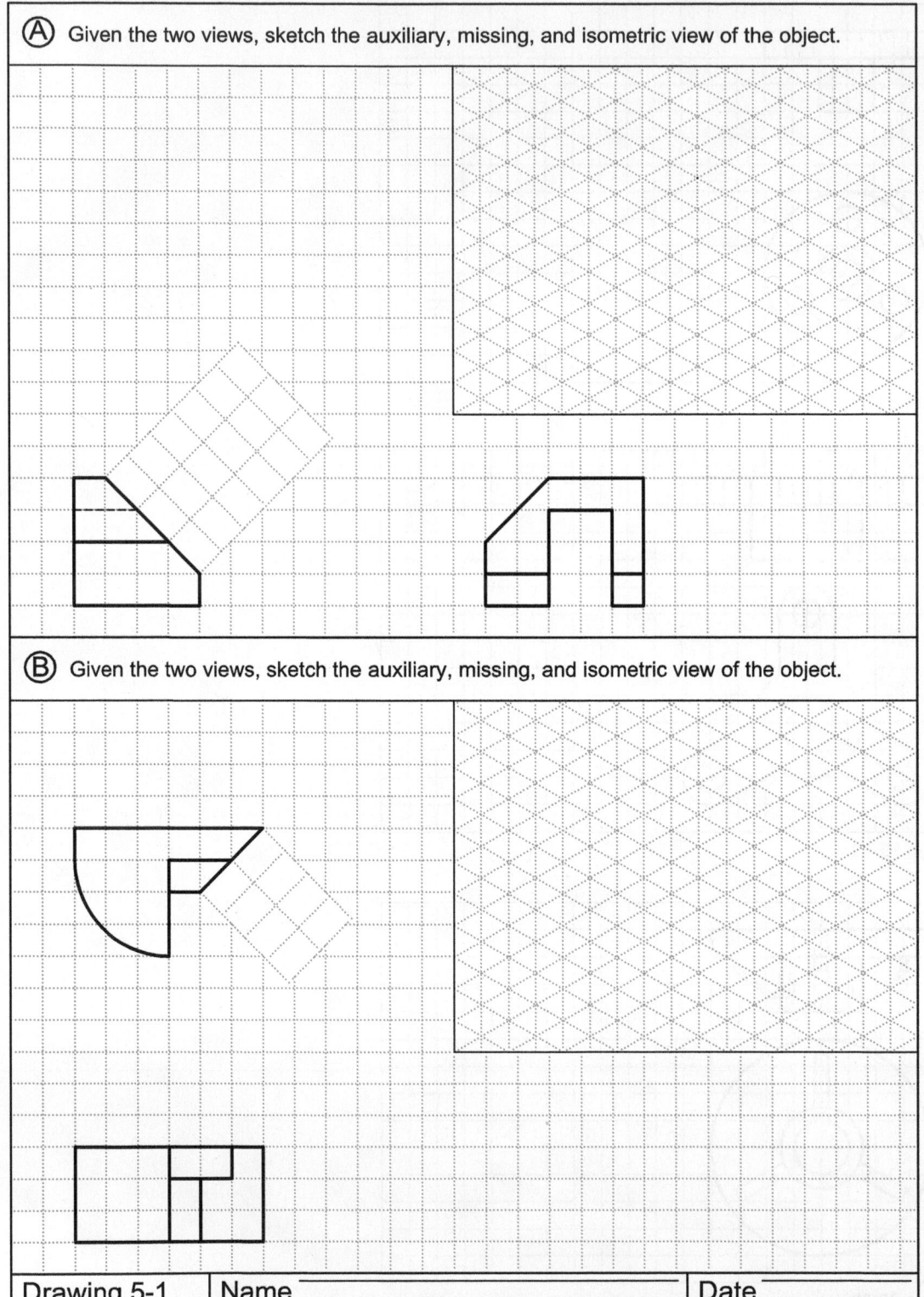

Ⓐ Given the two views, sketch the auxiliary, missing, and isometric view of the object.

Ⓑ Given the two views, sketch the auxiliary, missing, and isometric view of the object.

| Drawing 5-1 | Name _____ | Date _____ |

Ⓐ Given the two views, sketch the auxiliary, missing, and isometric view of the object.

Ⓑ Given the two views, sketch the auxiliary, missing, and isometric view of the object.

Drawing 5-2	Name	Date

Ⓐ Given the two views, sketch the auxiliary, missing, and isometric view of the object.

Ⓑ Given the two views, sketch the auxiliary, missing, and isometric view of the object.

| Drawing 5-3 | Name | Date |

Ⓐ Given the two views, sketch the auxiliary, missing, and isometric view of the object.

Ⓑ Given the two views, sketch the auxiliary, missing, and isometric view of the object.

| Drawing 5-4 | Name | Date |

(A) Given the two views, sketch the auxiliary, missing, and isometric view of the object.

(B) Given the two views, sketch the auxiliary, missing, and isometric view of the object.

| Drawing 5-5 | Name | Date |

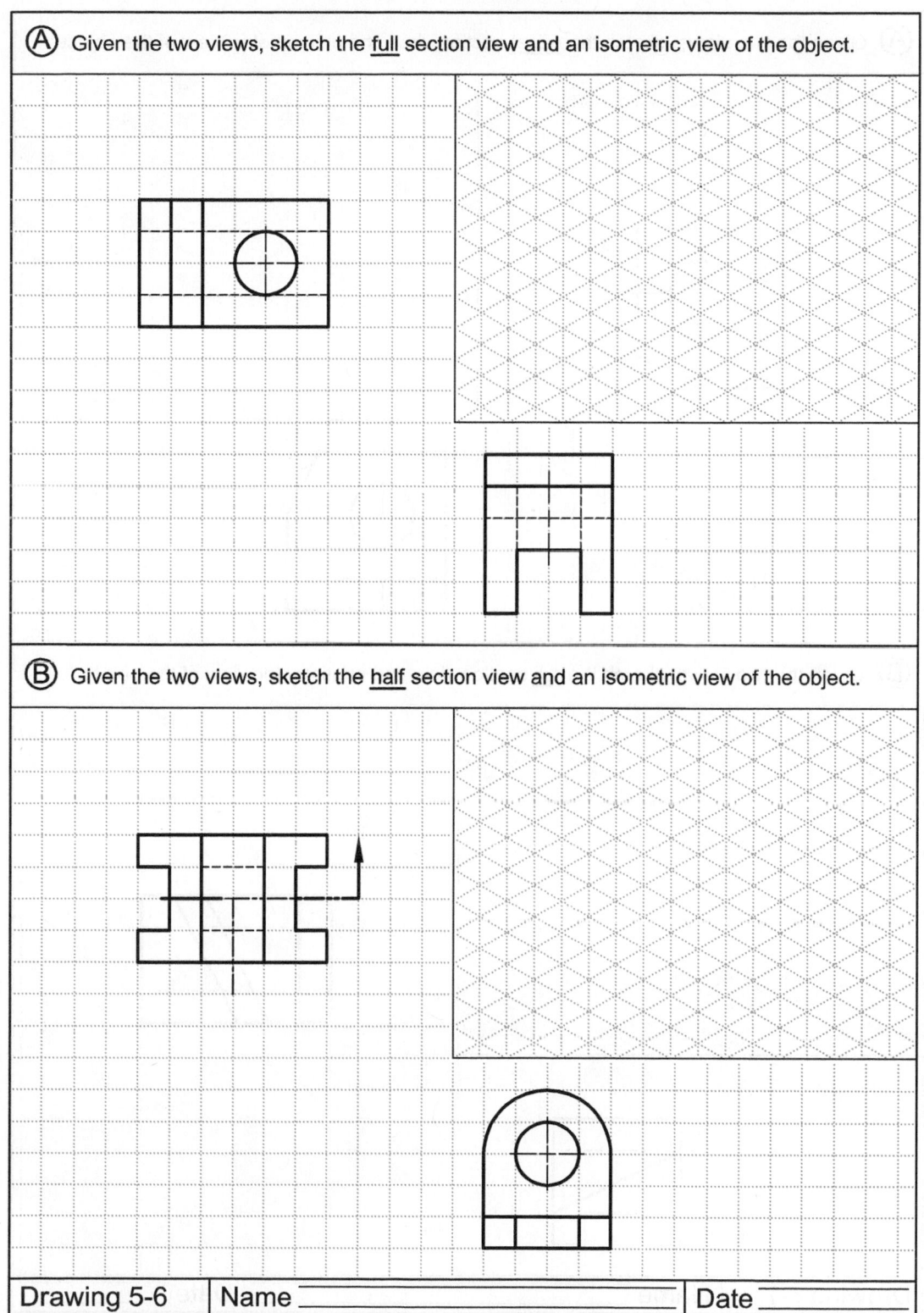

Ⓐ Given the two views, sketch the <u>full</u> section view and an isometric view of the object.

Ⓑ Given the two views, sketch the <u>half</u> section view and an isometric view of the object.

| Drawing 5-6 | Name | Date |

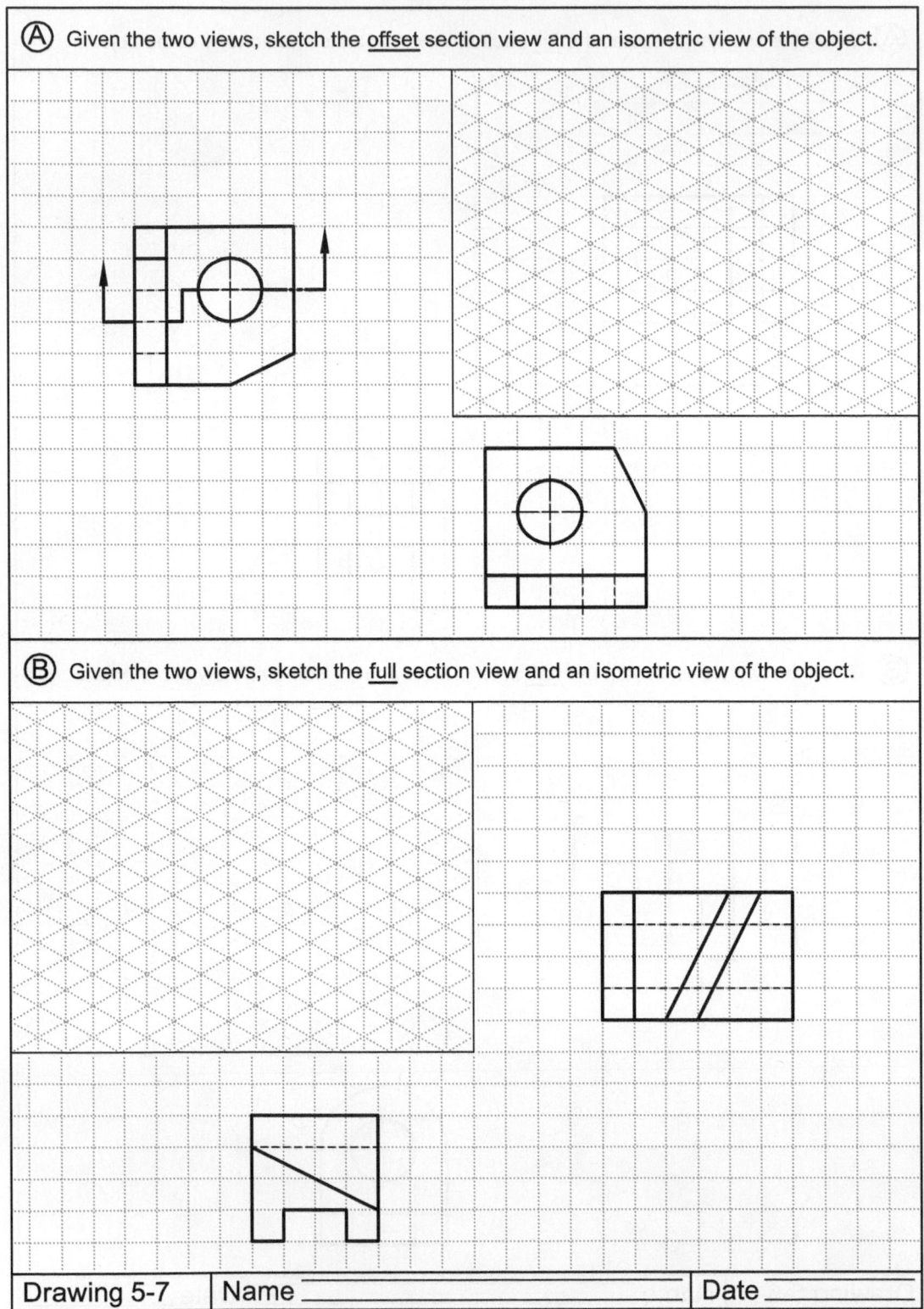

Ⓐ Given the two views, sketch the <u>offset</u> section view and an isometric view of the object.

Ⓑ Given the two views, sketch the <u>full</u> section view and an isometric view of the object.

Drawing 5-7 | Name | Date

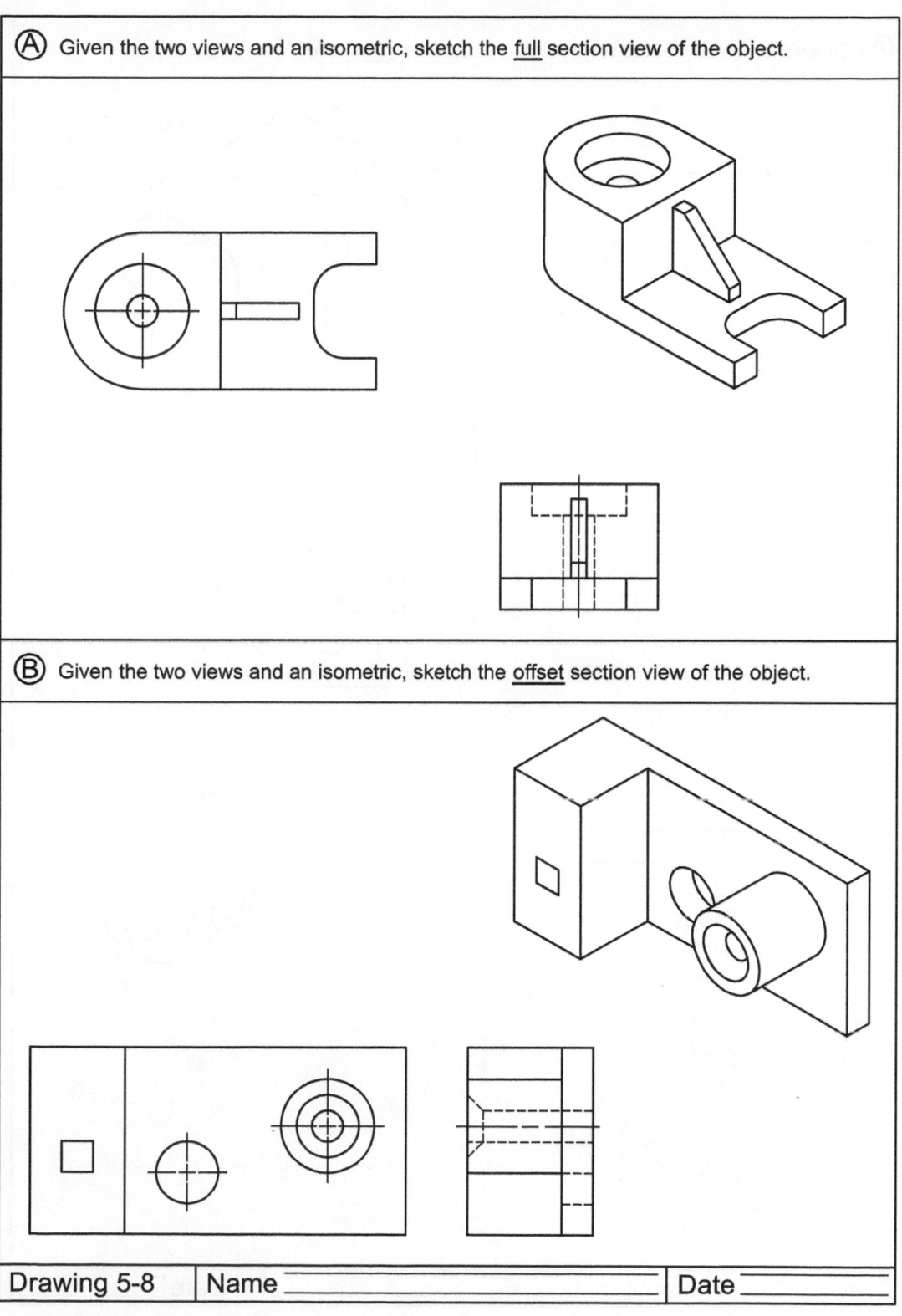

Ⓐ Given the two views and an isometric, sketch the __full__ section view of the object.

Ⓑ Given the two views and an isometric, sketch the __offset__ section view of the object.

Drawing 5-8	Name	Date

Ⓐ Given the two views and an isometric, sketch the **full** section view of the object.

Ⓑ Given the two views and an isometric, sketch the **aligned** section view of the object.

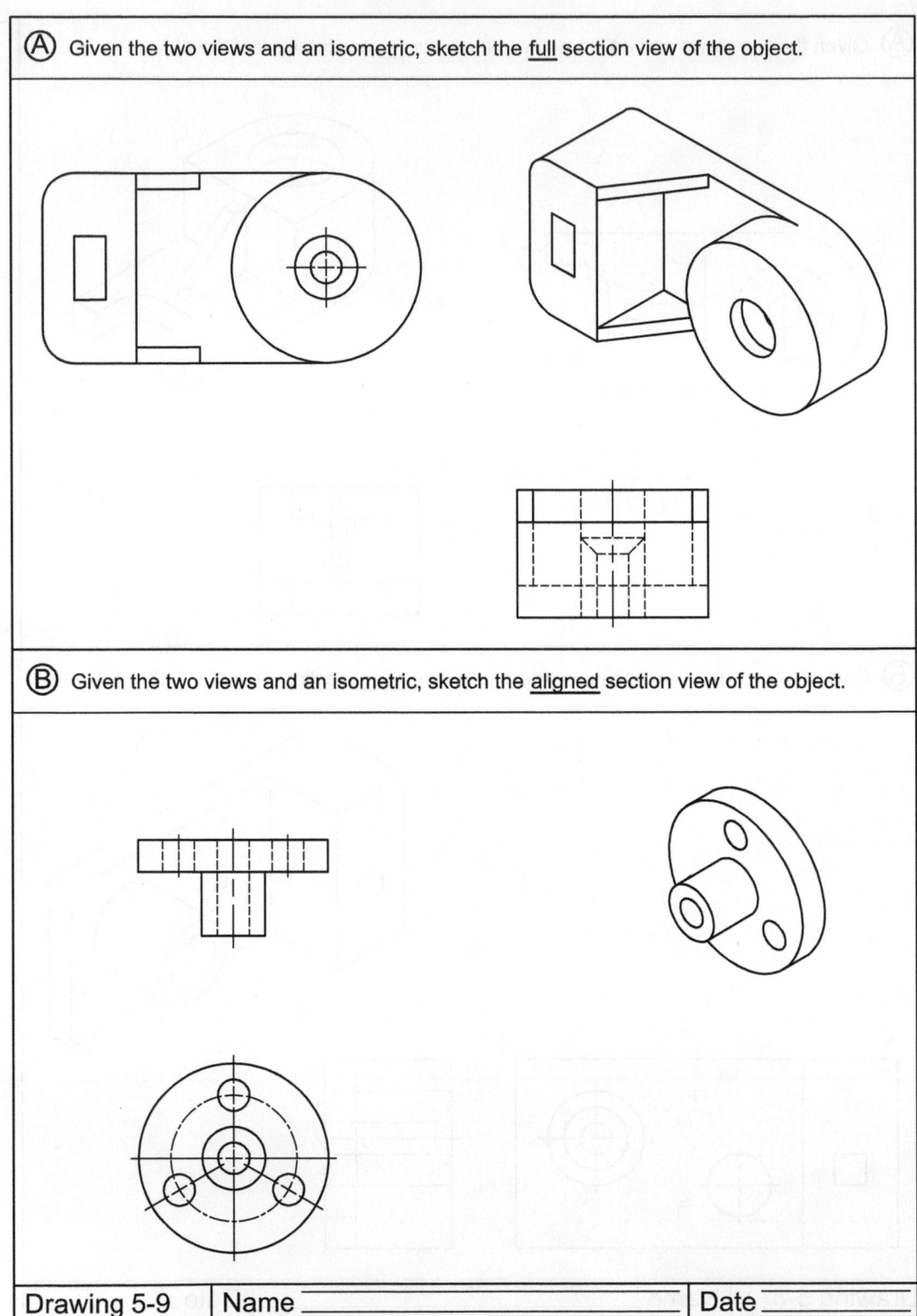

| Drawing 5-9 | Name | Date |

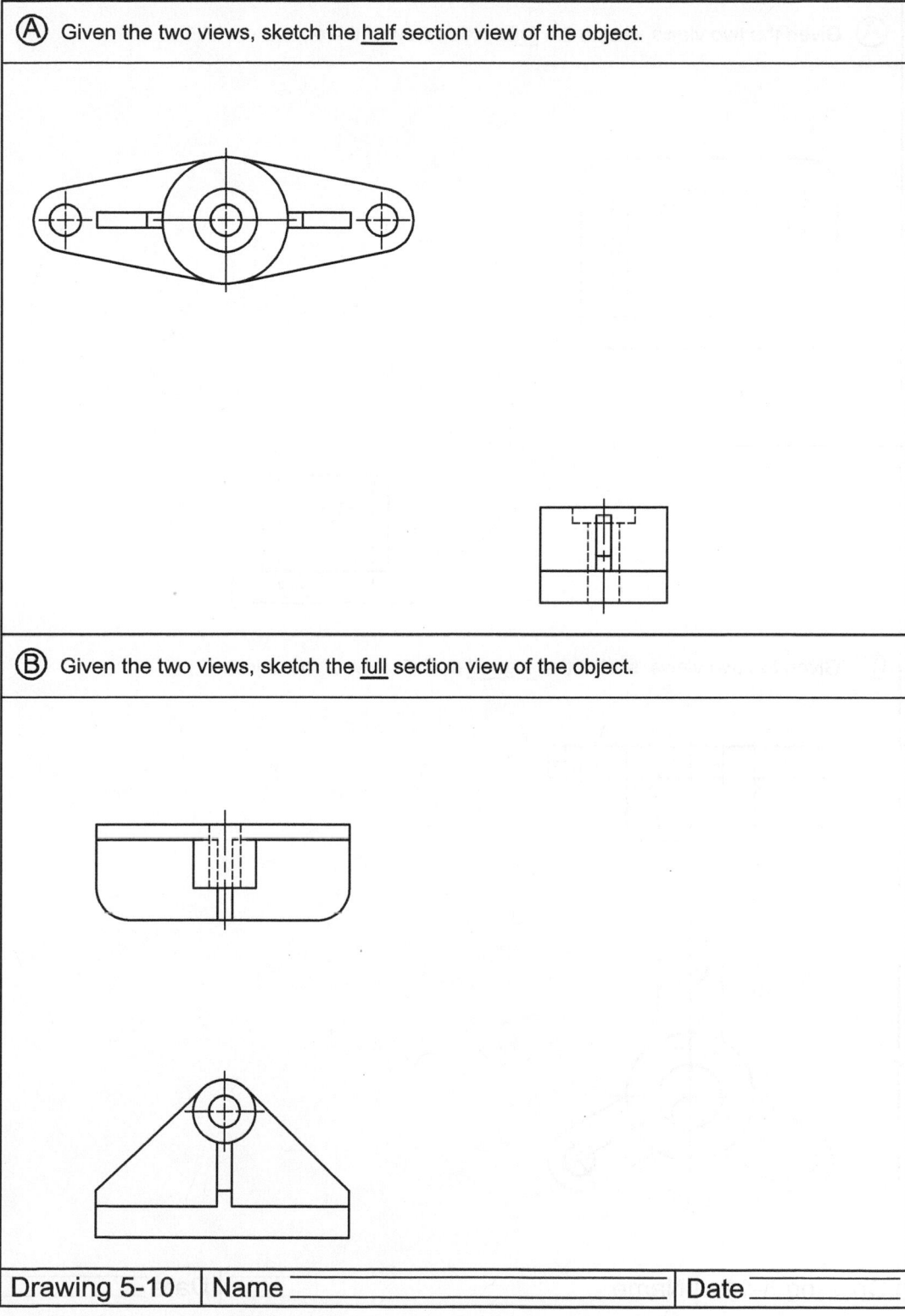

Ⓐ Given the two views, sketch the <u>half</u> section view of the object.

Ⓑ Given the two views, sketch the <u>full</u> section view of the object.

| Drawing 5-10 | Name | Date |

Ⓐ Given the two views, sketch the <u>full</u> section view of the object.

Ⓑ Given the two views, sketch the <u>aligned</u> section view of the object.

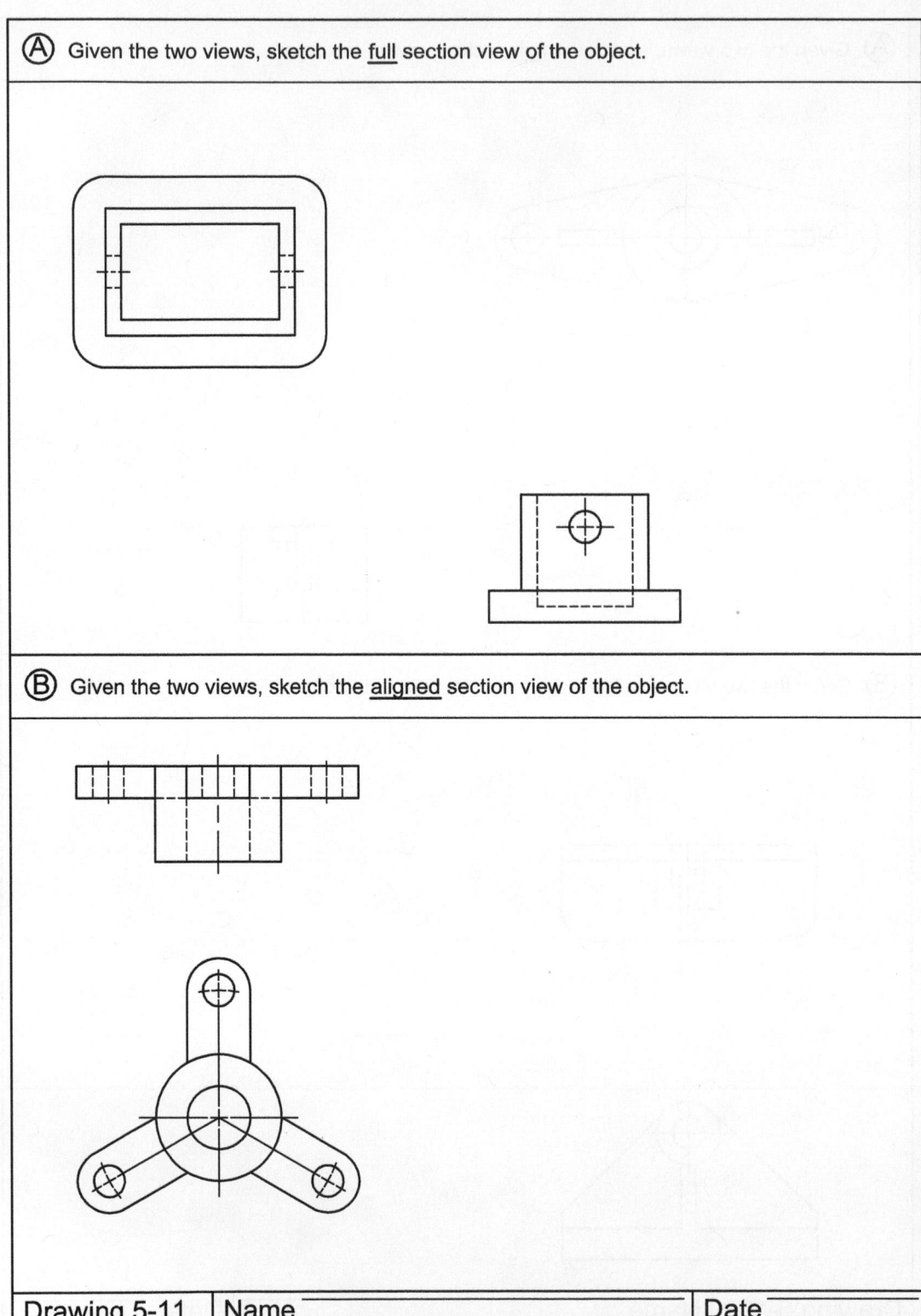

| Drawing 5-11 | Name | Date |

▌ DIMENSIONING

Introduction

An engineering drawing, once submitted to production for manufacture or construction, must include all of the information needed to build the part, assembly, or system. To this end, technical drawings include dimensions and general notes describing the size and location of part features, as well as details related to the construction or manufacture of the part.

A **dimension** is a numerical value used to define the size, location, geometric characteristic, or surface texture of a part or feature. The main goals of dimensioning (as laid out in ANSI/ASME Y14.5M, "Dimensioning and Tolerancing for Engineering Drawings"), are the following:

1. Use only the dimensions needed to completely define the part, nothing more.

2. Select and arrange dimensions to support the function and mating relationship of the part. It is important that the dimensioned part not be subject to differing interpretations.

3. In general, do not specify the manufacturing methods to be used in building the part. This is done both to leave options open to manufacturing and to avoid potential legal problems.

4. Arrange the dimensions for optimum readability. Dimensions should appear in true profile views and refer to visible object edges.

5. Unless otherwise stated, assume angles to be 90 degrees.

Units of Measurement

Drawings are typically dimensioned using either millimeters or decimal inches. Metric (Système Internationale) drawings normally employ millimeters specified as whole numbers, as shown in Figure 6-1a. In the English or Imperial system, the preferred units are inches expressed in decimal form, as seen in Figure 6-1b. In some disciplines, notably architecture and construction, fractional inches are still employed, but decimals are preferred because of the easier arithmetic and the greater precision they provide. Drawings in inches are typically specified using two-decimal-place accuracy. In both metric and English drawings, there is no need to specify the units of individual dimensions. Rather, a general note similar to "Unless otherwise stated, all dimensions are in millimeters (or inches)" appears.

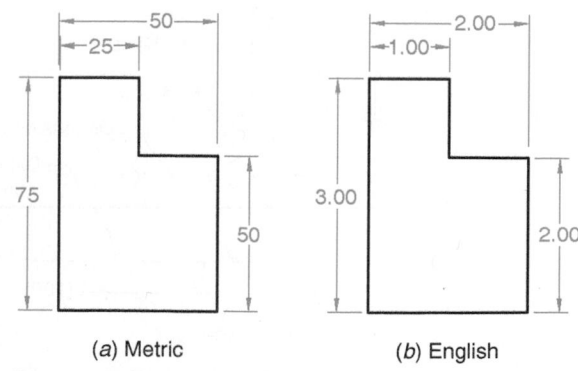

(a) Metric   (b) English

Figure 6-1 Metric versus English unit dimensions

Application of Dimensions

Dimensions are applied to a drawing through the use of dimension lines, extension lines, and leaders from a feature to a dimension or note. General notes are used to convey additional information.

Dimension lines are thin, dark lines used to show the direction and extent of a dimension. See Figure 6-2. A *dimension value* indicating the number of units of a measurement is associated with the dimension line. The height of the dimension value is typically 3 mm. By preference, dimension lines are broken to allow for the insertion of the dimension value. There is, however, an alternative dimension style that places the dimension value above an unbroken dimension line. Dimension lines are terminated by *arrows*, where the length of the arrowhead is equal to the dimension text height.

The dimension line of an angle being dimensioned is an arc drawn with its center at the apex of the angle.

As shown in Figure 6-2, thin, dark *extension lines* are typically drawn perpendicular to the associated dimension line. Extension lines extend from the view of an object feature to which they refer. A short, visible gap (1.5 mm) is included between the extension line and the view for clarity. In addition, the extension line extends 3 mm beyond its outermost related dimension line.

Leader lines are drawn from a feature to a note, dimension, or symbol. As shown in Figure 6-2, leaders are inclined straight lines, except for a small horizontal shoulder that extends to the mid-height of the first or last letter (or digit) of the note or dimension. These thin, dark lines start with an arrow at the feature being described. In certain circumstances, the feature is within an outline, in which case the arrow is replaced with a dot.

A *reference dimension* is used only for additional informational purposes. A reference dimension can be derived from other values shown on the drawing. It contains supplemental information and is not used for production or inspection purposes. Reference dimensions are

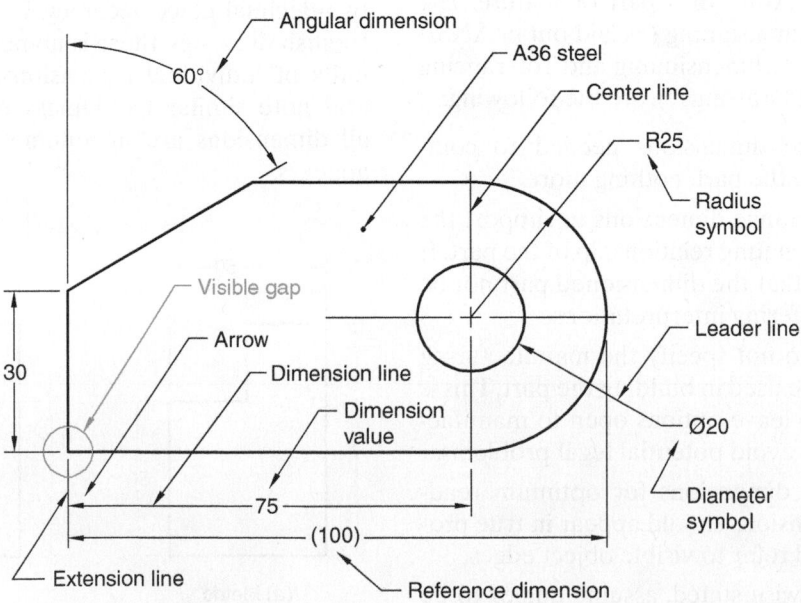

Figure 6-2 Terminology associated with dimensions

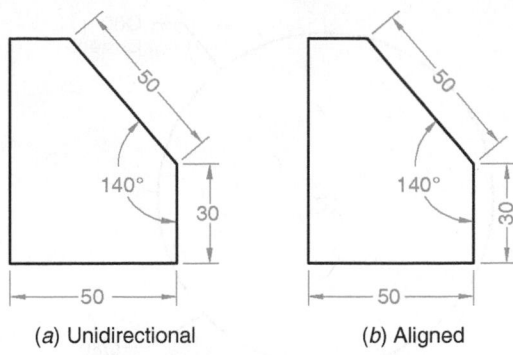

(a) Unidirectional   (b) Aligned

Figure 6-3 Unidirectional versus aligned dimensioning

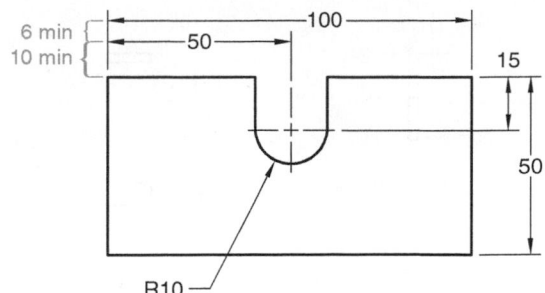

Figure 6-4 Spacing between parallel dimensions

easily identified because the associated dimensional value is placed in parentheses, as seen in Figure 6-2.

Thin, dark **center lines** also play a role in dimensioning, because they are used to locate the centers of cylindrical parts and holes.

READING DIRECTION FOR DIMENSIONAL VALUES

In *unidirectional dimensioning*, dimension values and text are oriented horizontally, as shown in Figure 6-3a. In an older dimensioning style called *aligned dimensioning*, dimensional values are oriented parallel to their dimension lines, as shown in Figure 6-3b. Aligned dimensioning is not recognized by ANSI.

ARRANGEMENT, PLACEMENT, AND SPACING OF DIMENSIONS

As was mentioned previously, dimensions are arranged for optimum readability. Several guidelines exist that govern the spacing, grouping, and staggering of parallel dimensions. There are also guidelines for dimensioning when space is limited.

A distance of at least 10 mm between the first dimension line and the part should be maintained. For succeeding parallel dimensions, this distance should be at least 6 mm (see Figure 6-4).

Parallel (i.e., either horizontal, vertical, or aligned) dimensions should be grouped and aligned, as shown in Figure 6-5, in order to present a uniform appearance.

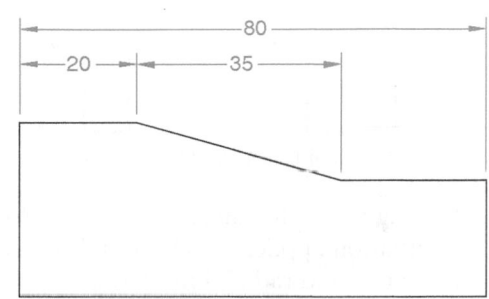

Figure 6-5 Grouping and alignment of parallel dimensions

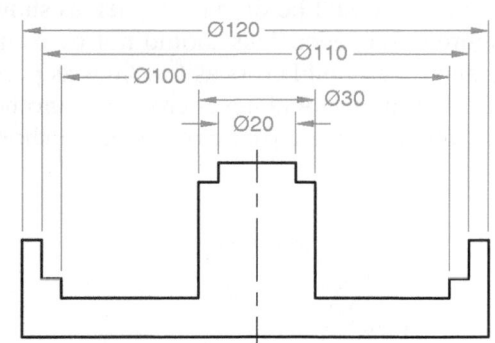

Figure 6-6 Staggering of parallel dimensions

The dimensional values of parallel dimensions should be staggered, as shown in Figure 6-6, to avoid crowding.

By preference, the dimensional value and arrows should appear inside the extension lines. Depending on the available space, however, it may be necessary to leave only the dimensional value inside, only the arrows inside, or nothing

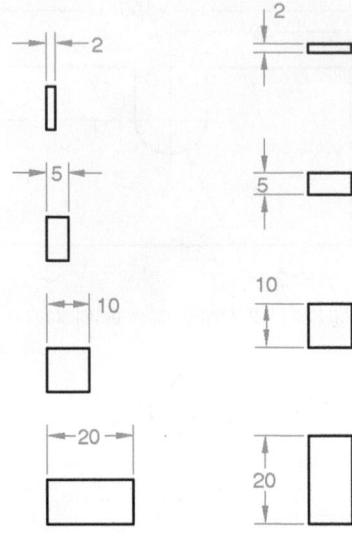

Figure 6-7 Placement of dimension text

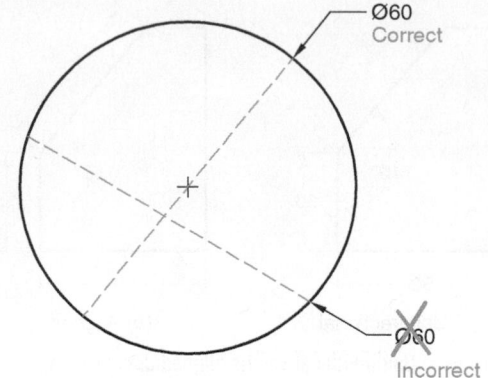

Figure 6-9 Leaders directed to circle or arc should pass through circle center when extended

inside. See Figure 6-7 for these possibilities. Note that this situation applies to horizontal, vertical, aligned, angular, and radial dimensions.

There are also guidelines for leaders that aim to improve the readability of a drawing. For instance, multiple leaders that are close to one another should be drawn parallel, as shown in Figure 6-8. Leader lines should not be overly long, and they should cross as few lines as possible. Leader lines should never cross one another. Finally, leaders directed to a circle or arc, as shown

in Figure 6-9, should be radial (i.e., pass through the center, if extended) to the hole or arc.

Using Dimensions to Specify Size and Locate Features

Dimensions are used to specify the size and location of features. Features are sized and located with linear (horizontal, vertical, or aligned), radial, diametric, and angular dimensions. Figure 6-10 provides a drawing with dimensions used to size the part features, and Figure 6-11, shows the same drawing, but with the dimensions used to locate part features now displayed.

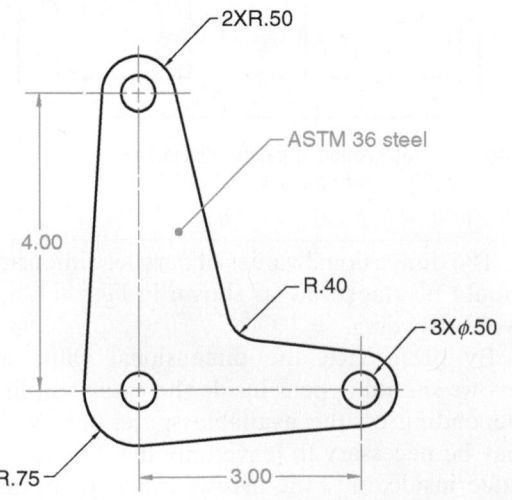

Figure 6-8 Multiple leaders in same vicinity should be parallel

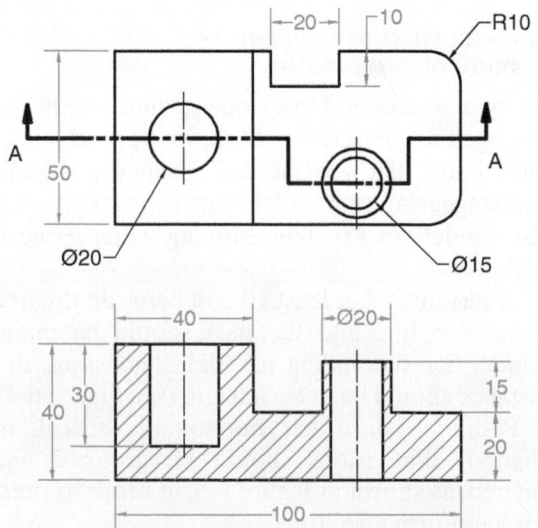

Figure 6-10 Dimensions used to size part features

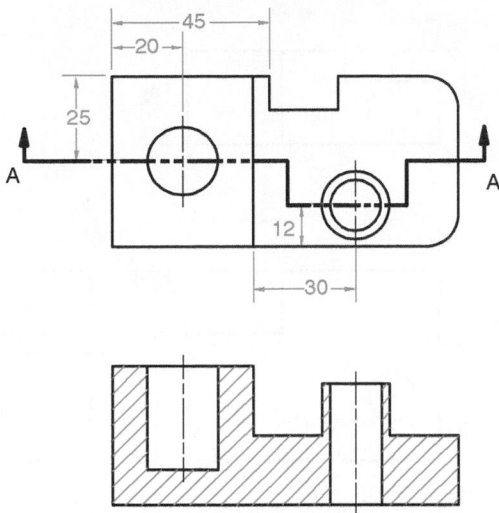

Figure 6-11 Dimensions used to locate part features

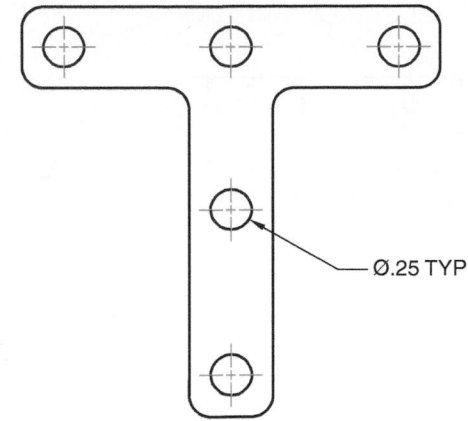

(*NOTE*: All fillets and rounds R.125)

Figure 6-13 General notes and abbreviations used in dimensioning

Symbols, Abbreviations, and General Notes

A number of symbols are employed in association with dimensioning. Figure 6-12 shows several of these symbols, including radius, diameter, spherical radius, spherical diameter, counterbore, countersink, deep, and times.

Whenever several features (e.g., holes, fillets, rounds) of the same type and size appear in a drawing, either a general note or the abbreviation TYP (for typical) may be used, as shown in Figure 6-13. Also note that the *times* symbol (i.e., 2 × R10) is also used to dimension multiple features of the same size (e.g., Figure 6-8).

Symbol Name	Symbol
Counterbore	⊔
Countersink	∨
Deep	▽
Diameter	Ø
Square	□
Places, Times	X
Radius	R
Spherical radius	SR
Spherical diameter	SØ

Figure 6-12 Dimensioning symbols

Dimensioning Rules and Guidelines

In this section several dimensioning rules or guidelines will be discussed. Note that on occasion these rules may be violated because of part complexity, lack of space, conflict with other rules, and the like. Rules concerning prismatic shapes will first be covered, followed by rules concerning cylinders and arcs.

PRISMS

1. *Do not repeat dimensions.* The depth of the object shown in Figure 6-14 on page 188 is 30. Although this dimension appears in the top view, it could just as well have been placed on the right view. In no case, however, should this same dimension appear in both views.

2. *Apply dimensions to a feature in its most descriptive view.* The object shown in Figure 6-14 contains a single (extruded "L" shape) feature. The most descriptive view of this feature is the front view. Note that four of the five dimensions needed to fully constrain the object appear on the front view.

3. *Dimension between views.* The object shown in Figure 6-14 contains two width dimensions (50 and 80) and two height dimensions (40 and 60). Because the width is shared

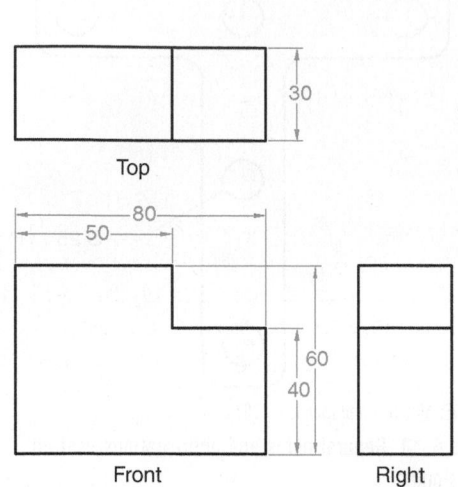

Figure 6-14 Rules and guidelines for dimensioning prisms

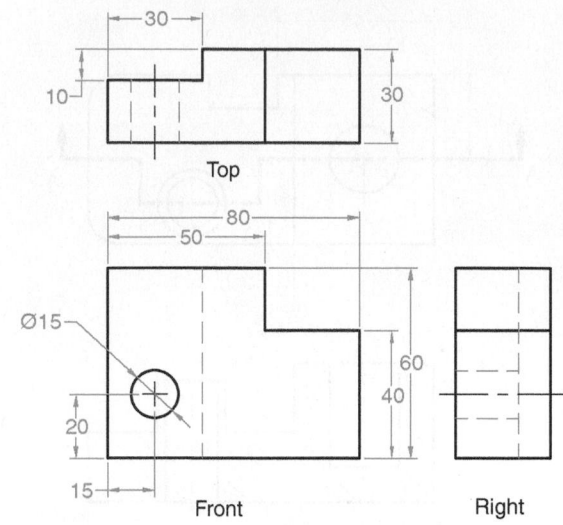

Figure 6-15 More rules and guidelines for dimensioning prisms

by the front and top views, these width dimensions are placed between these views. Likewise, the height dimensions are placed between the front and right views.

4. *Omit one (intermediate) dimension.* The object in Figure 6-14 contains two intermediate width dimensions, 50 and 30 (since 80 – 50 = 30). Only one of them is shown. Likewise, there are two intermediate height dimensions, 40 and 20 (since 60 – 40 = 20). Only one of them is shown. This is done to avoid cluttering the drawing with unnecessary dimensions, and also to avoid ambiguity in specifying tolerances (see the section on tolerance accumulation later in this chapter). As for which of the intermediate dimensions to include on the drawing, choose the one that is easiest to measure with calipers—in this case the width dimension of 50 and the height dimension of 40.

5. *Place smaller dimensions inside of larger dimensions.* Note that in Figure 6-14 the intermediate dimensions (50, 40) are placed closer to the view than the principal dimensions (80, 60). This practice helps to keep the drawing organized by avoiding the need

for extension lines that cross dimension lines.

Two additional features, an extruded cut and a hole, have been added to the object depicted in Figure 6-14 and can be seen in Figure 6-15.

6. *Dimension to visible object lines, not to hidden lines.* The dimensions of the newly added cut, 30 and 10, are placed on the view that best describes this feature (i.e., the top view), and not in either the front (30) or the right (10) view, where it would have been necessary to apply a dimension to a hidden line.

7. *Keep dimensions outside of the views.* The diameter of the through hole should be placed outside the view. If the drawing view is cluttered or the leader line needs to be extremely long to place the dimension outside the view, then this rule may be overridden.

8. *Extension lines may cross object lines and other extension lines.* In general it is desirable to avoid lines that cross; however, it is permissible for an extension line to cross an object line (e.g., extension lines for the dimensions of 20 and 15 locating the hole center) or another extension line (80 and 60).

Figure 6-16 Use of calipers to measure the diameter of a cylinder

CYLINDERS AND ARCS

As shown in Figure 6-16, calipers are used to directly measure the diameter of solid cylinders and round holes. It is for this practical reason that the diameter, rather than the radius, of circular features is specified on engineering drawings.

9. *Dimension the diameter of cylindrical parts in their rectangular view.* In Figure 6-17 the diameter of the boss (raised cylindrical) feature is dimensioned in its rectangular (i.e., front) view. Note that the diameter

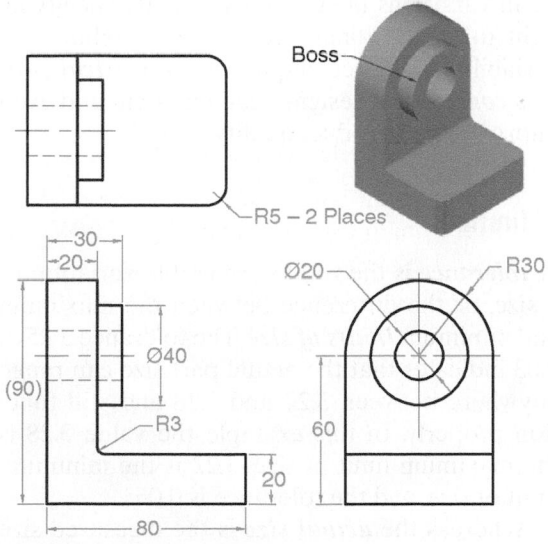

Figure 6-17 Rules and guidelines for dimensioning cylinders and arcs

symbol Ø (the Greek letter phi), precedes the dimension.

10. *Dimension the diameter of cylindrical holes in their circular view.* In Figure 6-17 the diameter of the through hole is dimensioned in the right view, where the hole appears as a circle.

11. *Dimension the radius of circular arcs in the view where their true shape is seen.* Note that the symbol R precedes the dimensional value of the radius. As seen in Figure 6-17, this rule applies to fillets (R3), rounds (R5), and other circular arcs (R30). For arcs less than or equal to 180 degrees, specify the radius. For arcs greater than 180 degrees, specify the diameter.

Some additional comments regarding Figure 6-17 are in order. First, note the use of the note Places (i.e., R5 – 2 Places) to eliminate the need for an additional R5 dimension. Similarly, the abbreviation TYP is common. Also note that the R30 radial dimension eliminates the need to dimension the overall depth (60) of the object. Similarly, the overall height (90) is not needed, although it is provided in the form of a reference dimension.

Finally, note that the overall width dimension (80) appears at the bottom of the front view, in apparent violation of Guideline 3, by which the width dimensions should appear between the front and top views. This is due to yet another guideline:

12. *Avoid overly long extension and leader lines.* Placing the 80 width dimension at the bottom, rather than the top, of the front view significantly reduces the length of the associated extension line.

Finish Marks

Parts formed by casting have rough external surfaces. When these cast parts are used in an assembly, surfaces in contact with other parts are machined or finished, in order to provide a smooth mating surface, reduce friction, and so on. In an engineering drawing, a finish mark symbol (√) is used to indicate that a surface is to be machined.

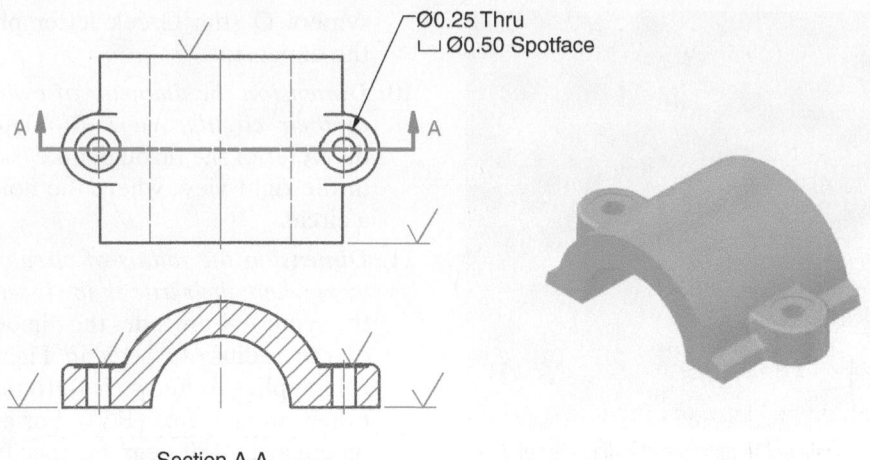

Ø0.25 Thru
⌴ Ø0.50 Spotface

Section A-A

Figure 6-18 Finish marks

As seen in Figure 6-18, finish marks are applied to all edge views, whether visible or hidden, of finished part surfaces.

■ TOLERANCING

Introduction

In manufacturing, the same process is typically employed to mass produce a single part. These parts are then combined with other, similarly mass-produced parts to create commercial products. Clearly these mass-produced parts must be interchangeable. However, when inspecting a batch of parts produced by the same manufacturing process, we would not find two parts that are exactly the same. In even the most precise manufacturing process, slight variations in part size are found.

Tolerancing is a dimensioning technique used to ensure part interchangeability by controlling the variance that exists in manufactured parts. This is accomplished by specifying a range within which a dimension is allowed to vary. As long as the size and location of part features fall within this tolerance zone, the part should function properly within an assembly.

Tolerancing is critical to the success of manufacturing. Beyond ensuring the interchangeability of parts, tolerancing directly influences both the cost and the quality of manufactured parts. Parts that are made to high accuracy are expensive. Depending on the type of product, extremely accurate parts may not be warranted. For example, the parts used to make a plastic toy do not need to be as accurate as automotive parts. As a general rule, tolerances should be stated as generously as possible, while still ensuring that the part will function properly. Doing so allows for the possibility of using a wider variety of processes to manufacture the part, and consequently it helps keep part costs low.

Manufacturing quality is primarily a function of part accuracy. High-quality parts exhibit small variations in size and shape. By specifying tight tolerance zones and then controlling part variability using techniques like *statistical process control*, the designer can maintain and even improve upon product quality.

Definitions

A *tolerance* is the total permissible variation of a size, or the difference between the maximum and minimum *limits of size*. The tolerance 3.25 ± 0.03 indicates that the actual part size can range anywhere between 3.22 and 3.28 and still function properly. In this example the value 3.28 is the maximum limit of size, 3.22 is the minimum limit of size, and the tolerance is 0.06.

Whereas the *actual size* is the measured size of a finished part, the *basic size* is the theoretical size from which a tolerance is assigned. In the example cited above, 3.25 is the basic size, and the actual size will fall between 3.22 and 3.28, assuming that it is within tolerance.

Tolerance Declaration

Tolerances may be expressed in different ways, including:

1. Direct tolerancing methods
2. General tolerance notes
3. Geometric tolerances

Direct tolerancing methods include (1) limit dimensioning and (2) plus-and-minus tolerancing. In limit dimensioning, the limits of size are directly represented as part of the dimension. The upper limit (maximum value) is placed above the lower limit (minimum value). When this is expressed in a single line, the lower limit precedes the upper limit. Figure 6-19 provides some examples of limit dimensioning.

In plus-and-minus tolerancing, the basic dimension is given first, followed by a plus-and-minus expression of the tolerance. Plus-and-minus tolerances may be either **unilateral** or **bilateral**. A unilateral tolerance is permitted to vary in only one direction from the basic size. Bilateral tolerances, on the other hand, may vary in either direction from the basic size. The variance of bilateral dimensions may be either equal or unequal. Figure 6-20 provides some examples of plus-and-minus tolerance dimensioning.

General notes such as "ALL DIMENSIONS HELD TO ±0.05" are sometimes used on engineering drawings. Such a note indicates that all dimensions that appear on the drawing should fall within 0.05 inch of the basic size.

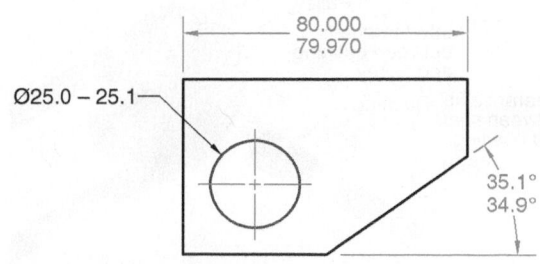

Figure 6-19 Limit dimensioning

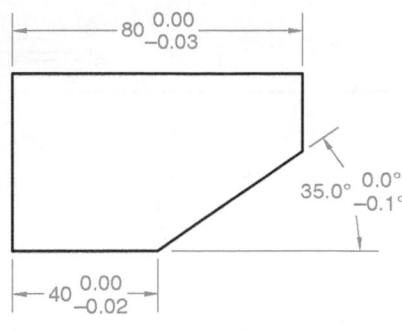

(*a*) Unilateral tolerancing

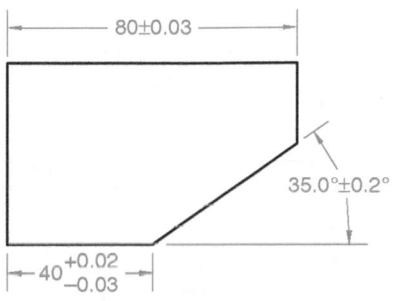

(*b*) Bilateral tolerancing

Figure 6-20 Plus-and-minus dimensioning

Figure 6-21 on page 192 provides an example of an object that has been dimensioned using geometric dimensioning and tolerancing (GD&T) techniques. GD&T is not covered in this text.

Tolerance Accumulation

When the location of a feature depends on more than one tolerance value, these tolerances will be cumulative. In the **chain dimensioning** technique employed in Figure 6-22*a* on page 192, for example, the tolerance accumulation between surfaces X and Y is ±0.03. The tendency for tolerances to stack up, or accumulate, can be reduced by using **base line dimensioning**, where all dimensions of a given type are specified from the same datum, as shown in Figure 6-22*b*. In this case, the tolerance variation between surfaces X and Y is reduced to ±0.02. If necessary, the tendency for tolerances to accumulate can be further controlled by **direct dimensioning**. The maximum variation between the directly dimensioned surfaces X and Y in Figure 6-22*c* is ±0.01.

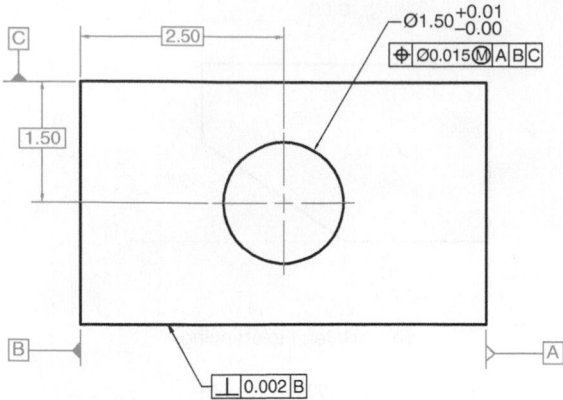

Figure 6-21 Geometric dimensioning and tolerancing

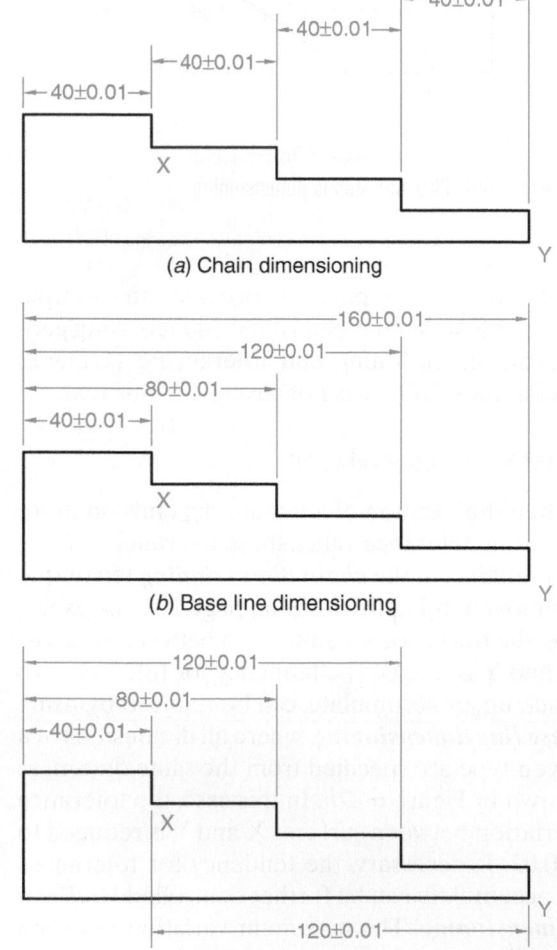

(a) Chain dimensioning

(b) Base line dimensioning

(c) Direct dimensioning

Figure 6-22 Accumulation of tolerances

Mated Parts

The tolerance of a single standalone part is of little importance. When a part is mated with other parts in an assembly, the true value of tolerancing becomes apparent. Mated parts must be toleranced as a system to fit within a prescribed degree of accuracy. Figure 6-23, for example, shows a shaft, bushing, and pulley assembly. The shaft must be able to turn freely within the bushing, while the bushing is force-fit into the pulley.

Fit refers to the degree of tightness or looseness between two mating parts. In a *clearance fit*, the internal member (e.g., shaft) is always smaller than the external member (e.g., hole). The shaft and bushing in Figure 6-23 have a clearance fit. The shaft is free to turn inside the hole.

In an *interference fit*, the internal member is always larger than the external member. An interference fit requires that the two parts be forced together, as is the case with the bushing and pulley shown in Figure 6-23. Note that an interference fit fastens two parts together without using adhesive or mechanical fasteners.

A *transition fit* ranges between a pure clearance fit and a pure interference fit. In a transition fit, either the internal shaft or the external hole may be larger, so that parts either slide or are forced together. If an assembly calls for a transition fit, the two sets (hole, shaft) of components can be measured and sorted into groups according to size (e.g., small, medium, large). The components are then assembled, with components from one group being mated with corresponding components from the matching group. This method, known as *selective assembly*, is a relatively inexpensive way to manufacture tight clearance or interference fits.

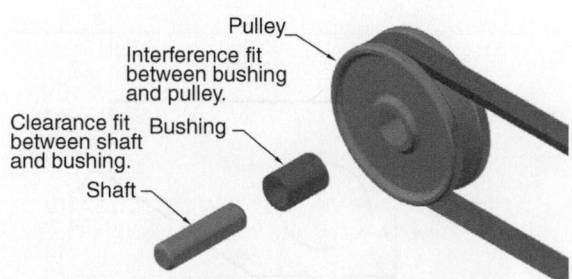

Figure 6-23 Pulley assembly tolerance fits

In a **_line fit_**, one of the limits on both the hole and the shaft are equal, which means that the shaft and hole may have the same size.

Figure 6-24 shows examples of clearance, interference, transition, and line fits, along with their upper and lower limits. Later in this chapter we will see that each of these classes of fit can be further categorized into subclasses.

The _allowance_ is the tightest possible fit between two mated parts. It is the difference between the smallest hole size and the largest shaft size. For a clearance fit, the allowance is positive and represents the minimum clearance between the two parts. For an interference fit, the allowance is negative and represents the maximum interference between the two parts. The allowance is calculated for the four cases shown in Figure 6-24.

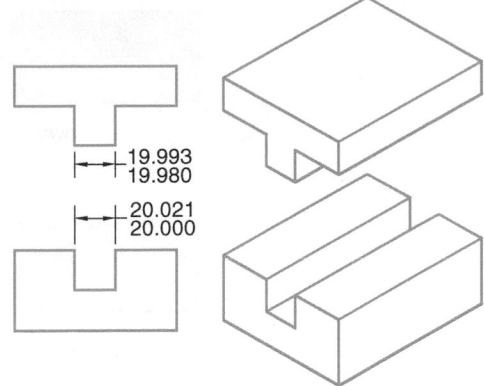

Figure 6-25 Fit of mated parts with parallel surfaces

Although these different types of fit typically refer to cylindrical features like shafts and holes, they also apply to parts with parallel surfaces that fit inside one another, as depicted in Figure 6-25.

With an understanding of tolerance, allowance, basic size, and types of fit, we can assign tolerances to a system of mated parts in order to achieve a particular type of fit (i.e., clearance, interference). All that is needed is a reference system, or method of calculation, for relating the tolerances and allowance to the basic size. In the following sections, two reference systems (hole and shaft) are discussed. Both English and metric units, though calculated differently, employ hole and shaft methods of calculation.

Basic Hole System: English Units

In the basic hole system, the minimum (i.e., lower limit) hole size is taken as the basic size. The allowance between the hole and shaft is then determined, and the tolerances are applied. The basic hole system is widely used because of the ready availability of tools (e.g., drills, reamers) capable of producing standard-size holes with precision. In effect, when using the basic hole system, we are choosing a standard drill size to create the hole and then turning down the shaft to fit that hole. Figure 6-26 on page 194 illustrates the basic hole system for both a clearance fit and an interference fit.

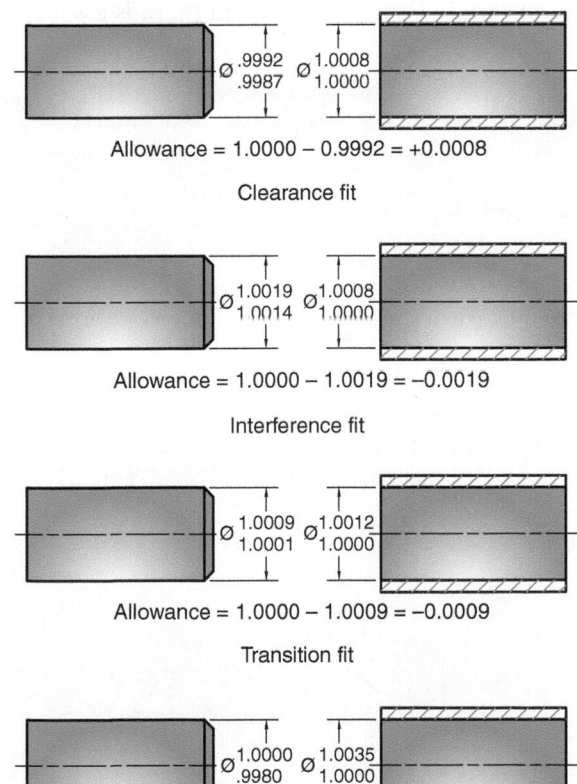

Allowance = 1.0000 − 0.9992 = +0.0008

Clearance fit

Allowance = 1.0000 − 1.0019 = −0.0019

Interference fit

Allowance = 1.0000 − 1.0009 = −0.0009

Transition fit

Allowance = 1.0000 − 1.0000 = +0.0000

Line fit

Figure 6-24 Comparison of different types of fits

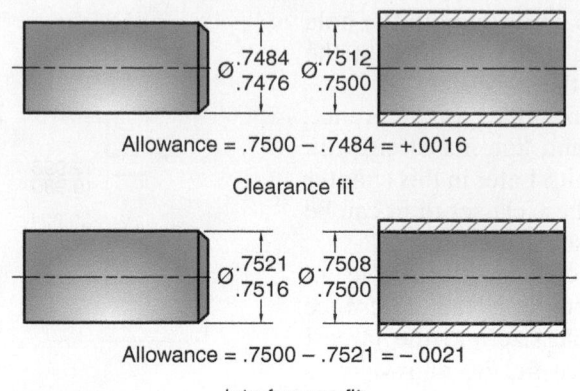

Allowance = .7500 − .7484 = +.0016

Clearance fit

Allowance = .7500 − .7521 = −.0021

Interference fit

Figure 6-26 Basic hole system for clearance and interference fits

Basic Shaft System: English Units

Less common than the basic hole system is the basic shaft system. In the basic shaft system, the maximum (i.e., upper limit) shaft is taken as the basic size. When using the basic shaft system, one selects a stock shaft and creates the mating hole to suit the shaft. The basic shaft system should be used only when there is a specific

---

Step-by-step tolerance calculation of a clearance fit using the basic hole system (see Figure 6-27)

Given:

- Basic size is .5000
- Allowance is +.0020
- Hole tolerance is .0016
- Shaft tolerance is .0010

1. Hole minimum = basic size = .5000

2. To find the upper limit on the shaft:

   Since
   Allowance = Hole minimum − Shaft maximum,
   Shaft maximum = Hole minimum − Allowance
   $$= .5000 − .0020 = .4980$$

3. To find the upper limit on the hole:

   Hole maximum = Hole minimum + Hole tolerance
   $$= .5000 + .0016 = .5016$$

4. To find the lower limit on the shaft:

   Shaft minimum = Shaft maximum − Shaft tolerance
   $$= .4980 − .0010 = .4970$$

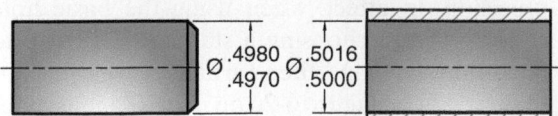

Figure 6-27 Step-by-step tolerance calculation: clearance, basic hole

Given:

- Basic size is 4.0000
- Allowance is –.0049
- Hole tolerance is .0014
- Shaft tolerance is .0009

1. Hole minimum = basic size = 4.0000

2. To find the upper limit on the shaft:

   Shaft maximum = Hole minimum – Allowance
   $$= 4.0000 - (-.0049) = 4.0049$$
   (because Allowance = Hole minimum – Shaft maximum)

3. To find the upper limit on the hole:

   Hole maximum = Hole minimum + Hole tolerance
   $$= 4.0000 + .0014 = 4.0014$$

4. To find the lower limit on the shaft:

   Shaft minimum = Shaft maximum – Shaft tolerance
   $$= 4.0049 - .0009 = 4.0040$$

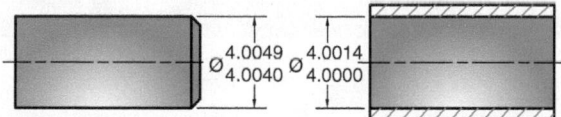

Figure 6-28 Step-by-step tolerance calculation; interference, basic hole

reason for using it, such as when a shaft cannot be easily machined to size, or when several parts requiring different fits must be mated to the same shaft. Figure 6-29 illustrates the basic shaft system for clearance and interference fits.

Preferred English Limits and Fits

In order to simplify the process of tolerancing mated parts, ANSI standards and accompanying tables have been developed for both English and metric units. The English unit standards are described in B4.1-1967 (R1994), "Preferred Limits and Fits for Cylindrical Parts." Although these standards are intended for holes, cylinders, and shafts, they can also be used for fits between parallel surfaces (see Figure 6-25 on page 193). The tables taken from B4.1 appear in Appendix A. To use these tables, the user

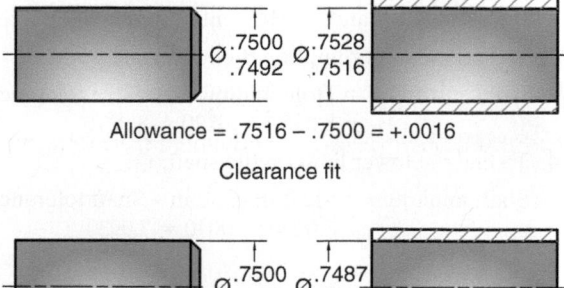

Figure 6-29 Basic shaft system for clearance and interference fits

TOLERANCING

195

provides the basic size and the type of fit. These tables employ the basic hole system.

B4.1 recognizes five different types of fit, as well as different classes within each type of fit. For any one class of fit (e.g., RC5), the fit produced between the mated parts results in the same fit characteristics, regardless of the basic size of the part features. The characteristics of the different types and classes of fit are described below.

RUNNING OR SLIDING CLEARANCE FIT (RC)

Clearance fits provide a similar running performance throughout a range of sizes. The clearances of the first two classes (RC1, RC2) are intended for use as slide fits, and the other classes (RC3 through RC9) are for free-running operation. RC1 and RC2 clearances increase more slowly with diameter than the other classes

to maintain accurate location at the expense of free relative motion. The other (free-running) classes range from precision (RC3) to loose (RC9).

LOCATIONAL CLEARANCE FIT (LC)

Tighter clearance fits than RC, intended for parts that are normally stationary but can be freely assembled and disassembled. They range from snug line fits for parts requiring accuracy of location, to looser fastener fits where freedom of assembly is of prime importance.

TRANSITION CLEARANCE OR INTERFERENCE FIT (LT)

Compromise between clearance and interference fits, for application where accuracy of location is important, but a small amount of either clearance or interference is permissible.

Step-by-step tolerance calculation of a clearance fit using the basic shaft system (see Figure 6-30)

Given:

- Basic size is 8.0000
- Allowance is +.0150
- Hole tolerance is .0120
- Shaft tolerance is .0070

1. Shaft maximum = basic size = 8.0000

2. To find the lower limit on the hole:

   Hole minimum = Shaft maximum + Allowance
   = 8.0000 + .0150 = 8.0150
   (because Allowance = Hole minimum − Shaft maximum)

3. To find the upper limit on the hole:

   Hole maximum = Hole minimum + Hole tolerance
   = 8.0150 + .0120 = 8.0270

4. To find the lower limit on the shaft:

   Shaft minimum = Shaft maximum − Shaft tolerance
   = 8.0000 − .0070 = 7.9930

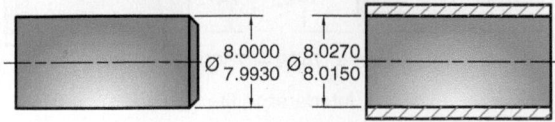

Figure 6-30 Step-by-step tolerance calculation: clearance, basic shaft

LOCATIONAL INTERFERENCE FIT (LN)

Force fit used where accuracy of location is of prime importance and for parts requiring rigidity and alignment with no special requirements for bore pressure. Not intended for parts designed to transmit frictional loads from one part to another based on tightness of fit.

FORCE OR SHRINK FIT (FN)

Force fit characterized by maintenance of constant bore pressures throughout the range of sizes. The interference varies almost directly with diameter, with the difference between its maximum and minimum values kept small to maintain the resulting pressures within reasonable limits.

Step-by-step tolerance calculation using English-unit fit tables, basic hole system (see Figure 6-31)

Given:

- Basic size = 2.0000
- Fit type is RC8
- Calculation method is basic hole

1. Using Appendix A, Running and Sliding Fits, with a basic size of 2.0000 and an RC8 fit type, the following information is extracted:

Nominal Size Range (Inches)	Class RC8		
Over To	Limits of Clearance	Standard Limits	
1.97 – 3.15		Hole H10	Shaft c9
	6	+4.5	−6.0
	13.5	0	−9.0

A note at the top of the table indicates that the limits are in thousandths of an inch. Thus the upper and lower limits on the hole, +4.5 and 0, are actually .0045 and 0, and the limits on the shaft, −6.0 and −9.0, are −.0060 and −.0090. These standard limits are added algebraically to the basic size to determine the actual tolerance limits. The limits of clearance give the tightest (0 − (−.0060) = +.0060) and the loosest (+.0045 − (−.0090) = .0135) possible fits. Recall that the tightest fit is the allowance.

2. Tolerance limits on the hole:

   Upper limit = 2.0000 + .0045 = 2.0045
   Lower limit = 2.0000 (= basic size)

3. Tolerance limits on the shaft:

   Upper limit = 2.0000 − .0060 = 1.9940
   Lower limit = 2.0000 − .0090 = 1.9910

4. Allowance = Hole minimum – Shaft maximum
              = 2.0000 − 1.9940 = .0060

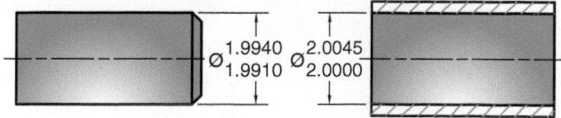

Figure 6-31   Step-by-step tolerance calculation: LC3, basic hole, preferred English-unit fit tables

Given:

- Basic size is 1.0000
- Fit type is LN2
- Calculation method is basic shaft

ANSI Standard B4.1 provides preferred limits and fits tables using only the Basic Hole System. To calculate tolerances of mated parts using the Basic Shaft System, the standard limits must be converted from basic hole to basic shaft.

1. Using Appendix A, Locational Interference Fits, with a basic size of 1.0000 and an LN2 fit type, the following information is extracted:

Nominal Size Range (Inches)	Class LN2		
Over To	Limits of Clearance	Standard Limits	
		Hole H7	Shaft p6
0.71 – 1.19	0	+0.8	+1.3
	1.3	−0	+0.8

2. These standard limits are using the Basic Hole System. In the Basic Shaft System, the upper limit on the shaft is taken as the basic size, meaning that the upper limit on the shaft should be 0, not +1.3. We can therefore convert the standard limits from basic hole to basic shaft by subtracting +1.3 from all of the standard limits. This has been done below:

	Hole	Shaft
Upper limit	+0.8 − 1.3 = −0.5	+1.3 − 1.3 = 0
Lower limit	−0 − 1.3 = −1.3	+0.8 − 1.3 = −0.5

3. Tolerance limits on the hole:

Upper limit = 1.0000 − .0005 = .9995
Lower limit = 1.0000 − .0013 = .9987

4. Tolerance limits on the shaft:

Upper limit = 1.0000 (= basic size)
Lower limit = 1.0000 − .0005 = .9995

5. Allowance = Hole minimum − Shaft maximum
= .9987 − 1.0000 = −.0013

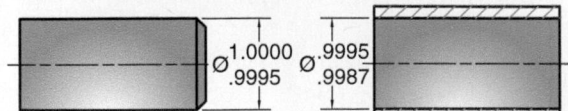

Figure 6-32 Step-by-step tolerance calculation: LN2, basic shaft, preferred English-unit fit tables

CHAPTER 6 DIMENSIONING AND TOLERANCING

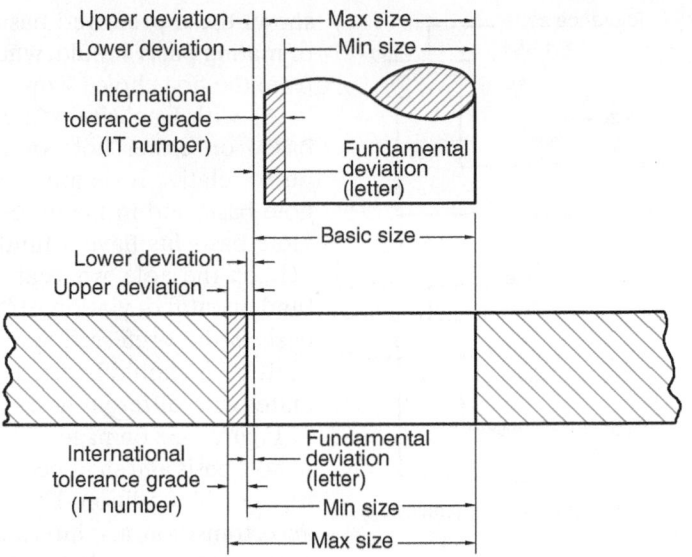

Figure 6-33 Illustration of definitions for metric limits and fits

Preferred Metric Limits and Fits

ANSI B4.2 – 1978 (1994), "Preferred Metric Limits and Fits," provides standards and tables for tolerancing fitted parts using metric units. Using this standard, a tolerance is specified using a special designation, for example 40H7. ANSI B4.2 begins with a series of definitions and an accompanying illustration similar to the one shown in Figure 6-33. These definitions include:

Basic size—The size to which limits or deviations are assigned. It is designated by the number 40 in 40H7.

Deviation—The algebraic difference between a size and the corresponding basic size.

Upper deviation—The algebraic difference between the maximum limit of size and the corresponding basic size.

Lower deviation—The algebraic difference between the minimum limit of size and the corresponding basic size.

Fundamental deviation—The deviation, upper or lower, that is closest to the basic size. It is designated by the letter H in 40H7.

Tolerance—The difference between the maximum and minimum size limits on a part.

Tolerance zone—A zone representing the tolerance and its position in relation to the basic size.

International tolerance grade (IT)—A group of tolerances that vary depending on the basic size, but that provide the same relative accuracy within a given grade. It is designated by the number 7 in 40H7 (IT7).

Hole basis—The system of fits where the minimum hole size is equal to the basic size. The fundamental deviation for a hole basis system is "H."

Shaft basis—The system of fits where the maximum shaft size is equal to the basic size. The fundamental deviation for a shaft basis system is "h."

Figure 6-34 on page 200 shows a toleranced size, along with the associated terminology. The International Tolerance grade establishes the magnitude of the tolerance zone (i.e., the amount of variation in part size that is allowed) for both internal (hole) and external (shaft) dimensions. It is expressed in grade numbers (e.g., IT7), with smaller grade numbers indicating a smaller tolerance zone.

The fundamental deviation establishes the position of the tolerance zone with respect to

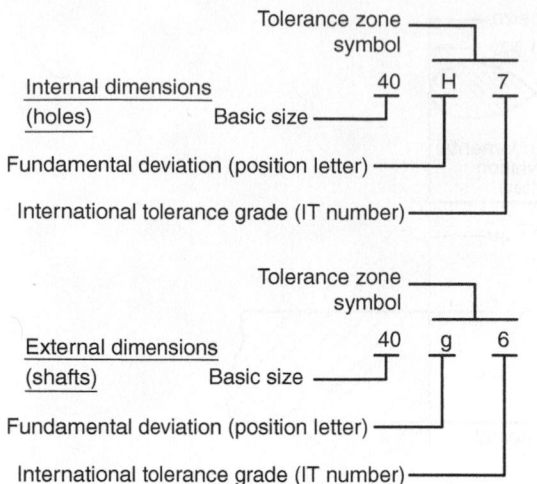

Figure 6-34 Metric-toleranced size with associated terminology

the basic size. It is expressed by "tolerance position letters"; upper-case letters (e.g., H) being used for hole dimensions, and lower-case letters (e.g., h) used for shaft dimensions.

A tolerance symbol (e.g., H7) is formed by combining the IT grade number and the tolerance position letter. The tolerance symbol identifies the actual maximum and minimum limits of the part. Toleranced sizes (e.g., 40H7) are determined by the basic size followed by a tolerance symbol.

A fit (e.g., 40 H8/f7) between mated parts is indicated by the basic size common to both components, followed by a symbol corresponding to each component, with the symbol for the internal (hole) part preceding the symbol for the external (shaft) part. Figure 6-35 shows a fit as designated using B4.2.

Standard, or preferred, sizes of round metal parts should be used whenever possible. Table 6-1

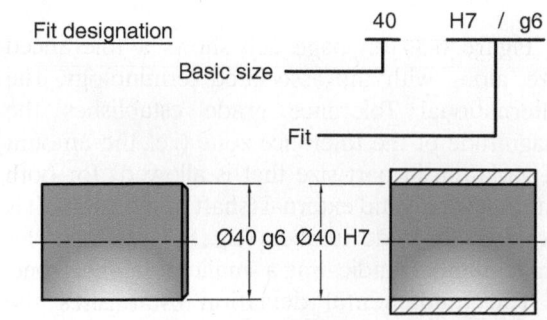

Figure 6-35 Metric-unit fit designation

shows these preferred basic sizes. The basic size of mating parts should, when possible, be chosen from the first choice sizes listed in this table.

As with English fits, metric preferred fits are based on either hole or shaft parts. Preferred fits to relative scale are shown in Figure 6-36 for hole basis and in Figure 6-37 for shaft basis fits. Hole basis fits have a fundamental deviation of "H" on the hole, whereas shaft basis fits have a fundamental deviation of "h" on the shaft. Hole basis is the preferred system in most cases, but shaft basis should be used when a common shaft mates with different holes.

Figure 6-38 on page 202 provides a description of hole basis and shaft basis fits that have the same relative fit condition. The limits and fits of clearance, transition, and interference fits are provided in the table appearing in Appendix A. Appendix B uses the fit types described in Figure 6-38 and the preferred sizes appearing in Table 6-1.

Table 6-1 **Preferred basic sizes**

Basic Size (mm)		Basic Size (mm)		Basic Size (mm)	
First Choice	Second Choice	First Choice	Second Choice	First Choice	Second Choice
1		1		100	
	1.1		11		110
1.2		12		120	
	1.4		14		140
1.6		16		160	
	1.8		18		180
2		20		200	
	2.2		22		220
2.5		25		250	
	2.8		28		280
3		30		300	
	3.5		35		350
4		40		400	
	4.5		45		450
5		50		500	
	5.5		55		550
6		60		600	
	7		70		700
8		80		800	
	9		90		900
				1000	

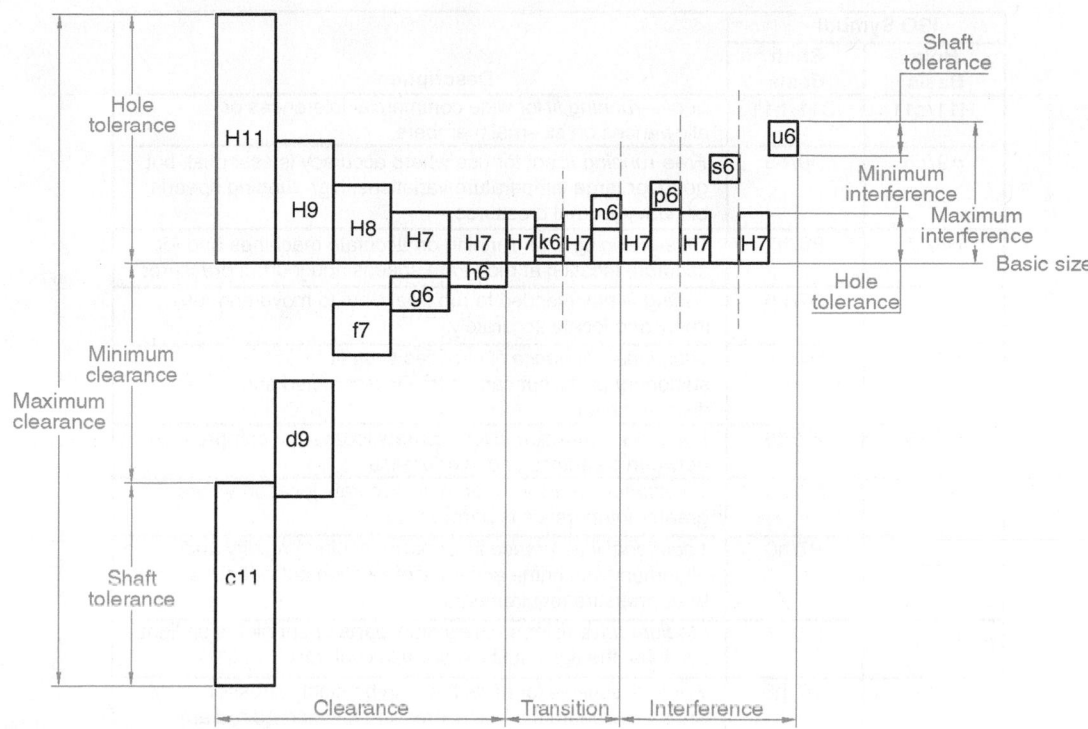

Figure 6-36 Preferred hole basis fits

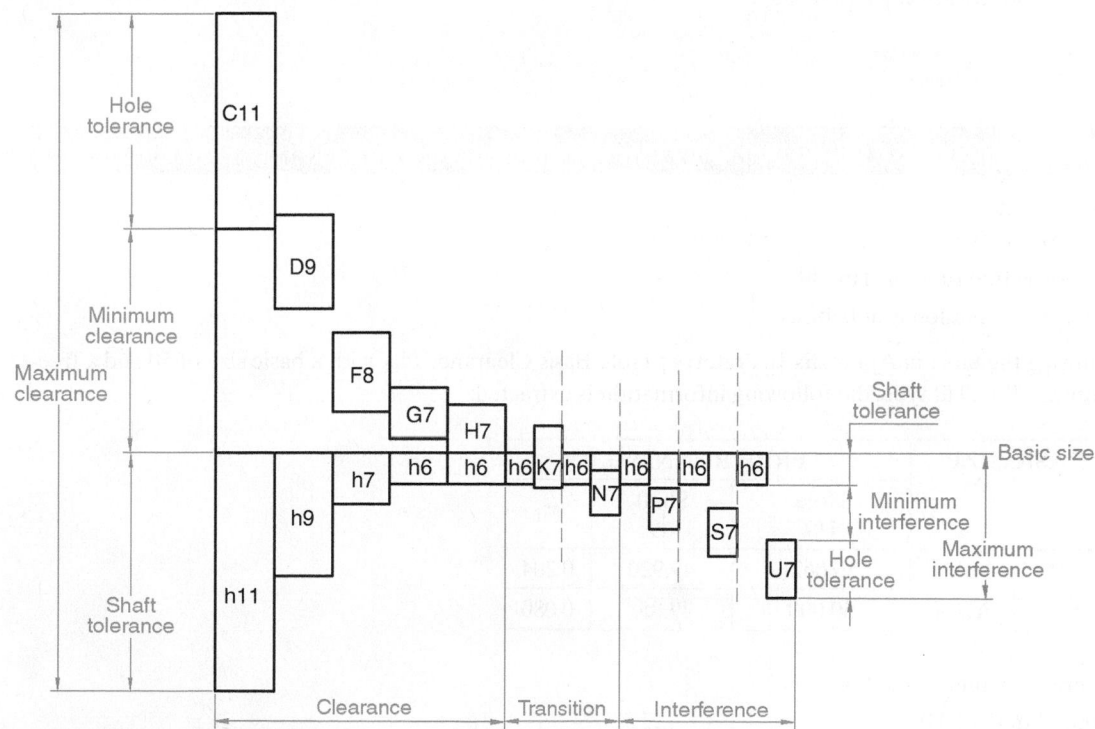

Figure 6-37 Preferred shaft basis fits

	ISO Symbol			
	Hole Basis	Shaft Basis	Description	
Clearance fits	H11/c11	C11/h11	*Loose-running fit* for wide commercial tolerances or allowances on external members.	More clearance
	H9/d9	D9/h9	*Free-running fit* not for use where accuracy is essential, but good for large temperature variations, high running speeds, or heavy journal pressures.	
	H8/f7	F8/h7	*Close-running fit* for running on accurate machines and for accurate location at moderate speeds and journal pressures.	
	H7/g6	G7/h6	*Sliding fit* not intended to run freely, but to move and turn freely and locate accurately.	
	H7/h6	H7/h6	*Locational clearance fit* provides snug fit for locating stationary parts, but can be freely assembled and disassembled	
Transition fits	H7/k6	K7/h6	*Locational transition fit* for accurate location, a compromise between clearance and interference.	
	H7/n6	N7/h6	*Locational transition fit* for more accurate location where greater interference is permissible.	
Interference fits	H7/p6	P7/h6	*Locational interference fit* for parts requiring rigidity and alignment with prime accuracy of location but without special bore pressure requirements.	
	H7/s6	S7/h6	*Medium drive fit* for ordinary steel parts or shrink fits on light sections, the tightest fit usable with cast iron.	
	H7/u6	U7/h6	*Force fit* suitable for parts that can be highly stressed or for shrink fits where the heavy pressing forces required are impractical.	More interference

Figure 6-38 Description of preferred fits

Step-by-step tolerance calculation using metric-unit fit tables, hole basis (see Figure 6-39)

Given:

- Basic size is 50
- Fit type is free running, H9/d9
- Calculation method is hole basis

1. Entering the table in Appendix B, Preferred Hole Basis Clearance Fits, with a basic size of 50 and a free-running H9/d9 fit type, the following information is extracted:

BASIC SIZE		FREE RUNNING		
		Hole H9	Shaft d9	Fit
50	MAX	50.062	49.920	0.204
	MIN	50.000	49.858	0.080

2. Tolerance limits on the hole:

   Upper limit = 50.062
   Lower limit = 50.000 (= basic size)

3. Tolerance limits on the shaft:

Upper limit = 49.920
Lower limit = 49.858

4. Allowance = Hole min − Shaft max
   = 50.000 − 49.920 = +0.080

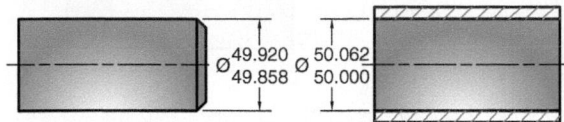

Figure 6-39 Step-by-step tolerance calculation: free-running, hole basis, preferred metric-unit fit tables

---

Step-by-step tolerance calculation using metric-unit fit tables, shaft basis (see Figure 6-40)

Given:

- Basic size is 30
- Fit type is medium drive, S7/h6
- Calculation method is shaft basis

1. Entering the table in Appendix B, Preferred Shaft Basis Interference Fits, with a basic size of 30 and a medium drive S7/h6 fit type, the following information is extracted:

BASIC SIZE		MEDIUM DRIVE		
		Hole S7	Shaft h6	Fit
30	MAX	29.973	30.000	−0.014
	MIN	29.952	29.987	−0.048

2. Tolerance limits on the hole:

Upper limit = 29.973
Lower limit = 29.952

3. Tolerance limits on the shaft:

Upper limit = 30.000 (= basic size)
Lower limit = 29.987

4. Allowance = Hole min − Shaft max = 29.952 − 30.000 = −0.048
   = 50.000 − 49.920 = 10.080

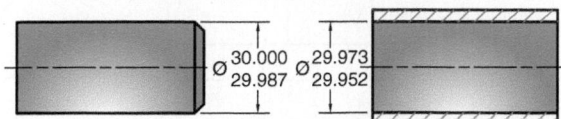

Figure 6-40 Step-by-step tolerance calculation: medium drive, shaft basis, preferred metric-unit fit tables

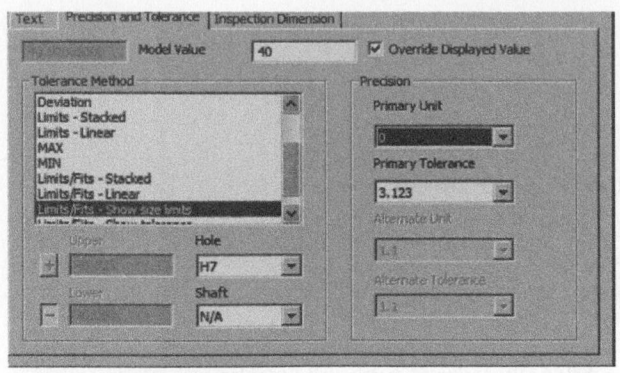

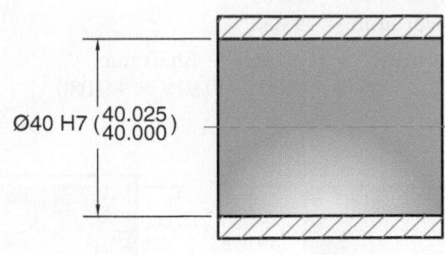

$$\varnothing 40 \ H7 \left(\begin{array}{c} 40.025 \\ 40.000 \end{array}\right)$$

Figure 6-41 Tolerance specification using CAD (Courtesy of Autodesk, Inc.)

Tolerancing in CAD

CAD programs typically provide a variety of tools for the specification of toleranced dimensions within a drawing. The dialog box shown on the left in Figure 6-41, for example, allows the designer to display tolerances in several ways. In Figure 6-41 the dialog box settings on the left result in the toleranced dimension shown on the right.

▌ QUESTIONS

DIMENSIONING

1. Select the dimensions A-Z, which represent good dimensioning (Figures P6-1 and P6-2). When finished, the object should be fully dimensioned. In cases where two dimensions locate or size the same feature, accept the one that best meets good dimensioning practice guidelines, while rejecting the other.

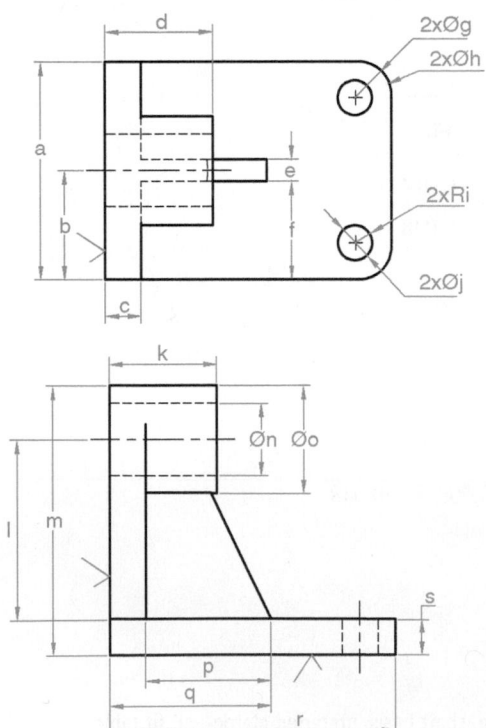

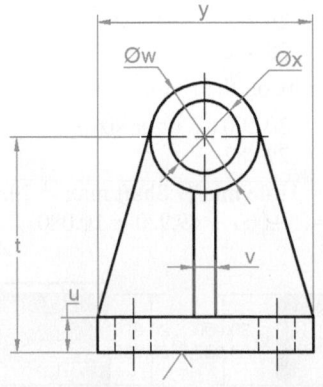

Figure P6-1

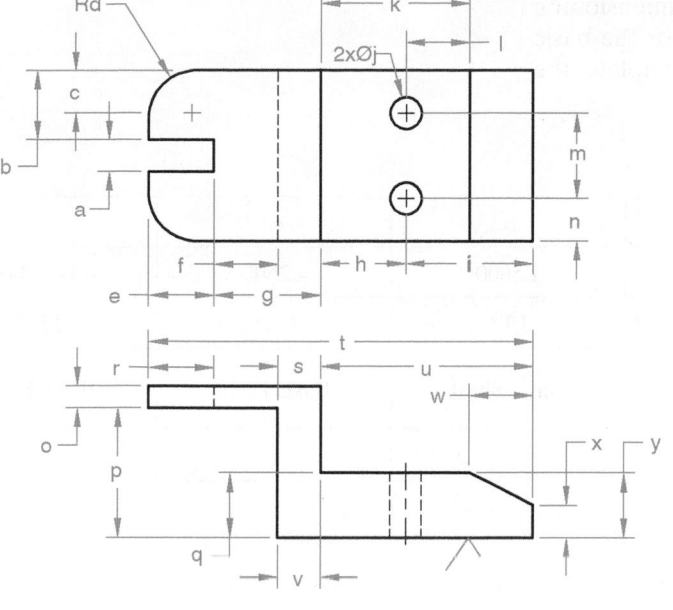

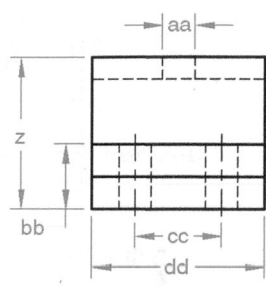

Figure P6-2

TOLERANCING

2. Given the method of calculation, basic size, fit (clearance, interference, or transition), tolerances, and the allowance, determine the limits on the hole and shaft.

	a.	b.	c.	d.	e.
Basic Size	10.000	1.5000	0.5000	30.000	16.000
Class of Fit	Clearance	Interference	Clearance	Interference	Transition
Method of Calculation	Shaft	Hole	Hole	Shaft	Hole
Limits of Hole Size					
Limits of Shaft Size					
Tolerance on Hole	0.200	0.0015	0.0010	0.016	0.025
Tolerance on Shaft	0.100	0.0014	0.0007	0.014	0.024
Allowance	0.150	−0.0030	0.0017	−0.032	−0.016

3. Using the appropriate limit dimensioning tables and either the basic hole or the basic shaft method of calculation, complete the table.

	a.	b.	c.	d.
Basic Size	2.5000	1.5000	4.2500	10.0000
Class of Fit	RC1	LT4	LN1	FN2
Method of Calculation	Basic Hole	Basic Shaft	Basic Hole	Basic Hole
Limits of Hole Size				
Limits of Shaft Size				
Tolerance on Hole				
Tolerance on Shaft				
Allowance				

	e.	f.	g.	h.
Basic Size				3.7500
Class of Fit	RC7	LC8	LN3	FN5
Method of Calculation	Basic Hole	Basic Hole	Basic Shaft	
Limits of Hole Size	0.7520	0.5028		3.7522
	0.7500	0.5000		3.7500
Limits of Shaft Size	0.7475		5.0000	
	0.7463		4.9990	
Tolerance on Hole				
Tolerance on Shaft		0.0016		0.0014
Allowance			−0.0035	

4. Using the appropriate limit dimensioning tables and either the basic hole or the basic shaft method of calculation, complete the table.

	a.	b.	c.	d.
Basic Size	10	250	25	4
Class of Fit	Loose Running H11/c11	Close Running H8/f7	Locational Clearance H7/h6	Medium Drive H7/s6
Method of Calculation	Preferred Hole Basis	Preferred Hole Basis	Preferred Shaft Basis	Preferred Hole Basis
Limits of Hole Size				
Limits of Shaft Size				
Tolerance on Hole				
Tolerance on Shaft				
Allowance				

	e.	f.	g.	h.
Basic Size	6			50
Class of Fit	Sliding H7/g6	Locational Transition N7/h6	Locational Interference H7/p6	Force U7/h6
Method of Calculation		Preferred Shaft Basis	Preferred Hole Basis	
Limits of Hole Size	6.012		80.030	
	6.000		80.000	
Limits of Shaft Size		160.000		50.000
		159.975		49.984
Tolerance on Hole		0.040		
Tolerance on Shaft	0.008		0.019	0.016
Allowance		−0.052		

Construct all extension lines, dimension lines, leaders, and arrowheads that are necessary to fully dimension the objects.

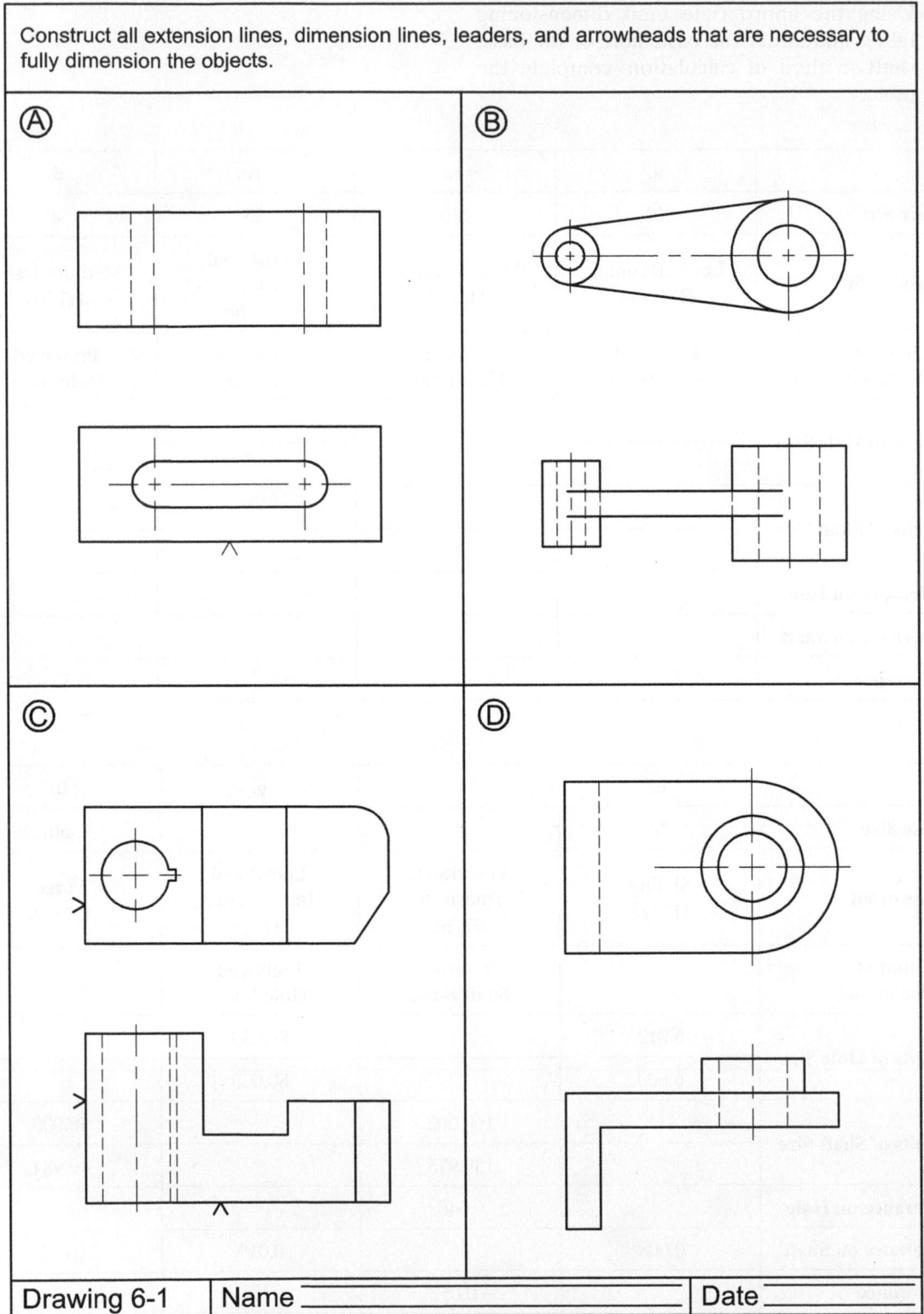

Drawing 6-1   Name _____   Date _____

Construct all extension lines, dimension lines, leaders, and arrowheads that are necessary to fully dimension the objects.

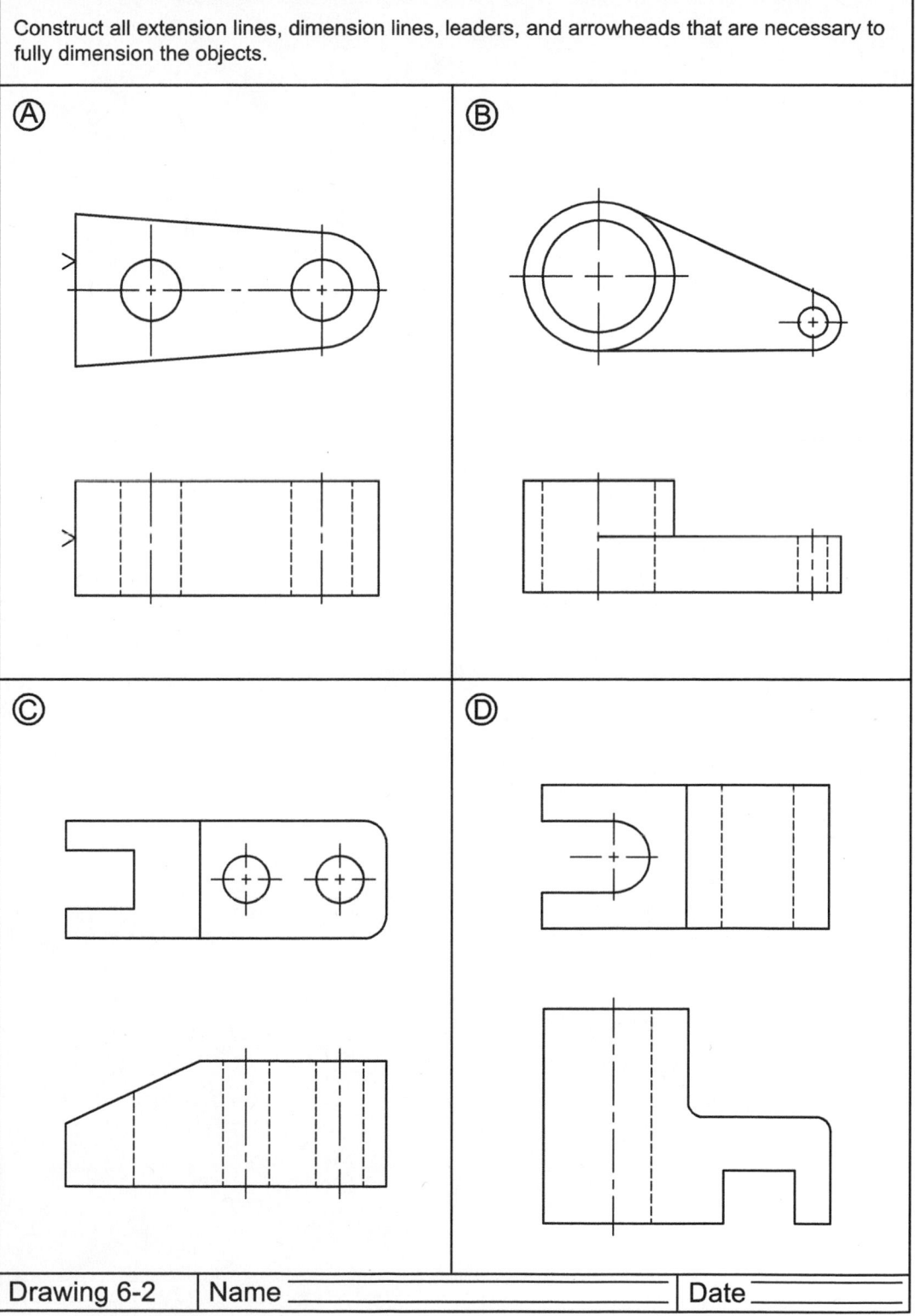

Drawing 6-2 | Name _____ | Date _____

Drawing 9-27   Name

CHAPTER

8

WORKING DRAWINGS

▌ INTRODUCTION

At the beginning of any project, there is no guarantee that a design will actually be executed. If a project is put out to bid, for example, only one of several competing preliminary designs will be selected. Even then, funding for the project may fall through. Similarly, for a company engaged in both research and development, management may decide to abandon the development of a new product.

Once the decision is made to build, however, the existing preliminary design must be further developed and detailed for production. The term *working drawing* is used to describe the complete set of drawing information needed for the manufacture and assembly of a product based on its design.

As discussed in the previous chapter, commercial products are almost always assemblies comprising several different parts. Perhaps the most recognizable element of a working drawing set is the *assembly drawing*. The purpose of the assembly drawing is to show how the different components fit together to form the product. An example of an assembly drawing is shown in Figure 8-1 on page 248.

An essential element of a working drawing is the parts list, or *bill of materials* (abbreviated BOM). The purpose of the BOM is to identify all parts, both standard and nonstandard, used in an assembly. The BOM for the assembly shown in Figure 8-1 is located in the upper-right corner of the drawing.

In addition to the assembly drawing, a set of working drawings also includes *detail drawings* of all nonstandard parts. As seen in Figure 8-2 on page 248, these individual part drawings contain multiple views, dimensions, notes, tolerance information, material specifications, and any other information necessary to the manufacture the part.

Working drawings are typically accompanied by written instructions called *specifications*, which serve to further clarify the details for manufacturing the product. An excerpt from the specifications describing the construction of a tugboat appears in Figure 1-14 in Chapter 1.

In addition to describing the details of a product's design and manufacture, working drawings and written specifications also serve as legal contracts. In the event that a problem arises with a design, working drawings may be called upon to help establish liability. Figure 8-3 on page 248 shows a professional engineer's stamp, taken from an engineering drawing. In stamping and signing the drawing, the engineer takes responsibility for the accuracy of the drawing's contents.

▌ THE IMPACT OF TECHNOLOGY ON WORKING DRAWINGS

As CAD technology has matured in the past 30 years, the elimination of the need for engineering drawings has frequently been predicted. This is perhaps unrealistic, but the importance of drawings, at least within some disciplines, has certainly diminished. In mechanical CAD (MCAD) for example, with its embrace of parametric solid modeling, the digital model has assumed the primary role in product definition,

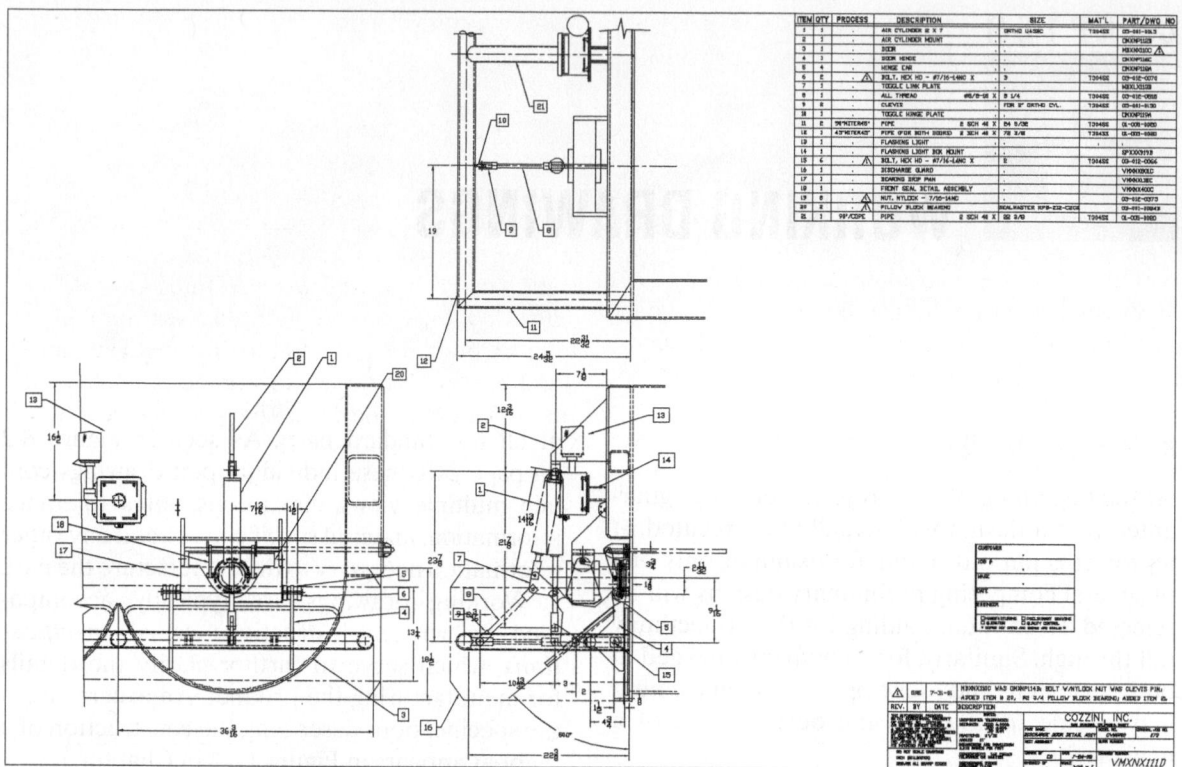

Figure 8-1 Typical assembly drawing (Courtesy of Cozzini, Inc.)

replacing the drawing. Later in this chapter we will see that it is fairly simple to extract 2D drawings from a 3D constraint-based model, either part or assembly. In the architecture, engineering, and construction (AEC) industry, recent strides have been made in the development of building

information modeling (BIM) software, showing that this ability to extract drawing information from a 3D model is not unique to the mechanical or aerospace industries.

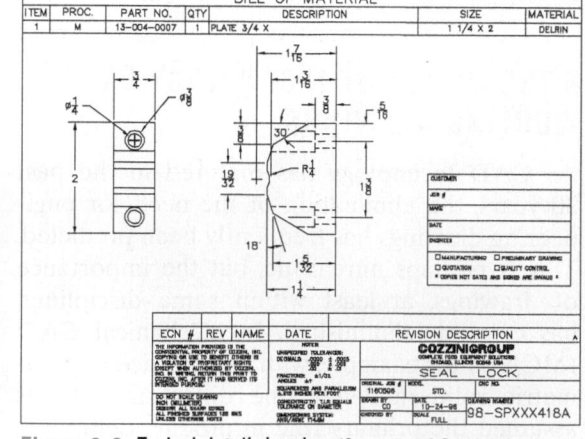

Figure 8-2 Typical detail drawing (Courtesy of Cozzini, Inc.)

Figure 8-3 Professional engineer's stamp (Courtesy of Jensen Maritime Consultants, Inc.)

ASME Y14.41-2003, "Digital Product Definition Practices," provides standards for the use of digital data sets as a substitute for working drawings. According to ASME Y14.41-2003, a product data definition set is a collection of one or more computer files that discloses, by means of graphical or textual presentations, the physical or functional requirements of an item. The Y14-41 standard supports two alternative methods for the development of the product definition data: (1) model only in digital format or (2) model and drawing in digital format.

CAD technology has affected the way working drawings are used by replacing traditional file storage (with its attendant risks[1] and demands for space) with electronic file storage. Computer networks and associated software enable authorized individuals to view the latest version of a working drawing at any time.

DETAIL DRAWINGS

A set of working drawings includes detail drawings of all nonstandard parts included in the product or assembly. Standard parts, either vendor-purchased or developed within the company, do not require individual part drawings. A detail drawing is a fully dimensioned multiview drawing that contains all of the information necessary to manufacture the part. Figure 8-4 on page 250 shows a detail drawing of a shaft part.

Part drawings are either developed directly in 2D or extracted from a 3D model. A typical part drawing includes the following information:

- Drawing views
- Dimensions
- Tolerances
- Material designation

- Surface finish
- Notes
- Title block
- Revision block

Although many parts are simple enough to be included together with other parts on the same drawing, there are good reasons to place each part on its own drawing. Doing so makes it possible to use a single identification number to identify both the part and the drawing. This same number can also be used to name the CAD file that documents the part. This practice considerably simplifies the task of keeping track of the ever-increasing number of parts, drawings, and files that even a small company must manage. It also makes it easy for the company to reuse parts in different assemblies.

In many companies, parts are identified by a unique code number. This code or identification number contains information about the part. Formal classification and coding techniques have been developed and are used to group similar parts on the basis of such characteristics as material, size, shape, function, process, or other information.

ASSEMBLY DRAWING VIEWS

The main purpose of an assembly drawing is to show all of the different parts in the assembly, and how these parts fit together to create the mechanism, device, component, or product. Using parametric modeling software, it is relatively easy to generate the views needed to accomplish this task. Two of the most commonly used assembly views are the section and the exploded view.

A section view of an assembly is used when the relationship between the different parts may not be apparent from an external view, as is the case with the ball valve shown in Figure 8-5 on page 251. In addition to the full section, half section and broken-out assembly views are also used, as seen in Figures 8-6 and 8-7 (on page 251), respectively, for a globe valve.

Certain conventions regarding section lines are to be followed when creating assembly section views:

[1] A fire destroyed all of the drawing files that belonged to an engineering firm the primary author worked for, Nickum, Spaulding & Associates, of Seattle, Washington. In another war story, a large set of original Mylar drawings was temporarily left by the loading dock at a large shipyard in the Pacific Northwest. The drawings were mistaken for trash and carted away. Happy ending: The drawings were recovered three days later at the local garbage dump.

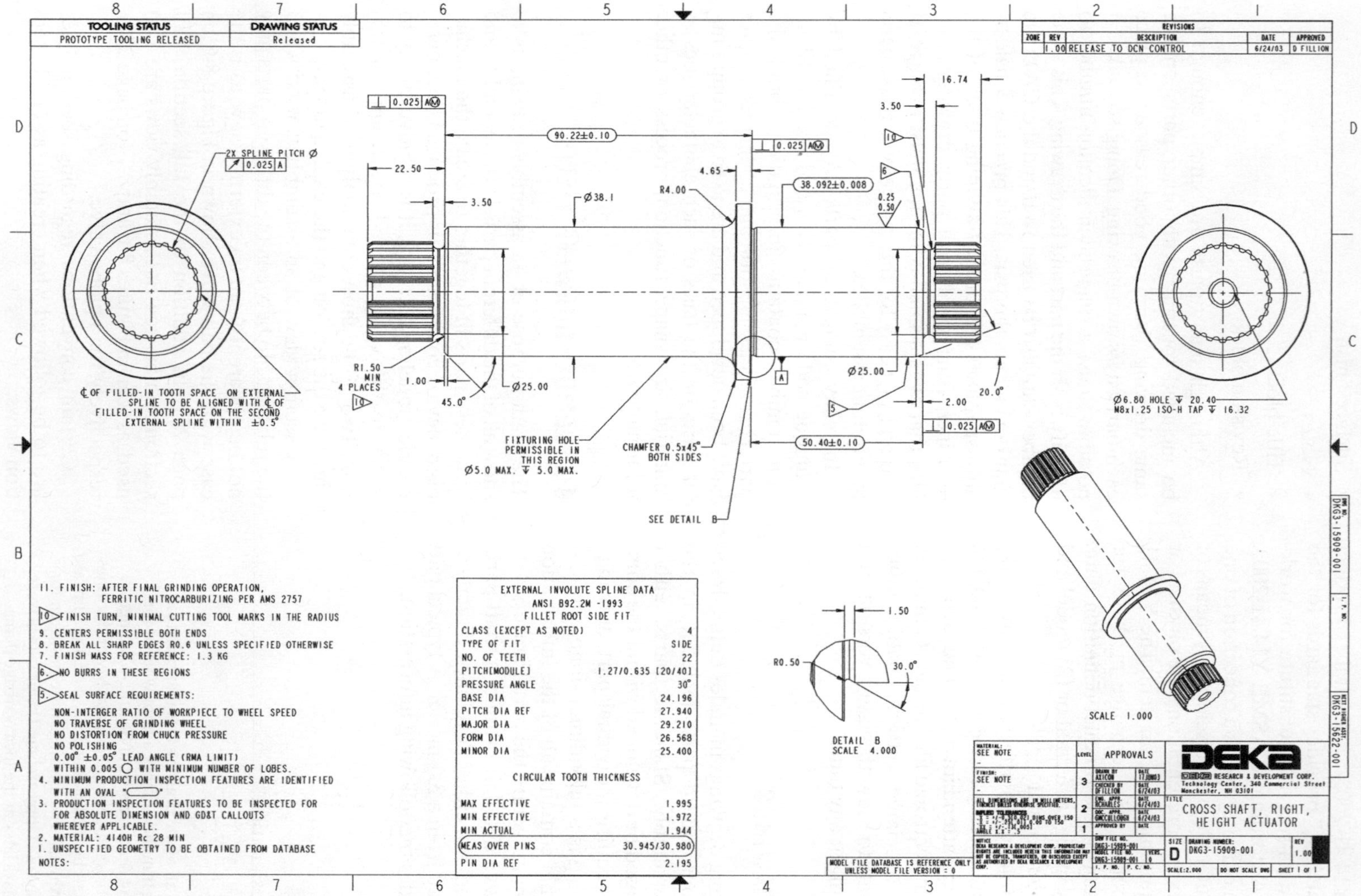

Figure 8-4 Detail drawing of a shaft part

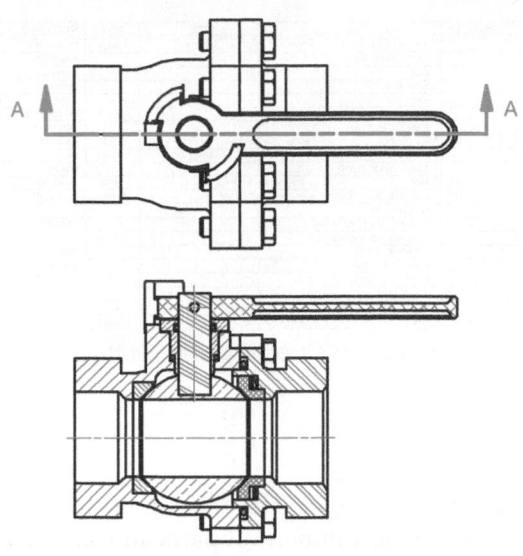

Section A-A

Figure 8-5 Assembly section view of a ball valve

- The section lining used for adjacent parts through which the cutting plane passes should be different. Either use different section line patterns for each material (as shown in Figure 8-5), or use the general-purpose (cast iron) section lining and vary the angle to help distinguish different parts (the approach used in Figures 8-6 and 8-7).

- Standard parts and solid parts with no internal detail are not sectioned. These parts include nuts, bolts, screws, shafts, pins, keys, bearings, spokes, and ribs. The shaft and nuts on the globe valve appearing in Figures 8-6 and 8-7 are unsectioned, for example.

- Extremely thin parts, such as gaskets and sheet metal parts, are shown solid, rather than sectioned.

An exploded view, like the one of the ball valve shown in Figure 8-8 on page 252, is also fairly easy to extract from a parametric assembly model. Because exploded views are easy to visualize, clearly showing the different parts and how they fit together to form an assembly, these views are commonly found in installation manuals and part catalogs.

When creating an assembly drawing, use the minimum number of views necessary to describe the part relationships. Often a single view is sufficient. It is not necessary to show

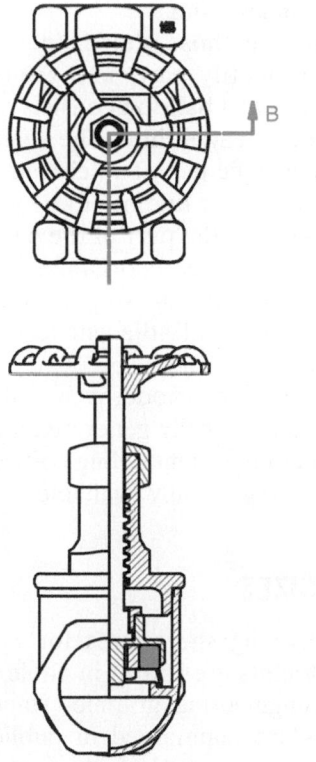

Section B-B

Figure 8-6 Half section view of a globe valve

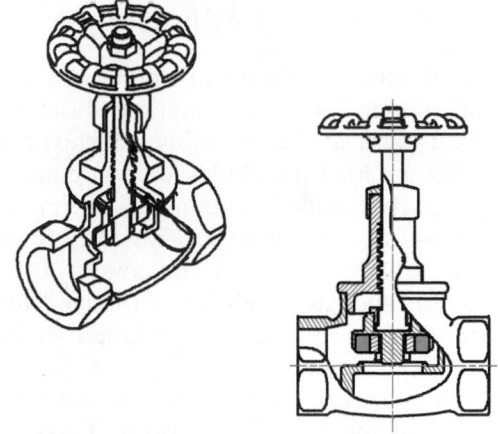

Figure 8-7 Broken-out section view of a globe valve

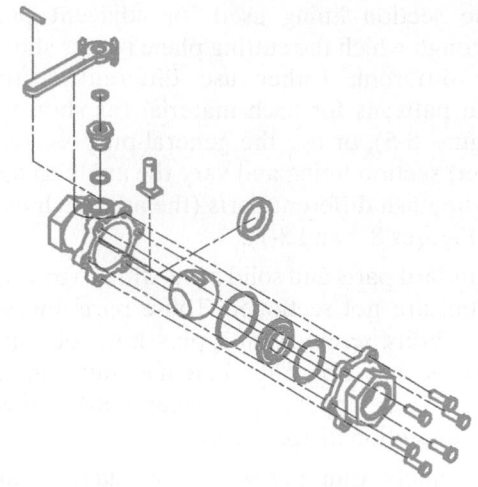

Figure 8-8 Exploded view of a ball valve

ITEM	QTY	PROCESS	DESCRIPTION	SIZE	MAT'L	PART/DWG NO
1	1	.	AIR CYLINDER 2 X 7	ORTHO U45BC	T304SS	05-001-0015
2	1	.	AIR CYLINDER MOUNT	.	.	DKXXP1123
3	1	.	DOOR	.	.	MBXXX010C ⚠
4	1	.	DOOR HINGE	.	.	DKXXP116C
5	4	.	HINGE EAR	.	.	DKXXP116A
6	2	. ⚠	BOLT, HEX HD − ⌀7/16-14NC X	3	T304SS	03-012-0070
7	1	.	TOGGLE LINK PLATE	.	.	DKXXP118A
8	1	.	ALL THREAD	#3/8-16 X	6 1/4	T304SS	03-012-0518
9	2	.	CLEVIS	FOR 2" ORTHO CYL.	T304SS	05-001-0150
10	1	.	TOGGLE HINGE PLATE	.	.	DKXXP119A
11	2	90°MITER45°	PIPE	2 SCH 40 X	24 5/32	T304SS	01-005-0020
12	1	45°MITER45°	PIPE (FOR BOTH DOORS)	2 SCH 40 X	72 3/8	T304SS	01-005-0020
13	1	.	FLASHING LIGHT	.	.	.
14	1	.	FLASHING LIGHT BOX MOUNT	.	.	SPXXX9193
15	6	. ⚠	BOLT, HEX HD − ⌀7/16-14NC X	2	T304SS	03-012-0066
16	1	.	DISCHARGE GUARD	.	.	VMXXX080C
17	1	.	BEARING DRIP PAN	.	.	VMXXX019C
18	1	.	FRONT SEAL DETAIL ASSEMBLY	.	.	VMXXX400C
19	8	. ⚠	NUT, NYLOCK − 7/16-14NC	.	.	03-012-0275
20	2	. ⚠	PILLOW BLOCK BEARING	SEALMASTER RPB-212-C2CR	.	03-001-0084B
21	1	90°/COPE	PIPE	2 SCH 40 X	22 3/8	T304SS	01-005-0020

Figure 8-9 Parts list for a specific product (Courtesy of Cozzini, Inc.)

individual parts in detail, because this is accomplished in the detail drawings. Standard parts are shown in the assembly views. Dimensions are not shown in an assembly drawing unless they pertain directly to the assembly as a whole. Hidden lines are typically not shown in an assembly drawing.

In addition to the assembly view showing all of the parts, the other main elements of an assembly drawing include a parts list providing information about these parts, and balloons used to identify the parts and relate them to the parts list. The parts list and balloons are discussed in the following section.

■ BILL OF MATERIALS AND BALLOONS

The bill of materials (BOM), or *parts list,* provides a tabular list of information about the individual parts in the assembly. As shown in Figure 8-9, a parts list includes an item number, a unique part identification number, a brief verbal description of the part, the quantity found in the assembly, the material, and possibly other information, such as weight or stock size. Note that standard parts are also included in the parts list.

The parts list generally appears on the right-hand side of the assembly drawing. Parts are listed in order of importance and/or size. If the column headings appear at the top of the parts list, then the most important parts appear at the top of the list. On some drawings the column headings appear at the bottom, in which case the most important parts appear at the bottom of the list. This practice makes it possible to add new, presumably less important parts without affecting the item numbering.

Balloons, like those shown in Figure 8-10, are used to identify parts and relate them to the parts list. Each balloon consists of a circled number and a leader line. The leader line is drawn to a specific part. The circled number is the item number (or find number) for the part; it is used to locate the part in the parts list. The balloons should be neatly organized, preferably in horizontal or vertical rows. Balloon leaders should not cross, and adjacent leaders should be parallel.

Once an assembly model is available, a parts list can be automatically generated and customized using parametric modeling software. A balloon tool is also generally available.

■ SHEET SIZES

Standard drawing sheet sizes for both metric and English units are shown in Table 8-1. These and other engineering graphics standards and conventions are maintained in publications by the American Society of Mechanical Engineers (ASME) and are approved by the American National Standards Institute (ANSI).

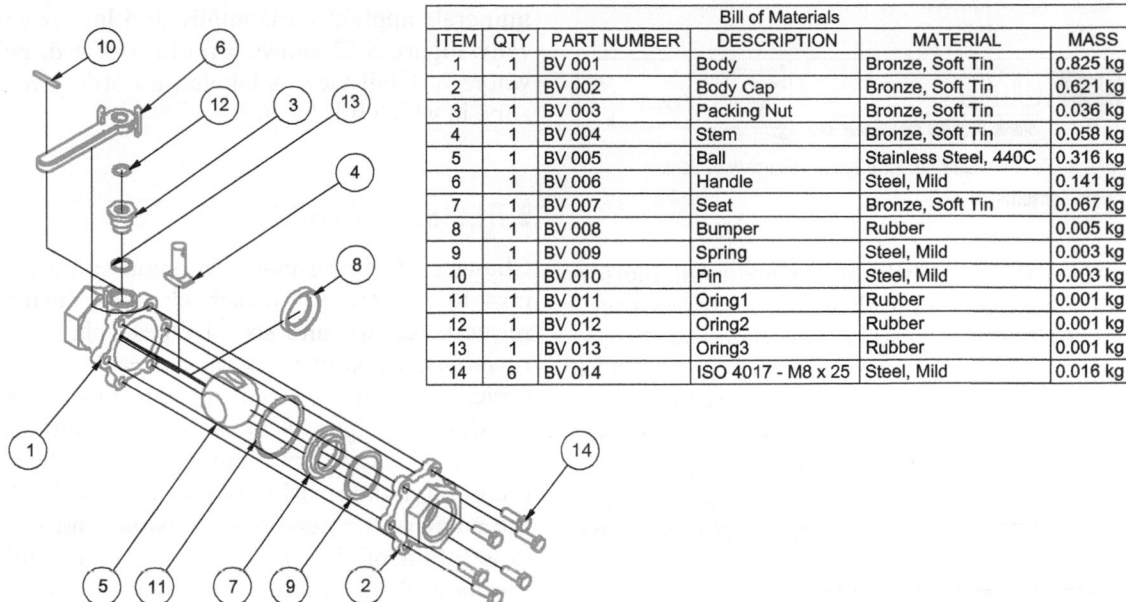

Bill of Materials					
ITEM	QTY	PART NUMBER	DESCRIPTION	MATERIAL	MASS
1	1	BV 001	Body	Bronze, Soft Tin	0.825 kg
2	1	BV 002	Body Cap	Bronze, Soft Tin	0.621 kg
3	1	BV 003	Packing Nut	Bronze, Soft Tin	0.036 kg
4	1	BV 004	Stem	Bronze, Soft Tin	0.058 kg
5	1	BV 005	Ball	Stainless Steel, 440C	0.316 kg
6	1	BV 006	Handle	Steel, Mild	0.141 kg
7	1	BV 007	Seat	Bronze, Soft Tin	0.067 kg
8	1	BV 008	Bumper	Rubber	0.005 kg
9	1	BV 009	Spring	Steel, Mild	0.003 kg
10	1	BV 010	Pin	Steel, Mild	0.003 kg
11	1	BV 011	Oring1	Rubber	0.001 kg
12	1	BV 012	Oring2	Rubber	0.001 kg
13	1	BV 013	Oring3	Rubber	0.001 kg
14	6	BV 014	ISO 4017 - M8 x 25	Steel, Mild	0.016 kg

Figure 8-10 BOM, balloons used to identify parts

Table 8-1 **Standard sheet sizes**

Metric (mm)		English (inches)	
A4	210 × 297	A	8.5 × 11.0
A3	297 × 420	B	11.0 × 17.0
A2	420 × 594	C	17.0 × 22.0
A1	594 × 841	D	22.0 × 34.0
A0	841 × 1189	E	34.0 × 44.0
ASME Y14.1M-1995		ASME Y14.1-1995	

Figure 8-11 Title block (Courtesy of Jensen Maritime Consultants, Inc.)

▮ TITLE BLOCKS

Every drawing contains a *title block* similar to the one shown in Figure 8-11. The purpose of the title block is to organize the information needed to identify the drawing. Additional information that is not given directly on the drawing is also included in the title block. Standard title block information includes the following:

- Company name and address
- Drawing title
- Drawing number
- Names, dates, and release signatures of designer, checker, and supervisor
- Revision number
- Sheet size
- Sheet number
- Scale of drawing
- Other information: project name, client name, material, general tolerance information, heat treatment, surface finish, hardness, weight estimate

The company logo is also frequently found on the title block. On most drawings the title block is located in the lower right-hand corner of the drawing.

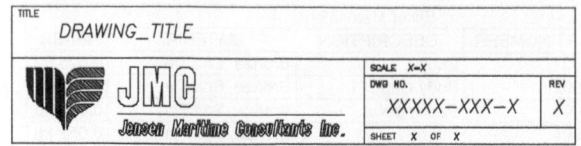

Figure 8-12 Continuation title block (Courtesy of Jensen Maritime Consultants, Inc.)

ANSI standard title blocks are available for both metric and English units in various sheet sizes. Companies tend to use their own standard title blocks. These standard title blocks can be made available from a CAD library or as a template, and they are easily inserted into a CAD file. In many CAD programs, the title block text fields are already completed or are set up for easy modification. In parametric solid-modeling software, the title block text fields are often associatively linked to the CAD database.

Drawings often contain more information than can be comfortably included on a single sheet. In this case a continuation title block can be used on all sheets after the first sheet. The continuation title block contains less information than the title block on the main sheet. Figure 8-12 shows an example of a continuation title block.

BORDERS AND ZONES

All drawings include a rectangular border. Some drawings also include *zones*. Zones are regular ruled intervals along the edges of the drawing border that assist users in finding information on a complicated drawing. These zones are similar to the zones found on highway road maps, with numerals applied horizontally and letters vertically. Figure 8-13 shows a portion of a drawing where a detail view is labeled according to the zone in which it is located.

REVISION BLOCKS

Changes are often made to engineering drawings to account for design changes, customer requests, errors, and so on. These changes, or *revisions*, are kept track of in a *revision block* typically located on the upper-right corner of the drawing. Figure 8-14 shows an example of a revision block. At a minimum, each modification recorded in the revision block includes the date, the name of the person responsible, and a brief description of the change. A revision number, normally placed in circle or triangle, is associated with each revision and placed both on the revision block and in the area on the drawing sheet where the revision has been made. In a zoned drawing, the zone of the affected area is listed in the revision block.

DRAWING SCALE

CAD models, whether 2D and 3D, should always be created full-size, without concern for how large or small the actual object may be. In this way the CAD data are based on the true size of the object. Added dimensions will be true-size, query commands will report true-size data, and the CAD database will be suitable for export to downstream analysis and manufacturing applications.

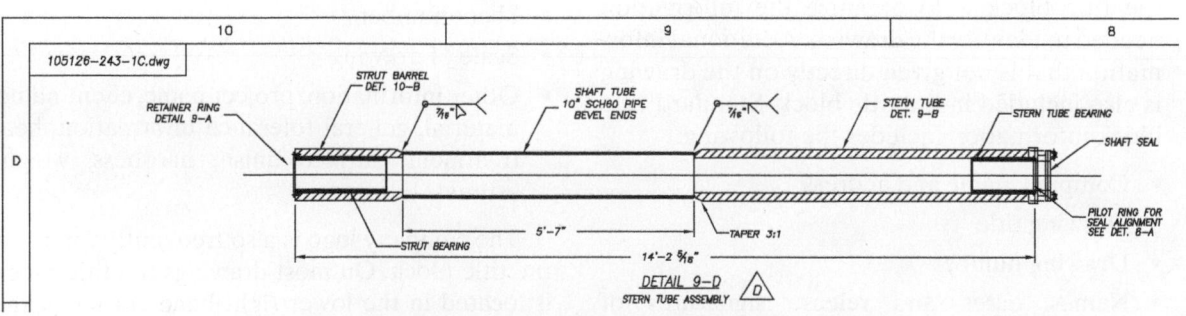

Figure 8-13 Drawing detail labeled by zone (Courtesy of Jensen Maritime Consultants, Inc.)

REVISIONS				
SYM	ZONE	DESCRIPTION	BY	DATE
PO	ALL	INITIAL ISSUE	YH	4/20 06
P1		1) OVERALL SHAFT LENGTH ADJUSTED PER UNDER WATER SHIPCHECK 4–26–06. 2) RUDDER AND EXISTING STRUT BEARING LOCATED. 3) NEW PROPELLER AND NOZZLE LOCATED TO CLEAR EXISTING RUDDER. 4) TRIMMING OFF 9-½" OF AFT EDGE OF SKEG PLATE SHOWN, TO SUIT THE LONGER CP PROPELLER HUB.	YH	4/29 06
P2		1) NUMBER AND LOCATION OF SHAFT BEARING REVISED. 2) ONE PILLOW BEARING REMOVED. 3) STERN TUBE AND BULKHEAD STUFFING BOXES REPLACED WITH SHAFT SEALS. 4) INDIVIDUAL LENGTHS OF TAIL SHAFT AND LINE SHAFT REVISED (OVERALL LENGTH REMAINS UNCHANGED) 5) SHAFT LINE MATERIALS CHANGED TO CARBON STEEL GR. 4 WITH BRONZE LINERS IN WAY OF STRUT AND STERN TUBE/SHAFT SEAL. 6) SHAFT DIAMETER AND STERN TUBE/STRUT DIAMETERS INCREASED. 7) RESERVATION NOTE #3 REMOVED.	YH	5/3 06
–		1) NEW ENGINE BLOCK RECEIVED FROM CUMMINS INSERTED IN DRAWING. 2) CRAFT MODEL S1 BCH 508 FL PILLOW BLOCK SHOWN. 3) SHAFT LINE ELEVATION (HEIGHT ABOVE BASE LINE) CORRECTED. 4) RESERVATION NOTE #2 REMOVED. GEN. NOTE #2 & #3 REVISED. 5) PRELIMINARY STATUS REMOVED.	YH	5/12 06
A		REVISED PROPELLER HUB FROM HUNDESTED INSERTED.	YH	5/19 06
B		1. LINE SHAFT BEARING RELOCATED PER HUNDESTED SUGGESTION. 2. PROPELLER NOZZLE TILTED 2 DEGREES.	YH	5/31 06
C		1. LINE SHAFT BEARING RELOCATED TO FRAME #39. 2. TAIL SHAFT LENGTH INCREASED, LINE SHAFT LENGTH DECREASED. 3. STERN TUBE AND STRUT DETAILS ADDED (SHT #2).	YH	6/12 06

Figure 8-14 Revision block (Courtesy of Jensen Maritime Consultants, Inc.)

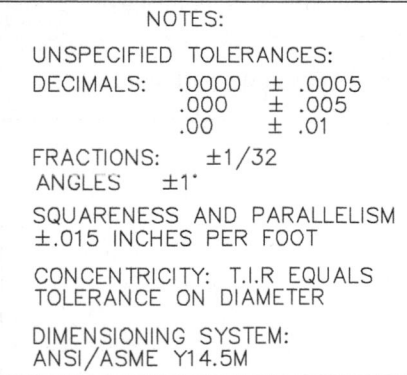

NOTES:

UNSPECIFIED TOLERANCES:

DECIMALS: .0000 ± .0005
.000 ± .005
.00 ± .01

FRACTIONS: ±1/32

ANGLES ±1°

SQUARENESS AND PARALLELISM
±.015 INCHES PER FOOT

CONCENTRICITY: T.I.R EQUALS
TOLERANCE ON DIAMETER

DIMENSIONING SYSTEM:
ANSI/ASME Y14.5M

Figure 8-15 General tolerance note (Courtesy of Cozzini, Inc.)

Table 8-2 Common drawing scales

1:1	1/8" = 1'- 0"
1:2	1/4" = 1'- 0"
1:4	3/8" = 1'- 0"
1:8	1/2" = 1'- 0"
1:10	1" = 1'- 0"
1:20	1" = 10'
1:30	1" = 20'
1:40	1/4" = 1"
1:50	1/2" = 1"
1:100	1" = 1"

Once the full-size model is complete, drawings are prepared and scaled to fit the designated drawing sheet size. Table 8-2 provides examples of some common scales used on engineering drawings.

The drawing scale is normally indicated on the title block. In the event that different views on the drawing sheet are drawn at different scales, the scale of the view is indicated on the view label.

■ TOLERANCE NOTES

Many companies include a general tolerance note similar to the one shown in Figure 8-15 on their detail drawings. These tolerance notes refer to any dimensions that have not been specifically toleranced. Tolerance notes are typically found in the lower-right corner of the drawing.

■ STANDARD PARTS

Standard parts may either be purchased from outside or produced within the company. Typical standard parts include threaded fasteners, bushings, bearings, pins, keys, pumps, and valves. Though standard parts do not require a detail drawing, they are shown on the assembly drawing and included in the parts list. Most 3D parametric modeling programs have a built-in standard parts library, from which standard part models can be directly inserted into an assembly model. Figure 7-29 in Chapter 7 shows a screen capture of a CAD library interface within a parametric modeling program.

■ WORKING DRAWING CREATION USING PARAMETRIC MODELING SOFTWARE

In the remaining pages of this chapter, various techniques to develop working drawings are described. These parametric modeling techniques include the creation of part drawings, assembly models, sectioned assembly and exploded views, as well as adding a parts list and balloons to an exploded view.

The steps required to generate a detail drawing of a part using parametric modeling software are described.

Step 1

Insert a base view of the part model.

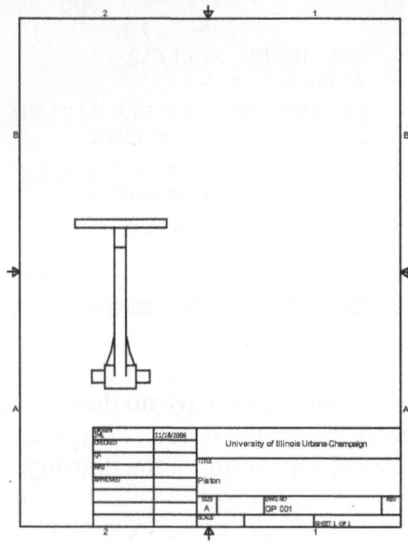

Step 2

Using the base view, project the other views.

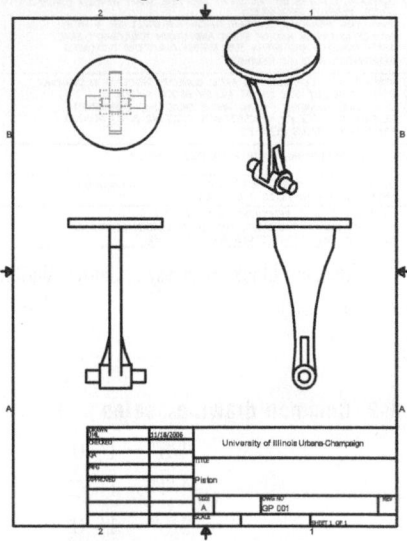

Step 3

Add centerlines.

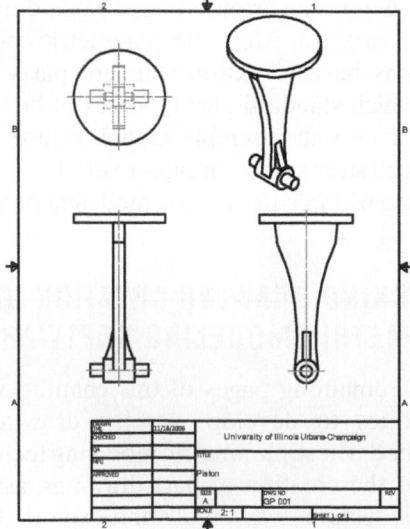

Step 4

Import parametric dimensions from part model. Note that these dimensions will probably need to be repositioned. In addition, at least some of these imported parametric dimensions will not be suitable for documentation purposes and will need to be replaced.

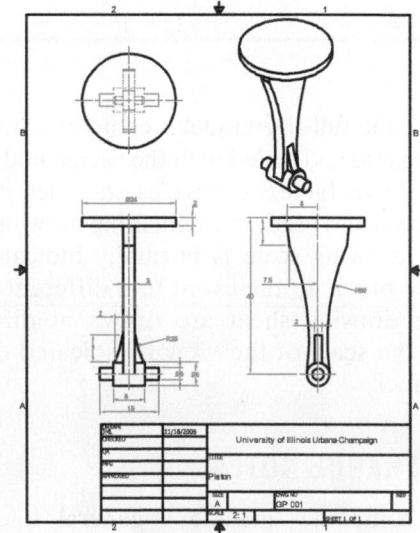

Figure 8-16 Extracting a detail drawing from a parametric model

The steps required to create a parametric assembly model are described, assuming that all of the part models have already been created.

Step 1
Import the base part into the assembly environment.

Step 2
Import the other component files into the assembly.

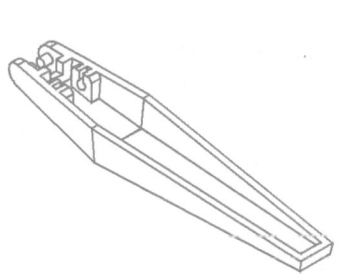

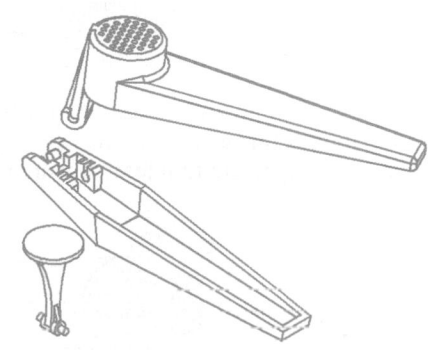

Step 3
Using assembly constraints like mate and insert, correctly position the components with respect to the base component and to one another.

Step 4
Properly constrained, moving parts will behave as they would in reality. The garlic press assembly model can be opened and closed, simulating the behavior of the actual device.

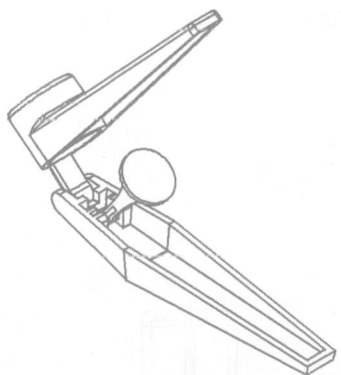

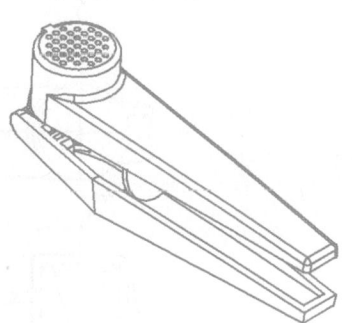

Figure 8-17 Creating an assembly model

This section shows the steps needed to create a sectioned assembly drawing from a parametric assembly model.

Step 1

Import a base view of an assembly into the drawing environment.

Step 2

Using a sectioning tool, create the section view(s). Note that the angle of the section lining (ANSI 31) differs for the different parts through which the cutting plane passes. Note also that some conventions (such as not applying section lining to cut thin features) are not automatically adhered to.

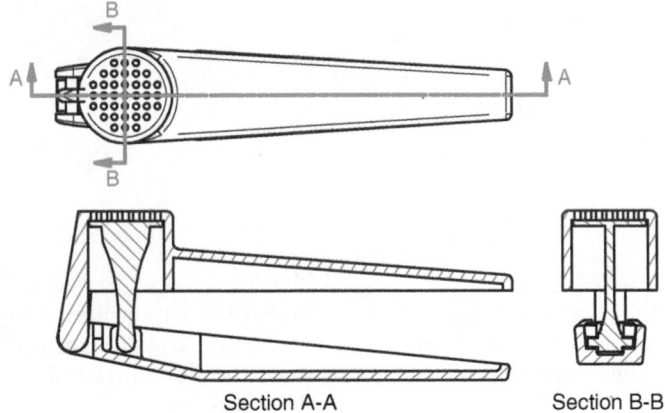

Section A-A   Section B-B

Step 3

Modify the hatch pattern to adhere to convention. Note also that the holes in the top of the cylindrical handle part have been suppressed for clarity at this stage.

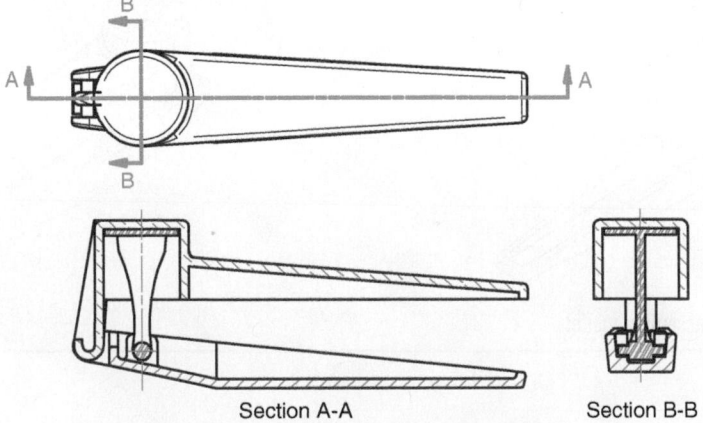

Section A-A   Section B-B

Figure 8-18 Extracting a sectioned assembly drawing from a parametric assembly model

This section shows the steps required to develop an exploded view of a parametric assembly model.

Step 1

Import an assembly model into the exploded-view creation environment.

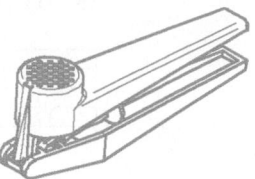

Step 2

Rotate open the cylindrical handle part.

Step 3

Translate the cylindrical handle part vertically.

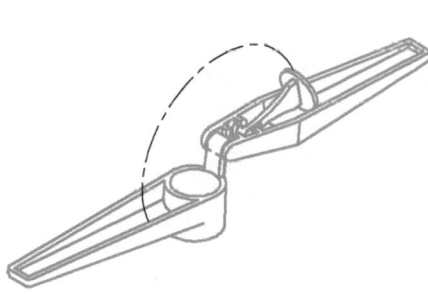

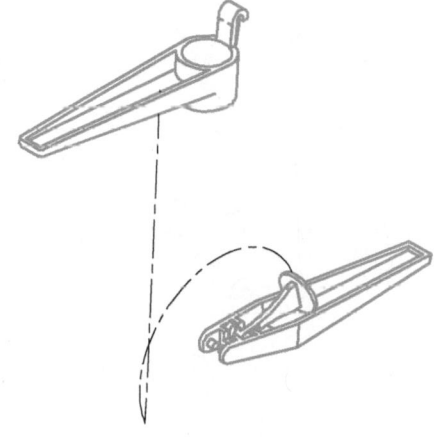

Step 4

Rotate the piston part to an upright position.

Step 5

Translate the piston part vertically.

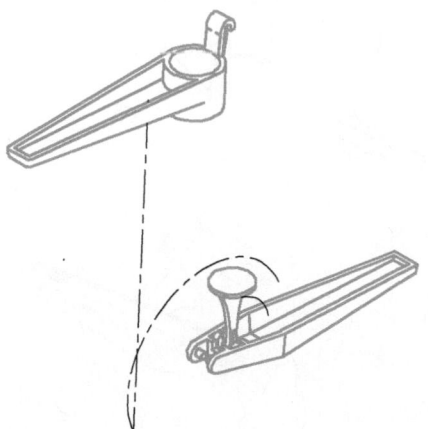

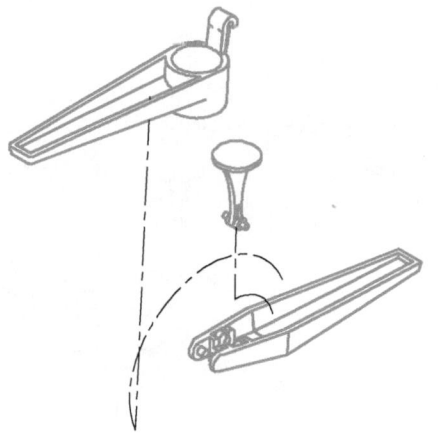

Figure 8-19 Creating an exploded view of a parametric assembly model

The process of generating an exploded-view drawing, parts list, and balloons is addressed, using parametric modeling software.

Step 1

Insert an exploded view into the drawing environment.

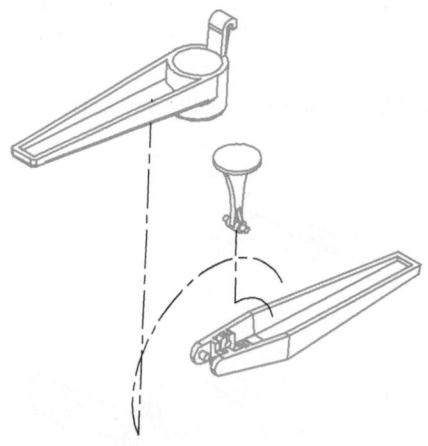

Step 2

Insert a parts list that is linked to the exploded view.

Parts List			
ITEM	QTY	PART NUMBER	DESCRIPTION
1	1	Handle	
2	1	Piston	
3	1	Cylinder_Handle	

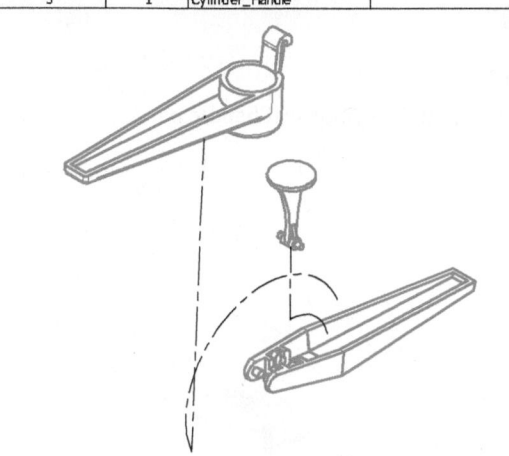

Step 3

Customize the parts list.

Bill of Materials					
ITEM	QTY	PART NUMBER	DESCRIPTION	MATERIAL	MASS
1	1	GP 001	Handle	Aluminum-6061	0.046 kg
2	1	GP 002	Piston	Aluminum-6061	0.006 kg
3	1	GP 003	Cylinder_Handle	Aluminum-6061	0.059 kg

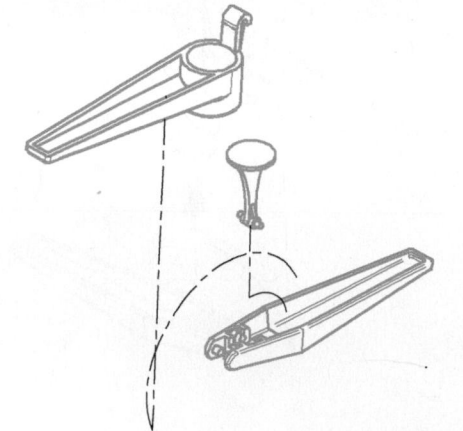

Step 4

Add balloons.

Bill of Materials					
ITEM	QTY	PART NUMBER	DESCRIPTION	MATERIAL	MASS
1	1	GP 001	Handle	Aluminum-6061	0.046 kg
2	1	GP 002	Piston	Aluminum-6061	0.006 kg
3	1	GP 003	Cylinder_Handle	Aluminum-6061	0.059 kg

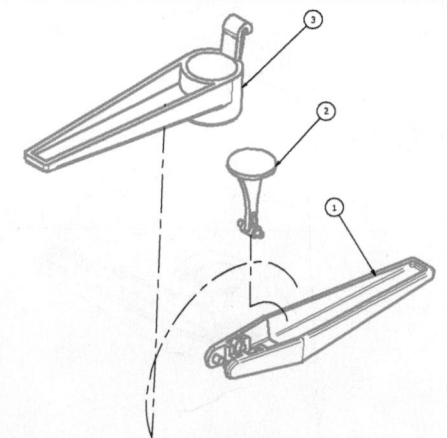

Figure 8-20 Creating an exploded-view drawing with parts list and balloons from a parametric assembly model

▌QUESTIONS

1. Given a dimensioned isometric view (Figures P7-1 through P7-24 in Chapter 7), create a dimensioned detail drawing of the part.
2. Given a dimensioned assembly view (Figures P8-1 through P8-5), create a complete set of working drawings of the assemblies.
   a. Create a solid model of each part.
   b. Create dimensioned detail drawings for each of the individual parts, including auxiliary views and section views when appropriate.
   c. Create an exploded pictorial assembly drawing, including balloons and parts list.
   d. Where appropriate, create a sectioned assembly drawing.

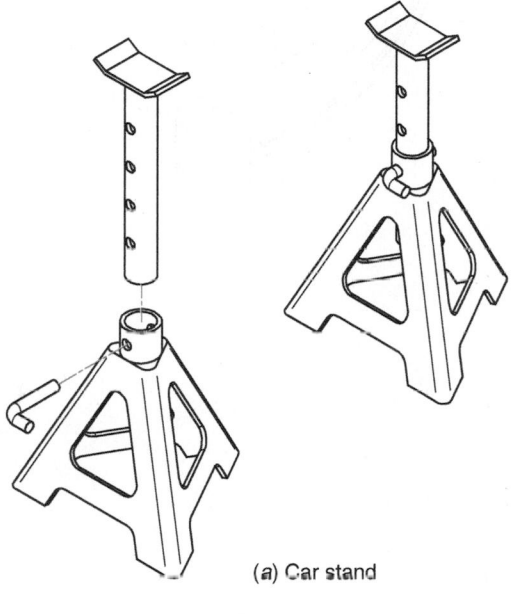

(a) Car stand

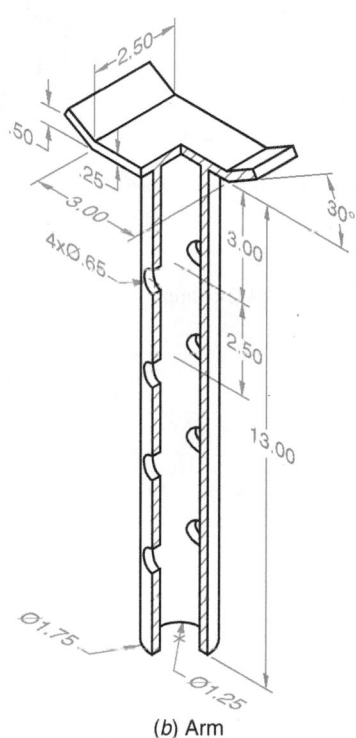

(b) Arm

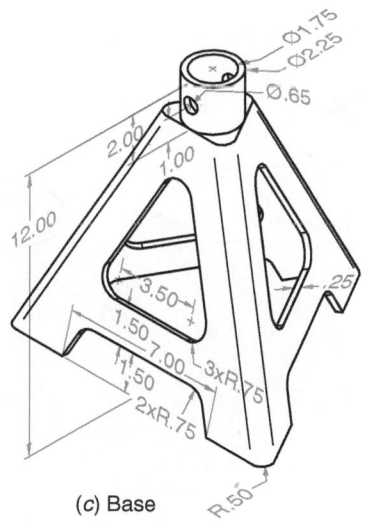

(c) Base

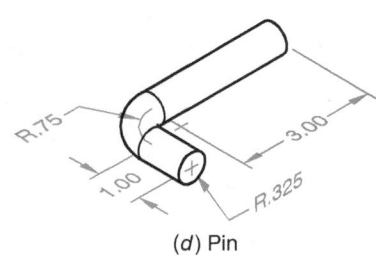

(d) Pin

Figure P8-1

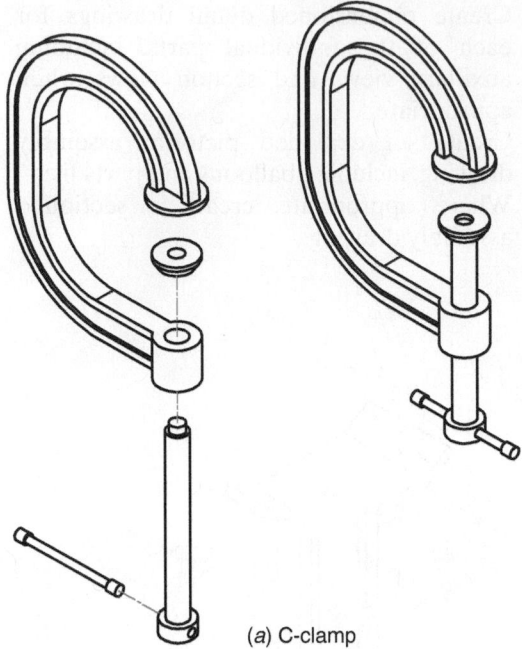

(a) C-clamp

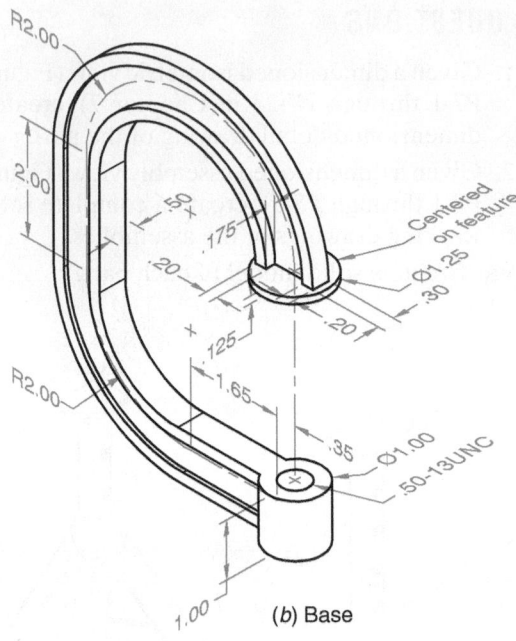

(b) Base

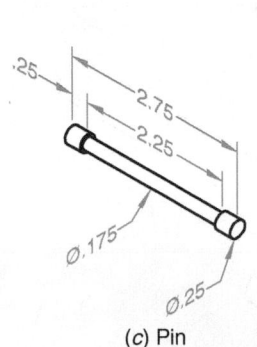

(c) Pin

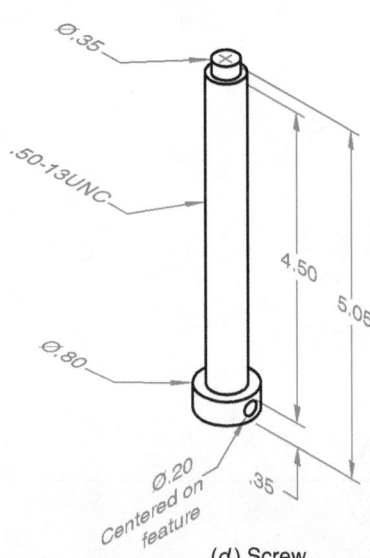

(d) Screw

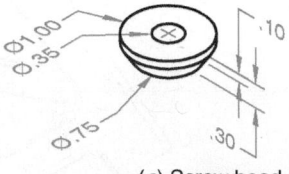

(e) Screw head

Figure P8-2

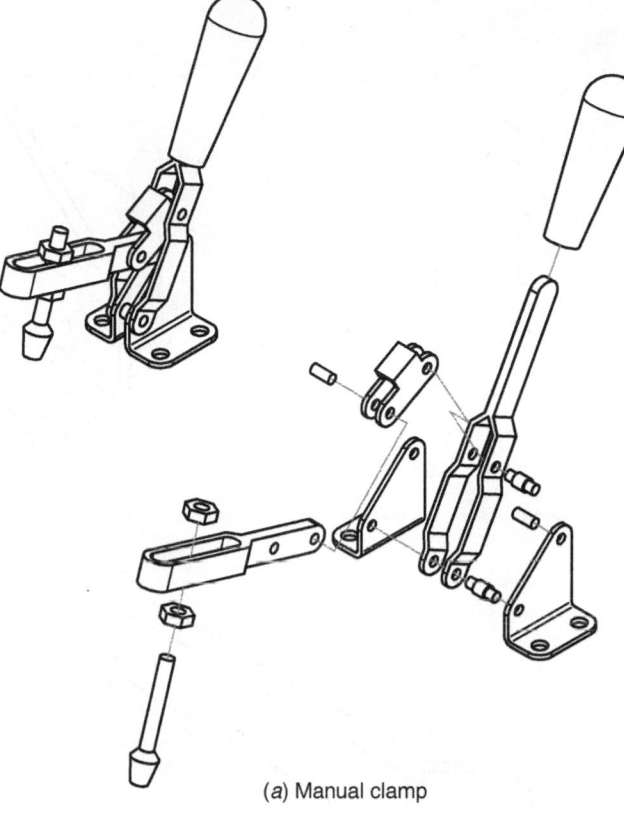

(a) Manual clamp

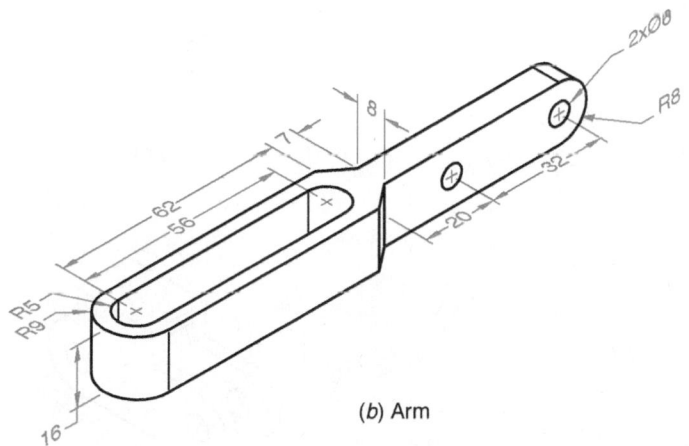

(b) Arm

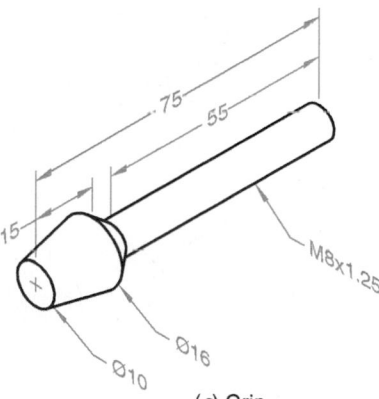

(c) Grip

Figure P8-3

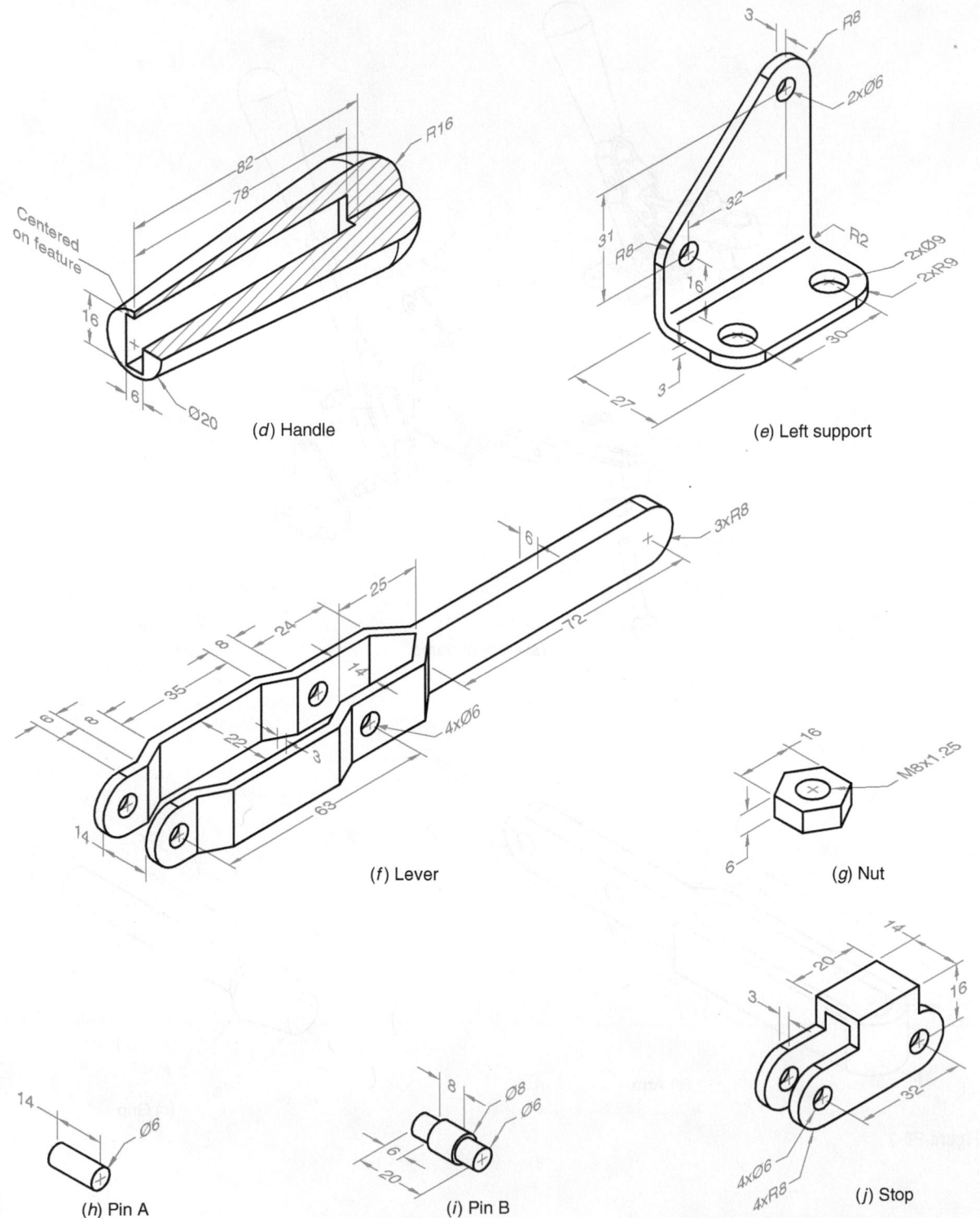

(d) Handle

(e) Left support

(f) Lever

(g) Nut

(h) Pin A

(i) Pin B

(j) Stop

Figure P8-3 (Continued)

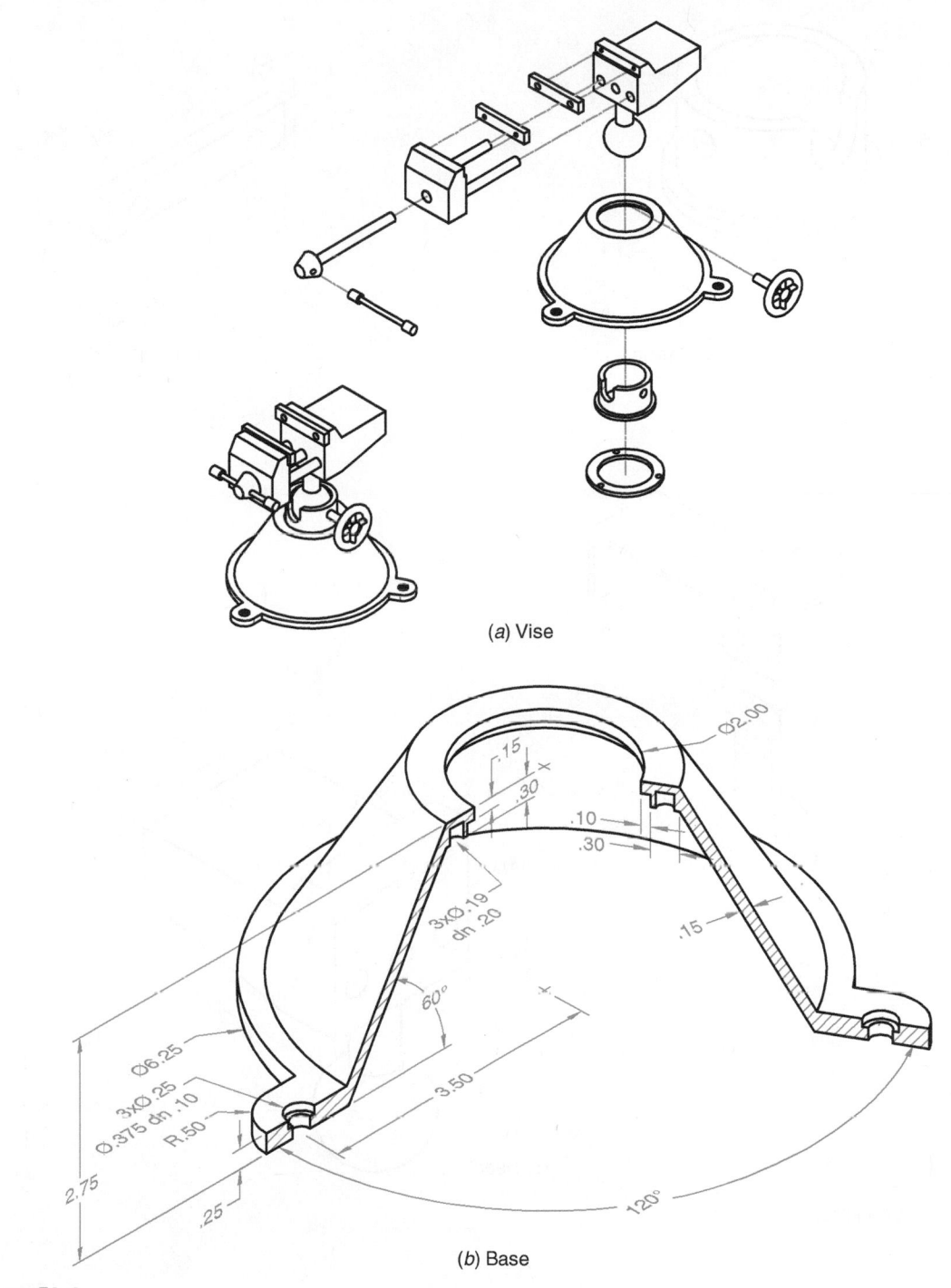

(a) Vise

(b) Base

Figure P8-4

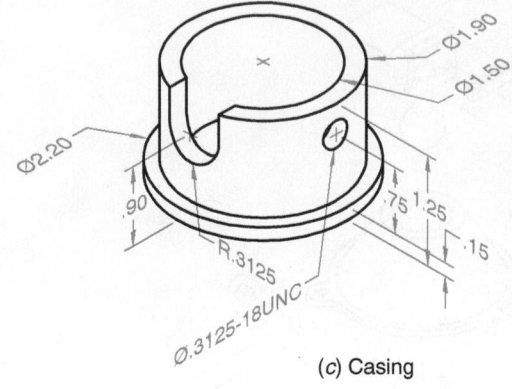

(c) Casing

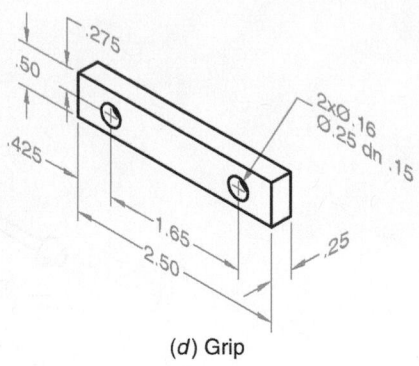

(d) Grip

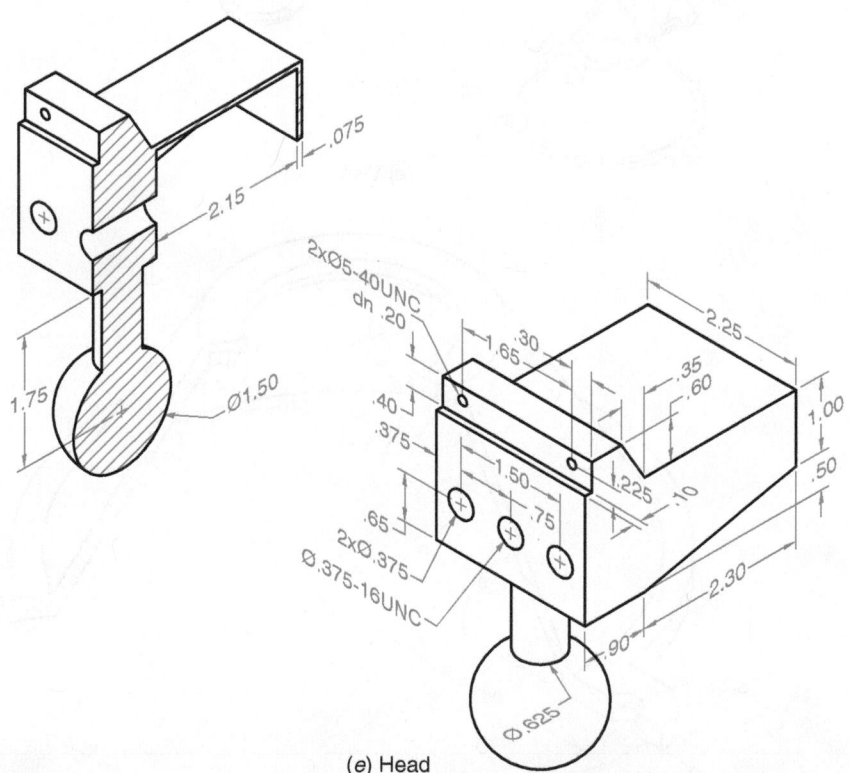

(e) Head

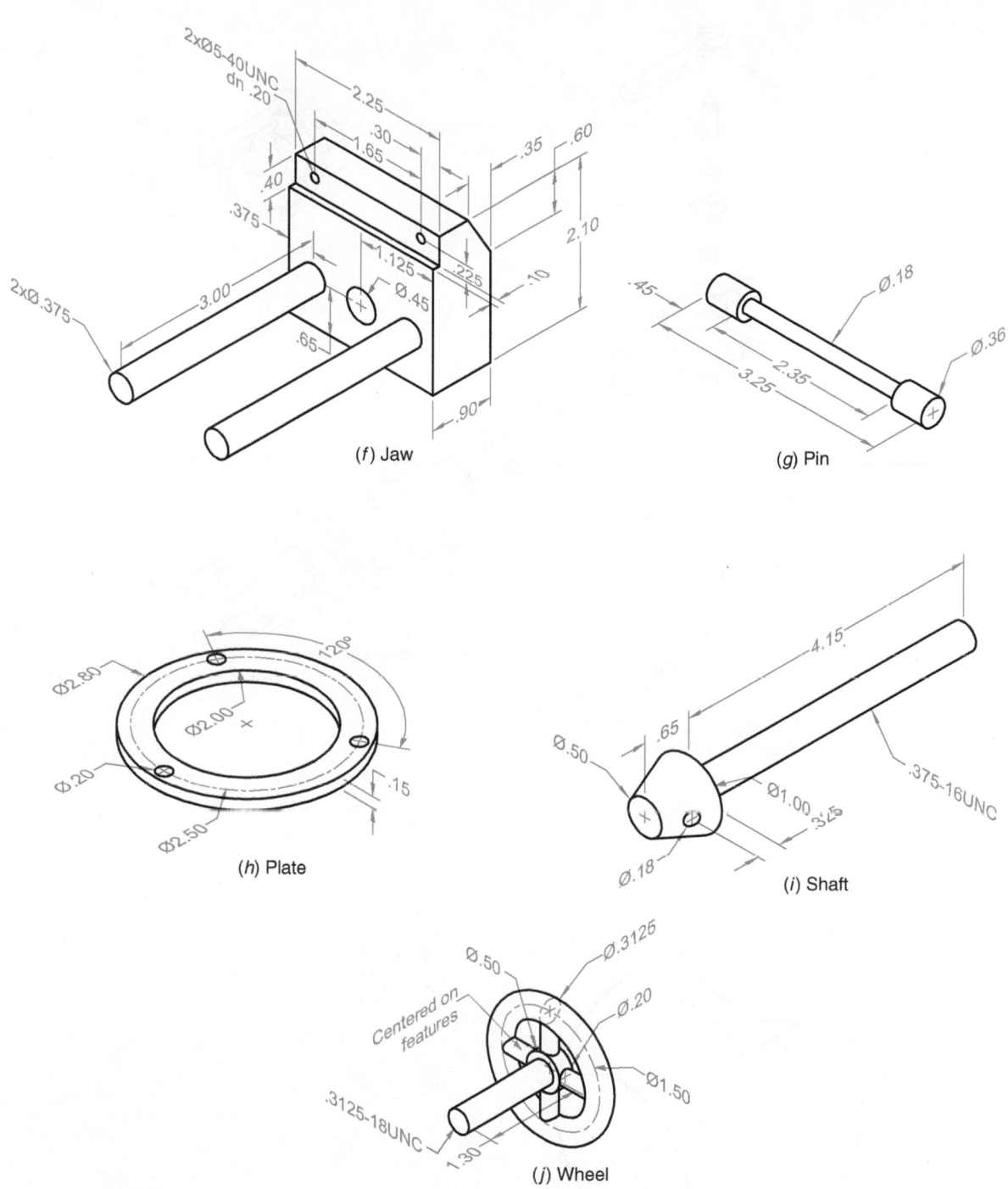

(f) Jaw

(g) Pin

(h) Plate

(i) Shaft

(j) Wheel

Figure P8-4 (Continued)

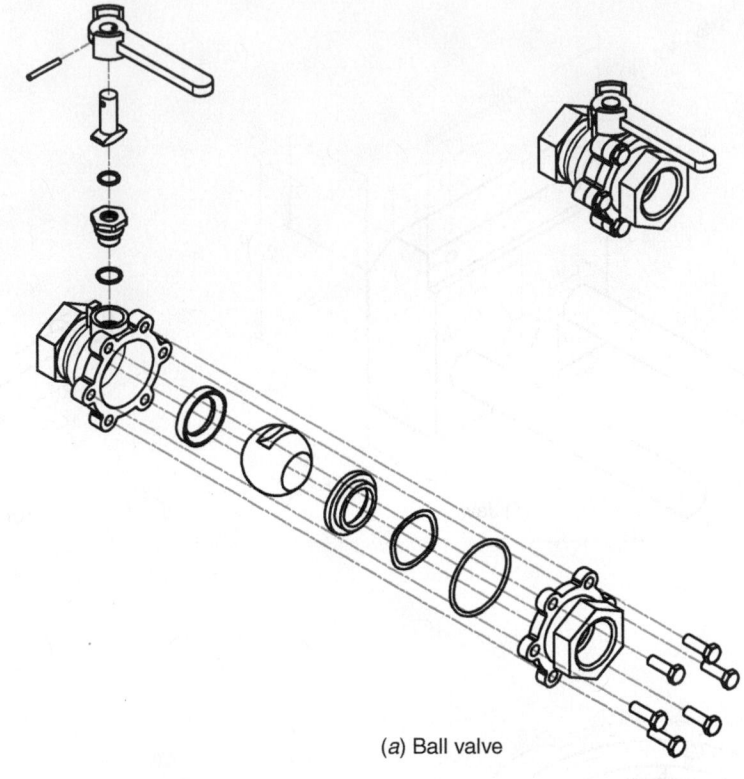

(a) Ball valve

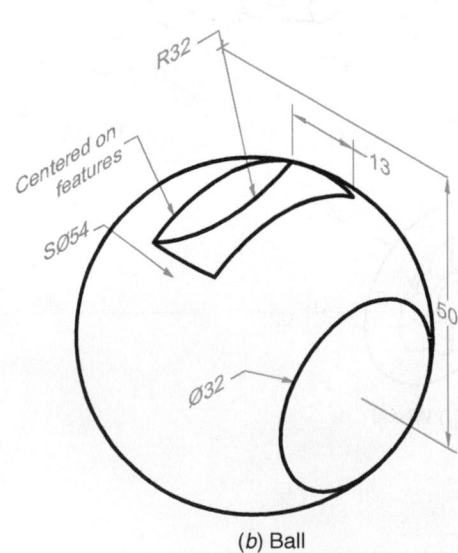

(b) Ball

Figure P8-5

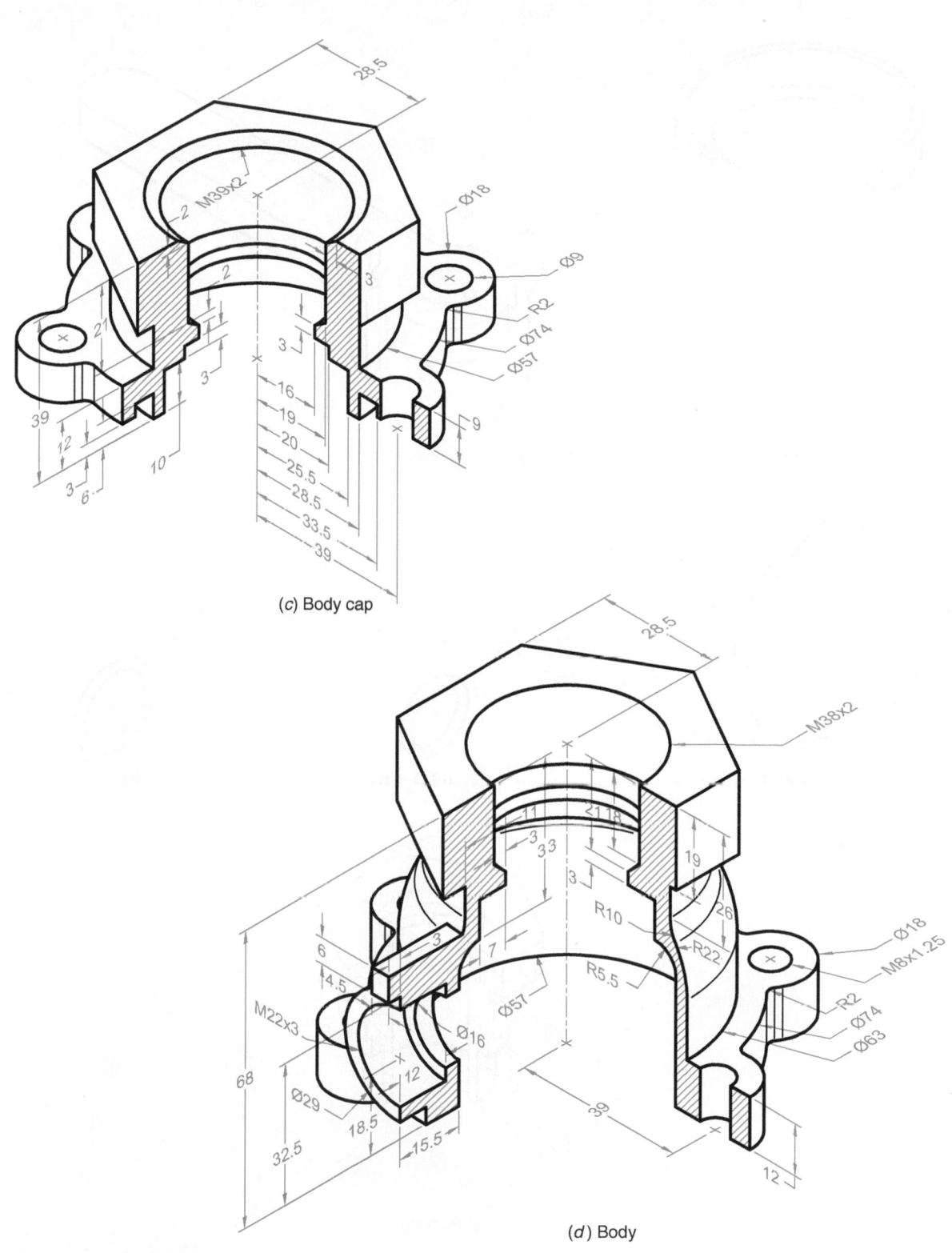

(c) Body cap

(d) Body

Figure P8-5   (Continued)

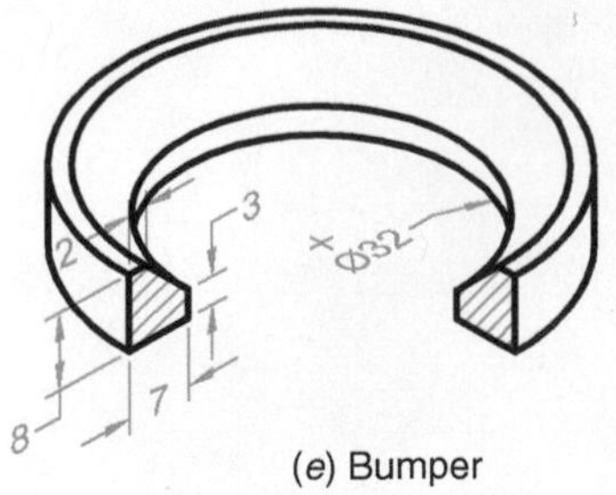

(e) Bumper

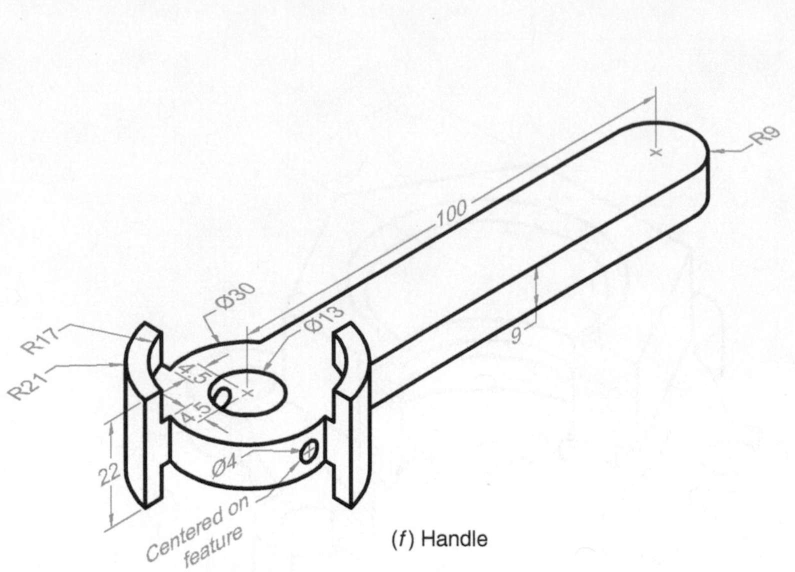

(f) Handle

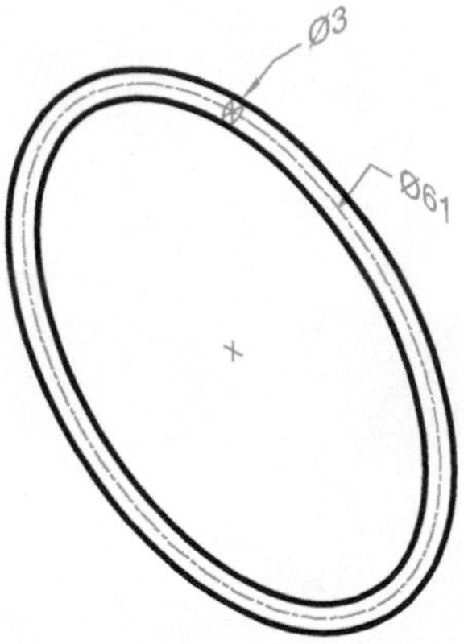

(g) O-ring 1

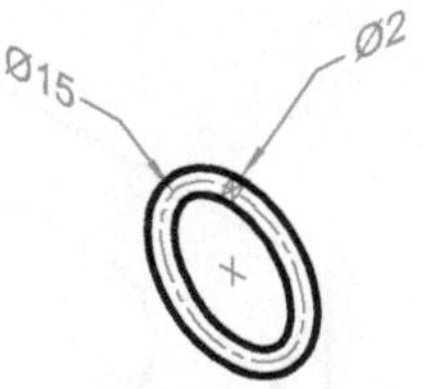

(h) O-ring 2

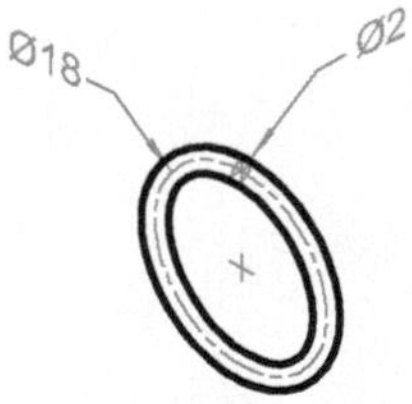

(i) O-ring 3

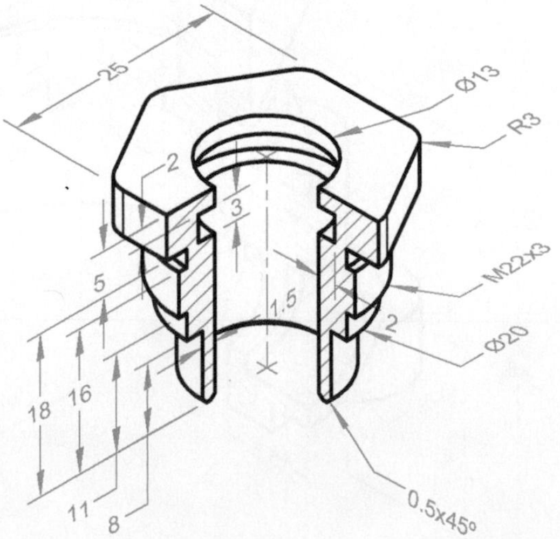

(j) Packing nut

Figure P8-5 (Continued)

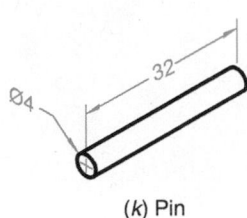

(*k*) Pin

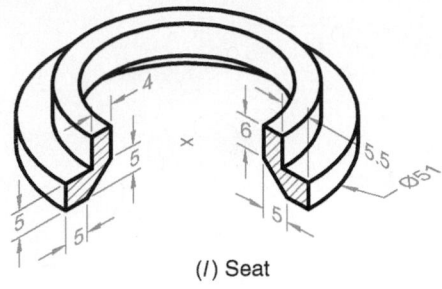

(*l*) Seat

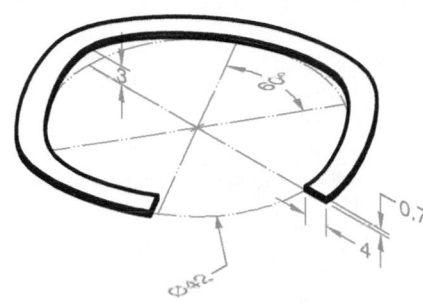

Uncompressed height is 7
compressed height is 3

(*m*) Spring

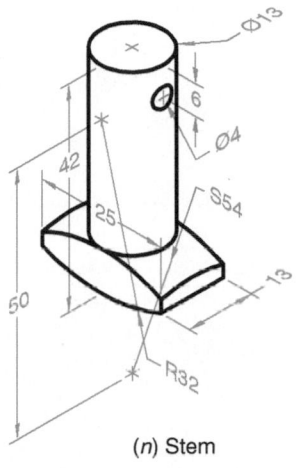

(*n*) Stem

Figure P8-5 (Continued)

13 PERSPECTIVE PROJECTIONS AND PERSPECTIVE SKETCHES

▌ PERSPECTIVE PROJECTION

Historical Development

Perhaps the single most important development in Renaissance art is the use of perspective.[1] Just prior to that time, paintings like those of Duccio di Buoninsegna (1255–1319) tended to be rather flat and two-dimensional (see Figure 13-1). Artists had yet to achieve an adequate understanding of human anatomy, and they had not developed techniques like shading and perspective to create an illusion of depth. Giotto (1267–1377), actually a contemporary of Duccio's, is generally considered the first Renaissance painter. In Figure 13-2, Giotto employs converging lines to suggest spatial depth, although these lines do not systematically converge to a single vanishing point.

In the work of later Italian Renaissance artists such as Leonardo da Vinci (1452–1519) and Raffaello Sanzio (1483–1520), we find paintings that employ one-point perspective to call the attention of the viewer to important details in the painting. See, for example, Figures 13-3 and 13-4 on page 328. The mathematical rules of perspective were developed and documented by people like the German Albrecht Dürer (1471–1528) and several Italian artists, including Brunelleshi and Alberti. Filippo Brunelleshi (1377–1446), a Florentine, invented a systematic method for determining perspective projections in the early

---

[1] The development of perspective during the Renaissance was recently chosen by *Time* magazine as one of the "100 Ideas That Changed the World."

Figure 13-1 Duccio di Buoninsegna, *Maesta*, Siena, 1308–1311 (Courtesy of The Bridgeman Art Library International)

Figure 13-2 Giotto, *Franciscan Rule Approved*, Assisi, Upper Basilica, c.1295–1300 (Courtesy of The Bridgeman Art Library International)

Figure 13-3 Leonardo da Vinci, *Last Supper*, 1498
(Courtesy of The Bridgeman Art Library International)

Figure 13-4 Raffaello Sanzio, *School of Athens*, Vatican, 1509
(Courtesy of The Bridgeman Art Library International)

1400's. Leon Battista Alberti (1404–1472) wrote the first treatise on perspective, *On Painting*.

Perspective Projection Characteristics

As we saw in Chapter 3, perspective differs from parallel projection in that in the former, the center of projection is a finite distance from the object. The projectors are therefore nonparallel rays that converge to the center of projection. As a consequence, when parallel object edges are not parallel to the projection plane, the edges converge to a *vanishing point* when projected. In addition, objects or features that are farther away from the projection plane are more *foreshortened* (i.e., smaller) than closer ones.

The principal advantage of perspective projection is that it produces a more realistic image. It closely approximates the view as seen by the human eye. A significant drawback of perspective

projection, however, is that it does a poor job in preserving the scale of the object. Consequently, dimensional information often cannot be extracted. In addition, perspective projections are generally more difficult to execute than parallel projections.

Classes of Perspective Projection

Perspective views are categorized according to the orientation of the object with respect to the projection plane. This orientation determines the number of principal axes (refer to Figure 3-4 in Chapter 3) that are parallel to the projection plane. If an axis is not parallel to the projection plane, then object edges parallel to this axis will not be parallel when projected. Rather, they will converge to a single point, called a vanishing point. There are three possible cases:

1. One-point perspective (one principal vanishing point)
2. Two-point perspective (two principal vanishing points)
3. Three-point perspective (three principal vanishing points)

In a top view looking down, Figure 13-5 illustrates the orientation of three identical cubes (more generally, three cube-shaped bounding boxes), with respect to a vertical projection plane. Note that, because the scene is viewed from above, the vertical projection plane appears as a line, since it is viewed on edge. Note also that the principal axes of each cube are also represented.

The cube on the left is oriented with one principal face parallel to the projection plane. If the principal axes of this principal enclosing box

Figure 13-5 Perspective classes

Table 13-1 Classes of perspective projection

Perspective Type	Principal Vanishing Points (PVPs)	Principal Axis Orientation
One-point	1	• One principal axis perpendicular to projection plane (PP) • Two principal axes parallel to PP
Two-point	2	• Two principal axes inclined to projection plane (PP) • One principal axis parallel to PP
Three-point	3	• All three principal axes inclined to PP • No principal axes parallel to PP

(PEB) are extended infinitely in both directions, only one axis intersects the projection plane; the other two axes (one vertical, one horizontal) are parallel to the projection plane. When the bounding box of an object is oriented in this way, a one-point perspective projection results.

The PEB in the middle has been rotated about a vertical axis so that only the vertical principal axis is parallel to the projection plane; the other two axes are inclined to the projection plane. In this orientation, a two-point perspective projection results.

Finally, imagine rotating the middle cube out of the plane of the paper about a horizontal axis. This is the position of the cube on the far right. Note that in this case, all three principal axes, when extended, intersect the projection plane. None of the three principal axes are parallel to the projection plane. In this orientation, a three-point perspective projection will result.

Vanishing Points

Before discussing these three cases in more detail, it is worth reiterating that, in a perspective projection, if parallel object edges are:

• Parallel to the projection plane, then the projected edges will also be parallel

• Inclined to the projection plane, then the projected edges will not be parallel; they will converge to a *vanishing point*

Returning to Figure 13-5, it is apparent that a one-point perspective of a box has one vanishing point, a two-point perspective of a box has two vanishing points, and a three-point perspective of a box has three vanishing points. These vanishing points are called *principal vanishing points*, since they are associated with the principal axes of the object. Table 13-1 provides a summary of the different perspective classes, along with the number of principal vanishing points of each.

Figure 13-6 on page 330 illustrates the process of locating principal vanishing points (PVPs) for a two-point perspective projection. Figure 13-6a shows a pictorial view of the object, projection plane, and center of projection (CP). Figure 13-6b shows the same elements when viewed from above. We can see that this orientation of the object with respect to the projection plane will result in a two-point perspective, since two principal axes are inclined to the projection plane. Two dashed construction lines are drawn through the CP, each parallel to an inclined principal axis, until they intersect the projection plane. Each point of intersection locates a principal vanishing point. Each construction line is parallel to a set of parallel edges on the object, with each edge set parallel to a principal axis. To summarize, taking a line parallel to an inclined object edge and passing it through the center of projection until the line pierces the projection plane locates the vanishing point for that object edge.

Figure 13-6c shows the projection plane and the resulting perspective projection. For clarity, the object is not shown. Note that the projected edges converge to the principal vanishing points.

Figures 13-6a and 13-6c show the ground line (GL) and the horizon line (HL). We will use these lines later in this chapter when constructing perspective sketches. The *ground line* is a horizontal line formed by the intersection of the

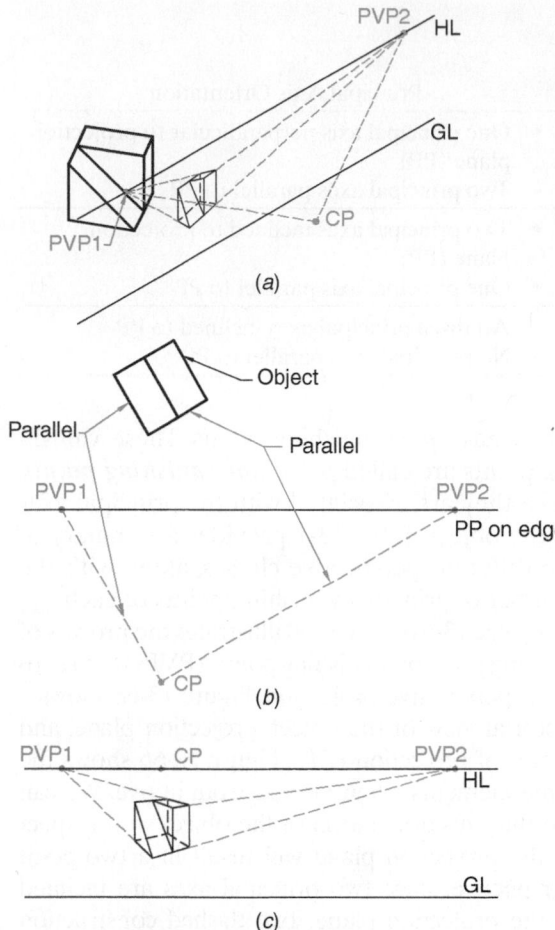

(a)

(b)

(c)

Figure 13-6 Locating principal vanishing points

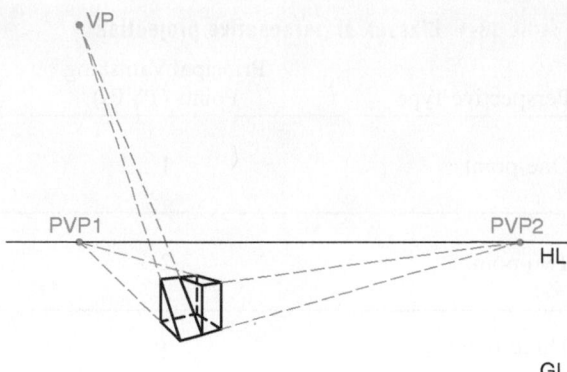

Figure 13-7 Another principal vanishing point

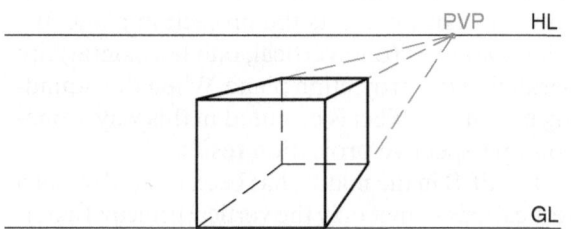

Figure 13-8 One-point perspective drawing of a cube

projection plane and the ground plane—that is, the plane on which the object rests. The *horizon line* represents the eye level of the observer. The horizon line is formed by the intersection of the projection plane and the horizontal plane that passes through the center of projection.

As seen in Figure 13-6c, both principal vanishing points lie on the horizon line. Each of these vanishing points results from the projection of a set of parallel object edges. Both parallel edge sets are themselves parallel to the ground plane. Vanishing points for edges parallel to a plane always lie along a straight line in the projection plane, with the line parallel to the plane. Here the plane is the ground plane, and the straight line is the horizon line.

Any parallel group of inclined object edges will converge to a vanishing point in a perspective projection, not just principal axis edges. For example, the object depicted in the two-point perspective in Figure 13-6c actually has three vanishing points; see Figure 13-7. The edges of the inclined surface on the actual object are inclined (i.e., not parallel) to the projection plane. Consequently, when projected, the inclined surface edges also converge to a vanishing point.

One-Point Perspective Projection

In a one-point perspective projection, one object face is parallel to the projection plane. One principal axis is perpendicular to the projection plane, and the other two principal axes (horizontal, vertical) are parallel to the projection plane.

Figure 13-8 shows a one-point perspective drawing of a cube. Note that the vertical edges of the projected cube are parallel to one another, as are the horizontal projected edges. Also note that the receding[2] edges of the cube are not parallel.

[2] Object edges not parallel to the projection plane will appear to recede back into space when projected.

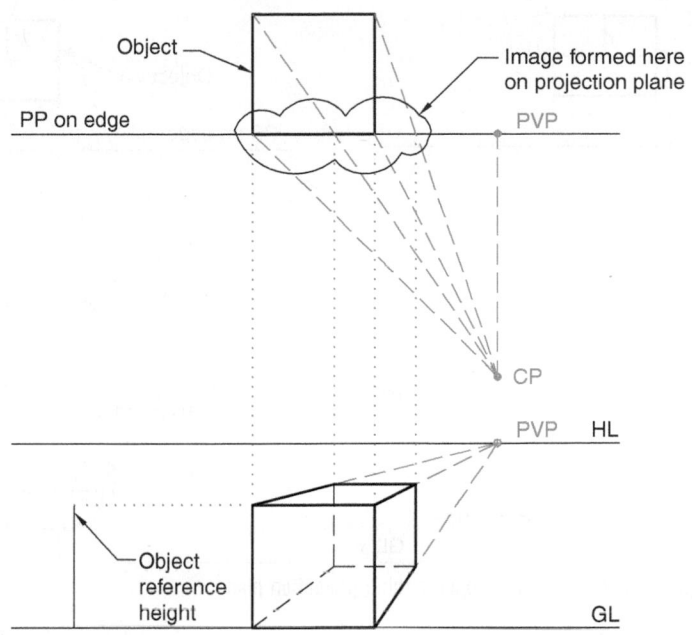

Figure 13-9 One-point perspective setup

Rather, they converge to a principal vanishing point.

Figure 13-9 shows the perspective arrangement used to obtain Figure 13-8. Once again, the top portion of the figure shows a view from above. The object, the projection plane, the center of projection, the projectors, and the construction lines used to locate the PVP are all depicted. Note that the front face of the object is coplanar with the projection plane. The projected image will appear within the encircled area on the projection plane.

The bottom half of Figure 13-9 shows the resulting one-point projection (object not shown for clarity). The dotted vertical lines connecting the two portions of the figure are used to locate the projected image on the projection plane. Because the front face of the cube lies in the projection plane, it will be projected true size.

Figure 13-10 on page 332 shows another example of a one-point perspective arrangement, this time with the object entirely behind the projection plane. Note that, because of this, the front face of the object is not projected true size.

Two-Point Perspective Projection

An example of a two-point perspective arrangement was shown in Figure 13-6. Let us review some of its characteristics:

- One set of principal edges (typically vertical) are parallel to the projection plane, causing the projected edges also to be vertical.

- The other two sets of principal edges, being inclined to the projection plane, will converge to vanishing points when projected. These principal vanishing points will lie on the horizon line.

- If the leading edge of the object lies behind the projection plane (as is the case in Figure 13-6), then none of the projected edges will appear true size.

Figure 13-11 on page 332 provides an example of a two-point perspective projection where the leading object edge lies in the projection plane. In this case, the leading edge is projected true size.

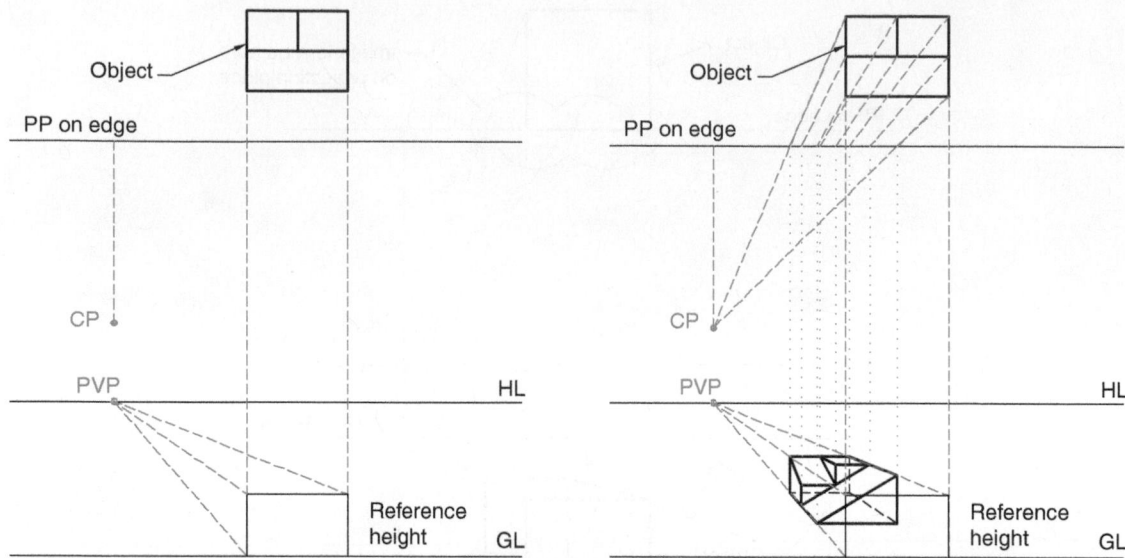

Figure 13-10 One-point perspective setup with object behind projection plane

Three-Point Perspective Projection

Because of the difficulty in their construction, three-point perspective drawings are rarely used. Figure 13-12 shows an example of a three-point perspective arrangement. Note that all three principal axes are inclined to the projection plane and that, when projected, all three sets of edges converge to principal vanishing points. Two of the PVPs lie on the horizon line; one does not.

Perspective Projection Variables

There are several variables that influence the appearance of a perspective projection. Some of these variables are discussed later.

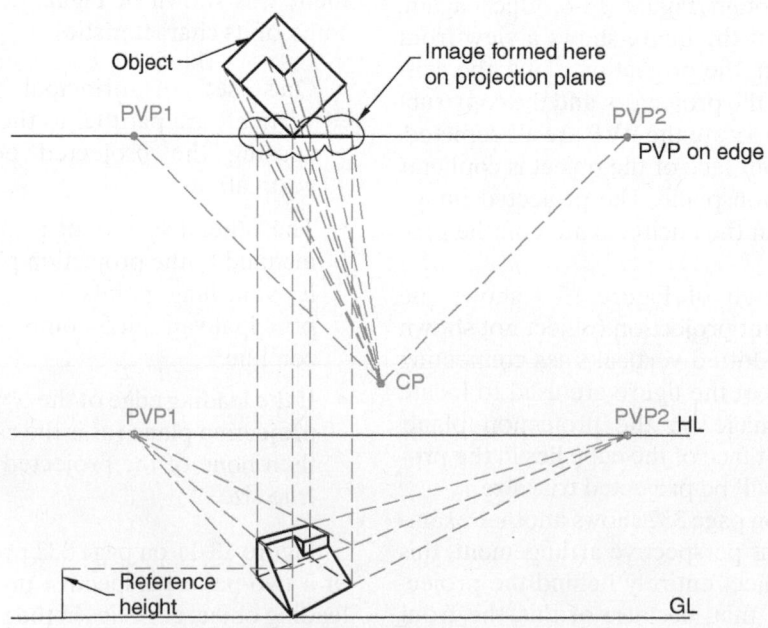

Figure 13-11 Two-point perspective projection setup; leading edge lies in projection plane

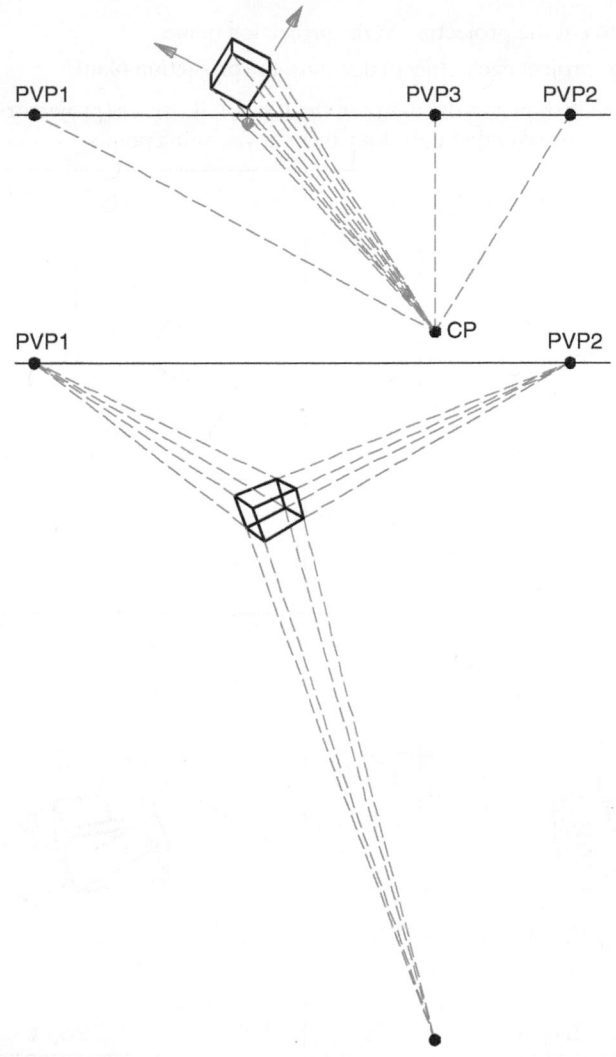

Figure 13-12 Three-point perspective setup

Perspective projection using a 3D CAD system

The procedure* described below may be used to create a perspective projection (or, for that matter, any of the planar projections described in Chapter 3) using a 3D CAD system like AutoCAD®. In this procedure, all of the common elements of a projection system (e.g., object, projection plane, center of projection, projectors), as well as the projection itself, are modeled.

1. Start by creating the object, the projection plane, and the center of projection. In Figure 13-13 on page 334:
   a. The object is modeled as a solid.
   b. The projection plane is represented using line segments to draw a vertically oriented rectangle.
   c. The center of projection is modeled as a solid sphere or point.

2. Use the line command and object snap settings to draw the projectors from the vertices of the object to the CP.

*This section is based on the work of Michael H. Pleck, who developed this technique at the University of Illinois at Urbana-Champaign.

(Continues)

3. Use the trim command to cut the projectors at the projection plane.

4. Use the line command to project each object edge onto the projection plane.

The completed projection is shown in the figure in the lower-left corner (projectors not shown). In the lower-right corner, projected edges are extended until they meet at vanishing points.

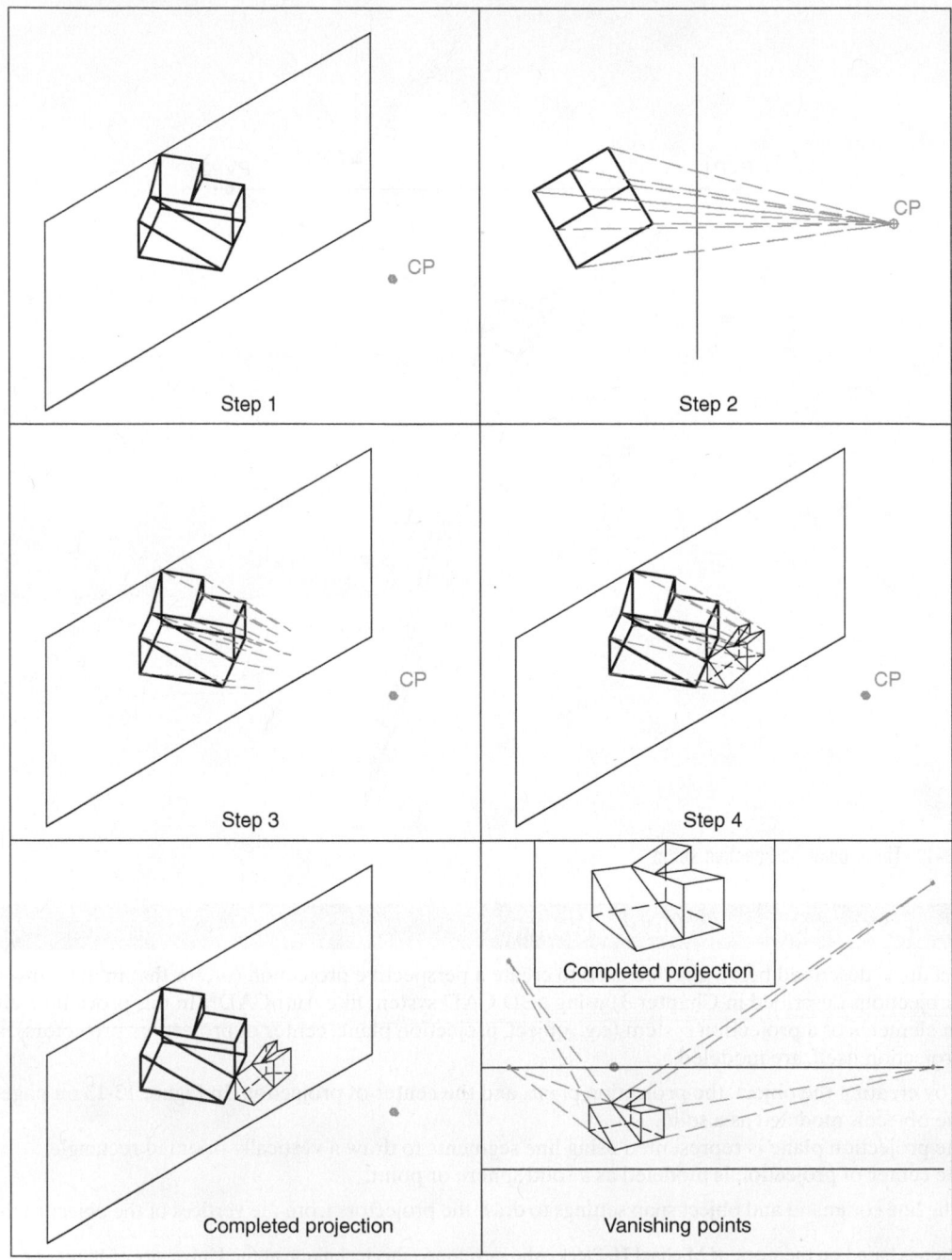

Figure 13-13 Perspective projection using a 3D CAD system

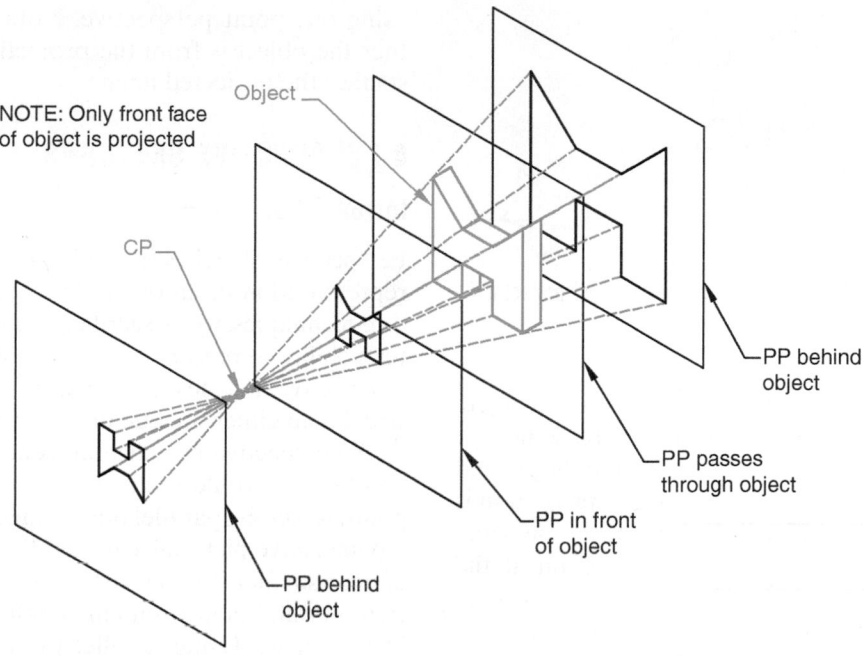

NOTE: Only front face of object is projected

Object

CP

PP behind object

PP passes through object

PP in front of object

PP behind object

Figure 13-14 Perspective projection plane projection

PROJECTION PLANE LOCATION

It has already been shown that in a perspective projection, the size of the projected image depends on the location of the projection plane with respect to the object and the center of projection. From Figure 13-14 (which is the same as Figure 3-13 in Chapter 3), it should be clear that the placement of the projection plane affects the size, and even the orientation, of the projected image. Here are the possibilities:

1. The projection place is behind the object, in which case the projected image is larger than the object.

2. The projection passes through the object, resulting in a projected image that is the same size as the object.

3. The projection plane is in front of the object, causing the projected image to be smaller than the object.

4. The projection plane is behind the center of projection. In this case, the projected image is inverted.

LATERAL MOVEMENT OF CP

If the center of projection is moved laterally with respect to the projection plane (or, equivalently, if the object is moved with respect to the center of projection), different projections will result, as shown in Figure 13-15. Generally speaking, it is recommended that the center of projection be placed in front of the object, slightly to one side.

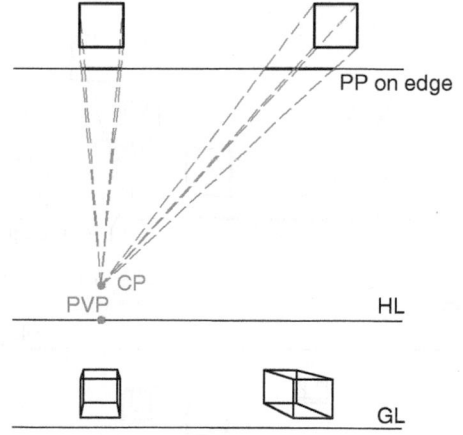

PP on edge

CP

PVP

HL

GL

Figure 13-15 Lateral movement of an object with the same center of projection

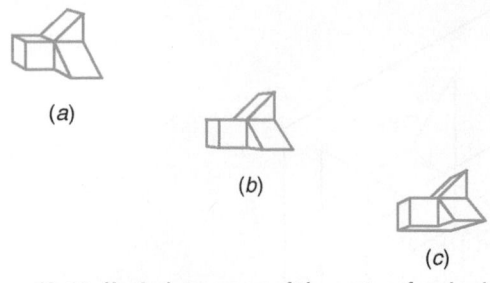

Figure 13-16 Vertical movement of the center of projection

VERTICAL MOVEMENT OF CP

Figure 13-16 shows how the projection of an object can change, depending on the vertical placement of the center of projection with respect to the ground plane. In Figure 13-16a, the center of projection is above the object. Figure 13-16b shows the same object, but with the center of projection at the same level as the object. Finally, in Figure 13-16c, the center of projection is below the object.

VARYING DISTANCE FROM CP

One of the strengths of perspective projection is that it results in a more realistic image than parallel projection. This is due to the fact that, much as in our own vision, the size of an image projected using perspective depends on the distance of the object from the projection plane. In Figure 13-17, cubes of the same size are projected

using one-point perspective. Note that the farther the object is from the projection plane, the smaller the projected image.

■ PERSPECTIVE SKETCHES

Introduction

Perspective sketches provide a more realistic representation of an object than parallel projection techniques, while sacrificing much of the latter's ability to preserve dimensional information. Perspective sketches are also more difficult to construct than either oblique or isometric sketches.

A perspective sketch represents an object as an observer would see it from a certain vantage point. Receding parallel object edges converge in a perspective pictorial, causing distant objects to appear smaller. In contrast, in a parallel projection, parallel edges remain parallel in the projected image. Using parallel projection, objects are projected as the same size, regardless of their distance from the projection plane.

Terminology

Key elements of a perspective sketch are shown in Figure 13-18. As we have already seen, these elements include the ground line, the horizon line, and vanishing points. The ground line represents the plane on which the object rests and is formed by the intersection of the ground plane with the projection plane. The horizon line represents the eye level of the observer.[3] A vanishing point is a position on the horizon to which depth projectors converge.[4]

One-Point Perspective Sketches

Recall that in a one-point perspective projection, one object face is parallel to the projection plane. This explains the similarity between a one-point

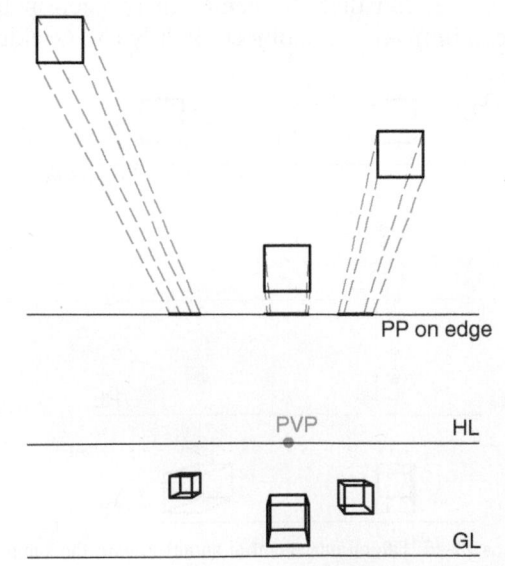

Figure 13-17 Varying object distance from projection plane

[3] Recall from the discussion of perspective projection earlier in the chapter that the horizon line is at the same height as the center of projection.

[4] In a perspective projection, parallel object edges inclined to the projection plane converge when projected. If these edges are also parallel to the horizontal ground plane, they will be projected along the horizon line.

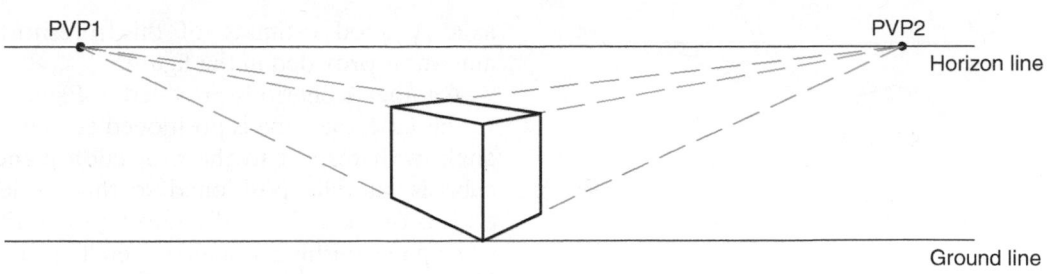

Figure 13-18 Perspective sketch terminology; two-point perspective

perspective and an oblique sketch, which is also oriented with two principal axes parallel to the projection plane. See Figure 13-19 for a comparison between one-point perspective and oblique pictorials.

The main difference between the two pictorials is that the receding edges are parallel in oblique projection, whereas in the one-point perspective the receding edges converge to a vanishing point.

In a one-point perspective projection, if the object's front face coincides with the projection plane, then the front face of the object is projected full scale (see Figure 13-14 on page 335). Otherwise, if it lies behind (or in front of) the projection plane, the projected front face will be smaller (or larger) than the actual. In practical terms, though, when constructing a perspective sketch, we first choose a vertical edge length representing the height of the front face. The horizontal width dimension is then scaled proportional to this vertical dimension.

Scaling along the receding depth axis involves some visual approximation. Figure 13-20 shows two one-point perspective sketches of the same cube. On the left, the cube has been laid out using the same distance L along the horizontal, vertical, and converging axes. Clearly, the resulting pictorial appears to be too long in the receding

axis direction. This distortion occurs because foreshortening along the depth axis has not been accounted for. On the right, the depth dimension has been reduced, resulting in an improved representation of a cube. Although the amount of foreshortening depends on several variables, a good rule of thumb is to foreshorten the converging axis dimension on a one-point perspective sketch to approximately two-thirds of the actual.

Two-Point Perspective Sketches

In a two-point perspective sketch, the object is oriented so that only one principal axis, typically vertical, is parallel to the projection plane. The other two principal axes are inclined to the projection plane. As a consequence, vertical object edges remain parallel when projected, while the two other sets of principal edges converge to different vanishing points. Both of these principal vanishing points lie on the horizon line (see Figure 13-21 on page 338).

If the projection plane passes through the leading vertical edge of the object, this edge will be projected full size (see Figure 13-14 on page 335).

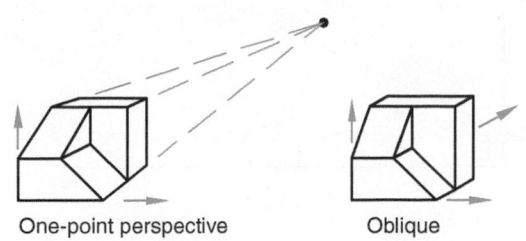

One-point perspective   Oblique

Figure 13-19 Comparison of one-point perspective and oblique sketches

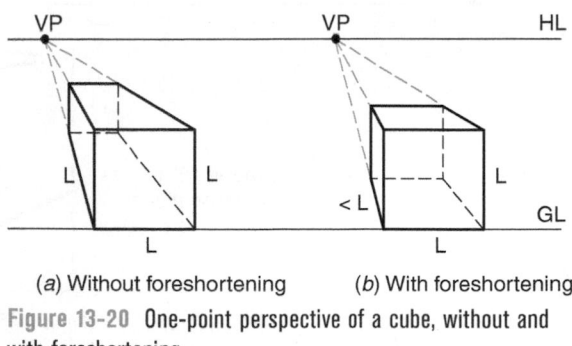

(a) Without foreshortening   (b) With foreshortening

Figure 13-20 One-point perspective of a cube, without and with foreshortening

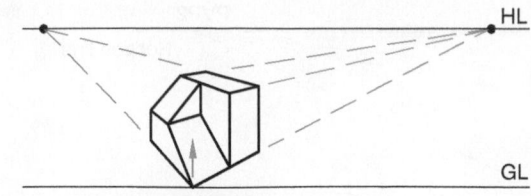

Figure 13-21 Two-point perspective sketch

Otherwise, if the edge is behind (or in front of) the projection plane, the projected vertical will be smaller (or larger) than the true length. When constructing a two-point perspective sketch, though, one simply chooses a vertical edge length representing the height of the leading edge of the bounding box, without regard for the location of this edge with respect to the projection plane. Convergence lines are then drawn from the leading edge end points to both principal vanishing points.

As was the case with one-point perspective, the amount of foreshortening along the receding axis must be estimated in order to create a well-proportioned sketch. In the case of a two-point perspective, however, there are two receding axes.

Figure 13-22 shows a two-point perspective sketch of a cube, where the cube is placed at a 45-degree angle to the projection plane. If the horizontal distances from the leading edge to each principal vanishing point are equal, as is the case in Figure 13-22, then the amount of foreshortening will be the same along each receding

axis. A good estimate of this foreshortening amount is provided in the figure.

Another scenario is provided in Figure 13-23. In this case, the cube is positioned at a 30-degree angle with respect to the projection plane. The cube is laterally positioned so that its leading edge is one-fourth the distance between the two principal vanishing points. Given this scenario, Figure 13-23 provides reasonable foreshortening estimates along both receding axes. Note that the closer the PVP is to leading edge distance, the greater the amount of foreshortening.[5]

Proportioning Techniques

A useful proportioning technique when constructing perspective sketches is to sketch the diagonals of a receding face in order to locate the midpoint of that face. This point can then be projected to an adjacent edge in order to locate other key vertices. Figure 13-24 illustrates this technique for both one-point and two-point perspective sketches.

In Figure 13-25 this technique is extended to allow for partitioning of a trapezoidal area into thirds and quarters. See the section on partitioning lines in Chapter 2 for additional information.

[5] Figures 13-22 and 13-23 are taken from *Graphics for Engineers*, 2nd Edition by Jerry Dobrovolny and David O'Bryant, John Wiley & Sons, 1984.

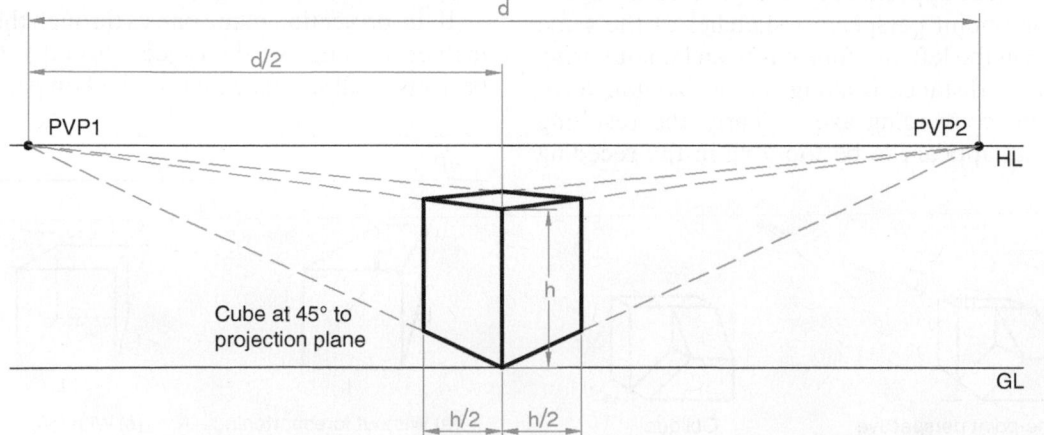

Figure 13-22 Two-point perspective pictorial of a cube at a 45-degree angle to the projection plane (Taken from Jerry Dobrovolny and David O'Bryant, *Graphics for Engineers, 2nd Edition,* John Wiley & Sons, 1984)

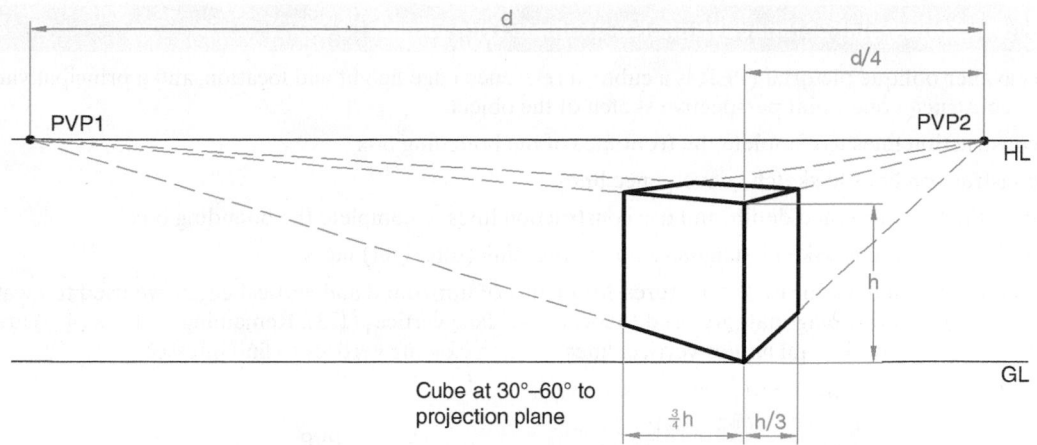

Figure 13-23 Two-point perspective pictorial of a cube at a 30-degree angle to the projection plane (Taken from Jerry Dobrovolny and David O'Bryant, *Graphics for Engineers, 2nd Edition,* John Wiley & Sons, 1984)

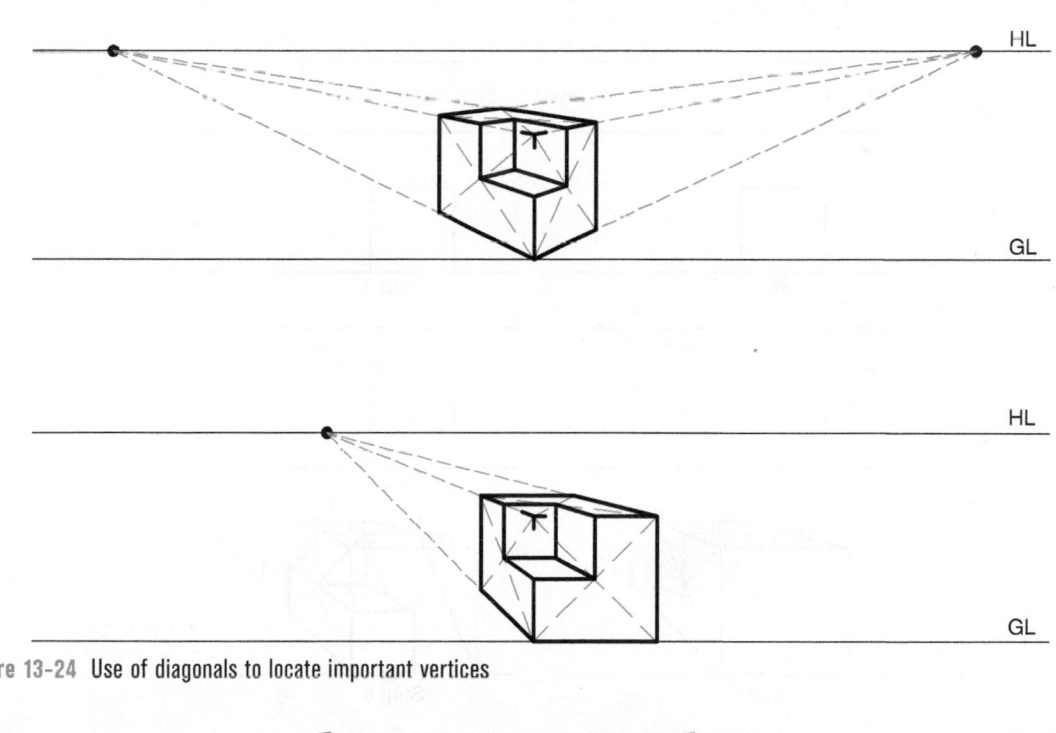

Figure 13-24 Use of diagonals to locate important vertices

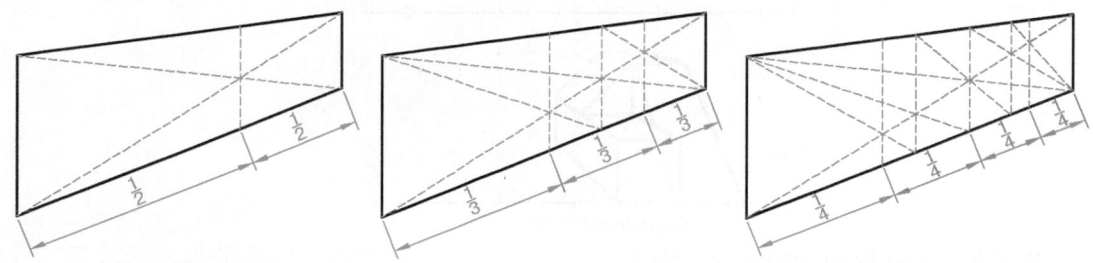

Figure 13-25 Partitioning trapezoidal areas

Given a cavalier oblique pictorial (PEB is a cube), a reference edge height and location, and a principal vanishing point, construct a one-point perspective sketch of the object.

1. Use construction lines to complete the front face of the bounding box.

2. Use construction lines to sketch convergence lines.

3. Estimate the foreshortened depth, and use construction lines to complete the bounding box.

4. Using construction lines, sketch diagonals on the receding (top, right) faces.

5. Using construction lines, locate key features. Midpoints of horizontal and vertical edges are used to locate 2, 5, and 7. Intersecting diagonals are used to locate mid-face vertices (1, 3). Remaining vertices (4, 6) are located by passing horizontal and/or vertical lines through existing vertices to find intersections.

6. Go bold.

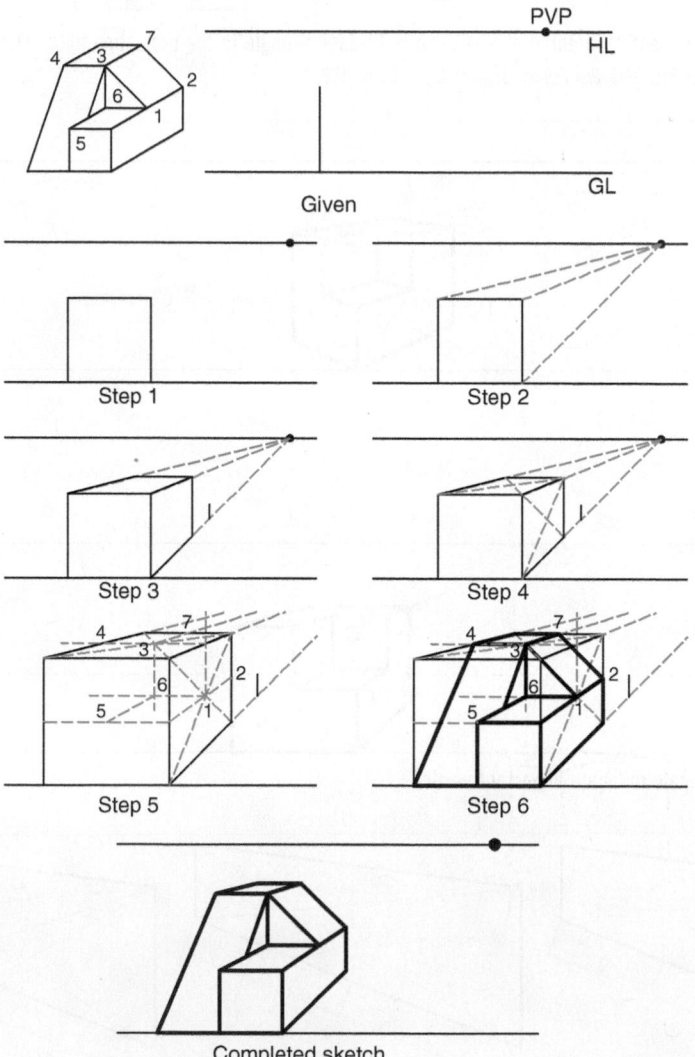

Figure 13-26 Multiple steps for one-point perspective sketch

Given a cavalier oblique, the reference edge height and location, and the location of the principal vanishing points:

1. Use construction lines to sketch convergence lines to PVP1 and PVP2. Also lay out the unforeshortened dimensions of the object's PEB.

2. After estimating the foreshortened depths, complete the bounding box (use construction lines).

3. On the front face, sketch the diagonals of the face. Also pass a vertical line through the point of intersection of the diagonals.

4. In order to partition the front face into three segments, sketch the diagonal lines shown in Step 4.

5. Sketch two more vertical lines on the front face, each one passing through the intersection formed by the diagonal lines created in Steps 3 and 4.

6. Sketch the diagonals on the left and top faces.

7. Sketch a vertical line from the intersection of the left face diagonal to the upper-left edge of the bounding box, and then sketch a line from this intersection point to the intersection of the top face diagonals. Finally, extend this line until it intersects the right edge of the top face.

8. Go bold.

Given

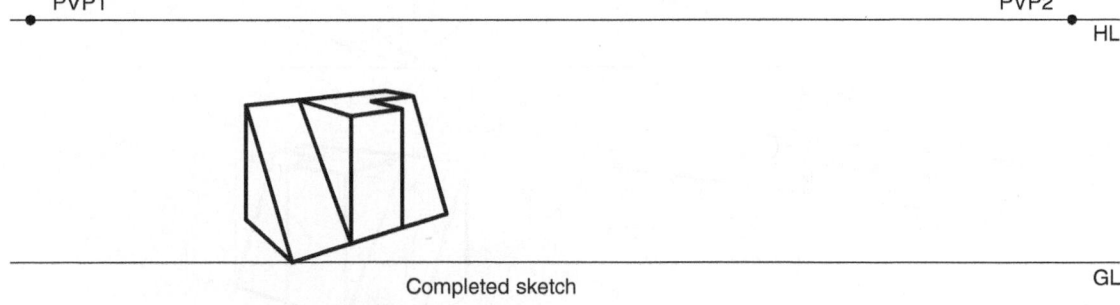

Completed sketch

(Continues)

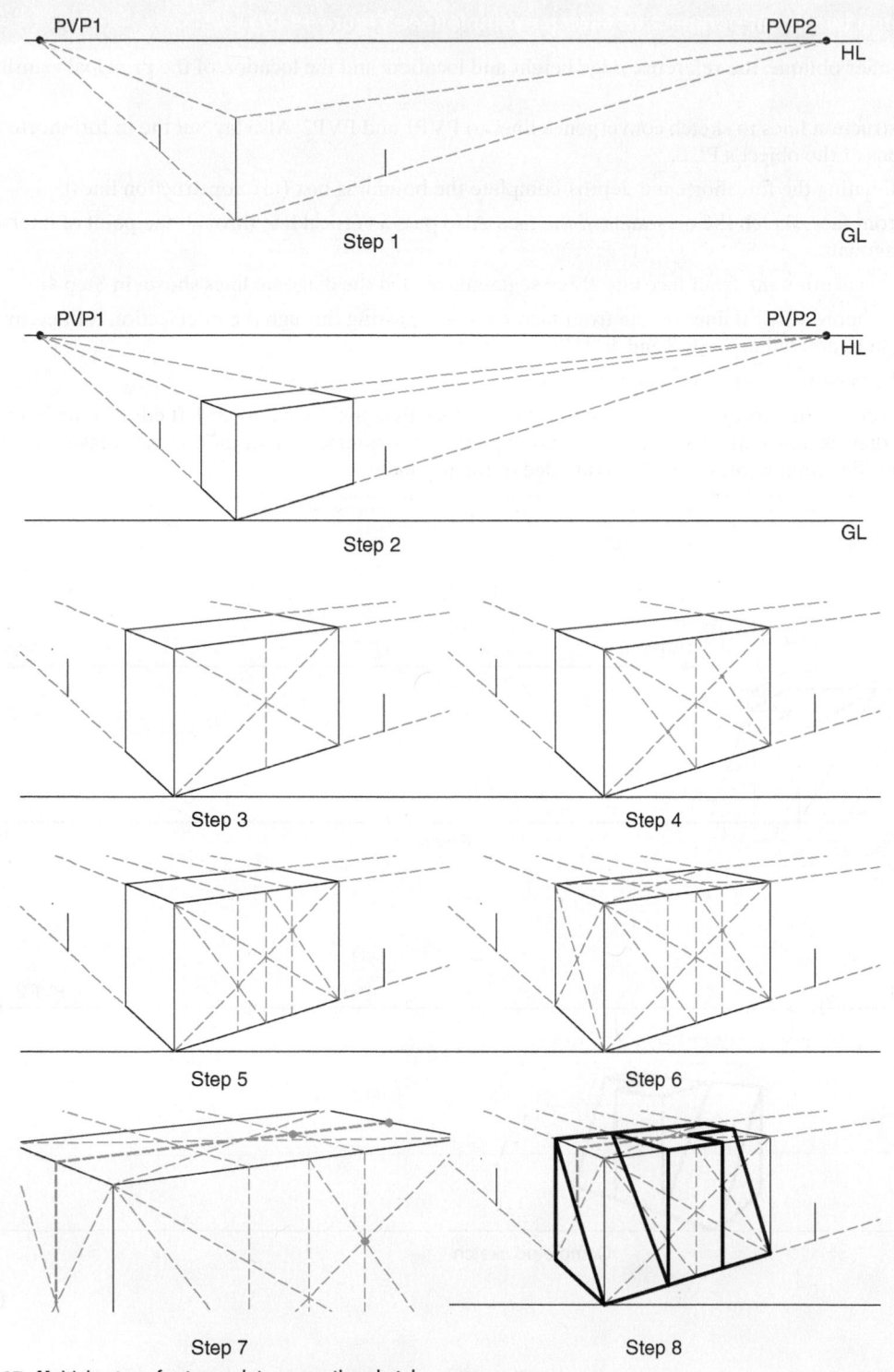

Step 1

Step 2

Step 3

Step 4

Step 5

Step 6

Step 7

Step 8

Figure 13-27 Multiple steps for two-point perspective sketch

(a) Oblique
- All three sets of PEB edges (horizontal, vertical, receding) remain parallel.

(b) Isometric
- All three sets of PEB edges (vertical, 30° to right, 30° to left) remain parallel.

(c) One-Point Perspective
- Two sets of PEB edges (horizontal, vertical) remain parallel.
- One set converges to PVP.

(d) Two-Point Perspective
- One set of PEB edges (vertical) remains parallel.
- Two sets converge to PVPs.

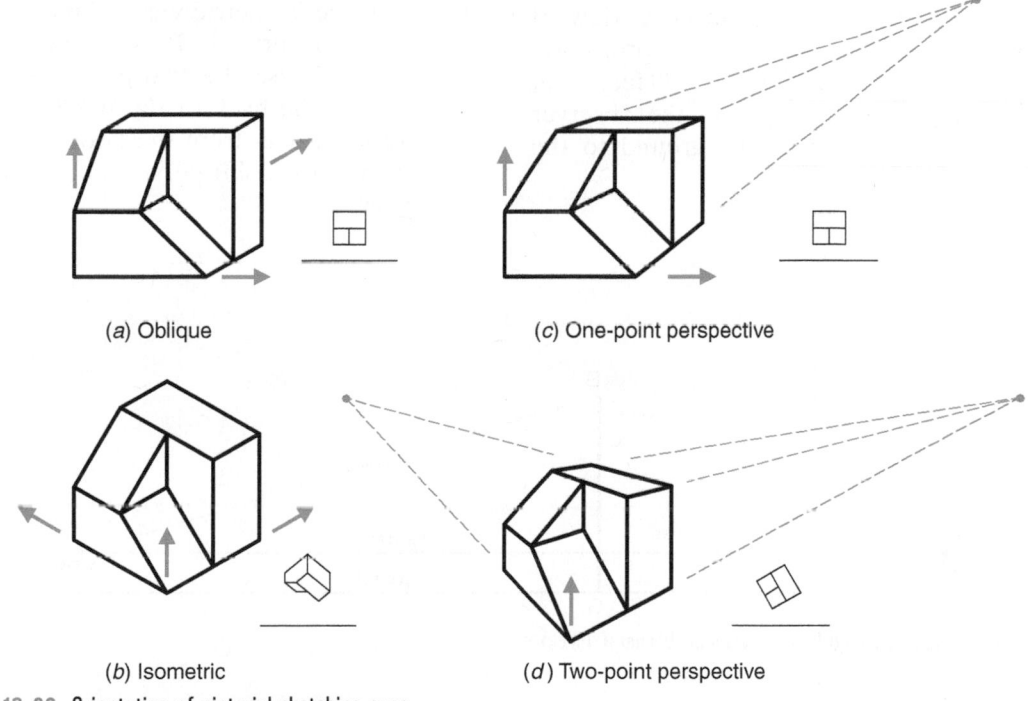

(a) Oblique   (c) One-point perspective

(b) Isometric   (d) Two-point perspective

Figure 13-28   Orientation of pictorial sketching axes

▌QUESTIONS

TRUE OR FALSE

1. A two-point perspective has two principal axes parallel to the projection plane.
2. Two-point perspective and oblique projections have the same number of principal axes inclined to the projection plane.
3. In a perspective projection, if the projection plane is located in front of the object, the projected image will be smaller than the object.

MULTIPLE CHOICE

4. Figure P13-1 shows a perspective view of a vertical pole projected onto a projection plane. If the length of the pole is 30 feet, what is the approximate height of the observer (i.e., the distance from the ground to the observer's eye level)?
   a. 0 feet
   b. 3 feet
   c. 6 feet
   d. 12 feet
   e. 15 feet
   f. 30 feet
   g. Not determinable

5. Given the isometric view of the cut block objects appearing in P3-4 through P3-65 in Chapter 3, use the one-point perspective set up in the back of the book (or download worksheet from the book website) to sketch a one-point perspective view of the object.

6. Given the isometric view of the cut block objects appearing in P3-4 through P3-65 in Chapter 3, use the two-point perspective set up in the back of the book (or download worksheet from the book website) to sketch a two-point perspective view of the object.

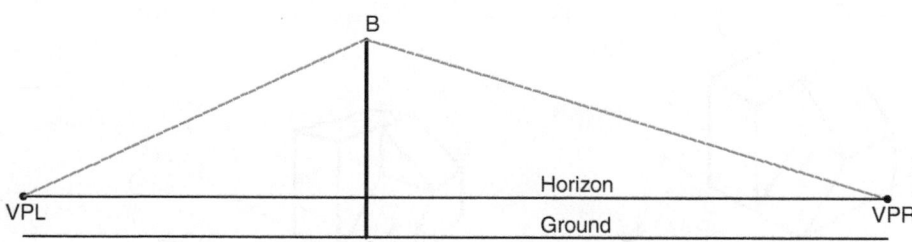

Figure P13-1 (Figure adapted from the work of Michael H. Pleck)

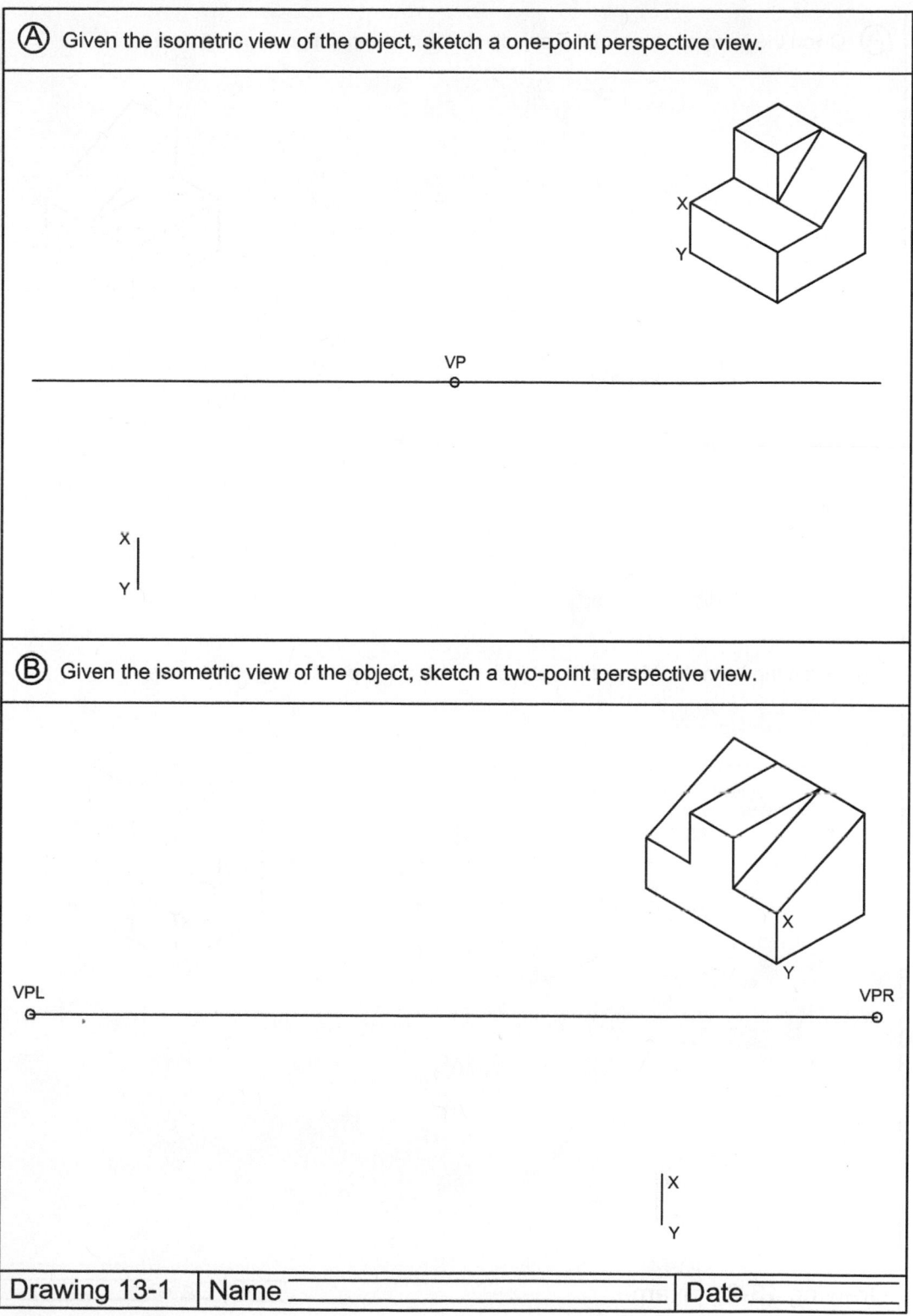

Ⓐ Given the isometric view of the object, sketch a one-point perspective view.

X
Y

VP

X
Y

Ⓑ Given the isometric view of the object, sketch a two-point perspective view.

X
Y

VPL

VPR

X
Y

| Drawing 13-1 | Name | Date |

Ⓐ Given the isometric view of the object, sketch a one-point perspective view.

VP

X
Y

Ⓑ Given the isometric view of the object, sketch a two-point perspective view.

VPL   VPR

X
Y

Drawing 13-2 | Name _____ | Date _____

A ANSI PREFERRED ENGLISH LIMITS AND FITS

APPENDIX A ANSI PREFERRED ENGLISH LIMITS AND FITS

ANSI RUNNING AND SLIDING FITS (RC)–ENGLISH UNITS

American National Standard Running and Sliding Fits (ANSI B4. 1–1967, R1979)

Tolerance limits given in body of table are added to or subtracted from basic size (as indicated by + or − sign) to obtain maximum and minimum sizes of mating parts.

Values shown below are in thousandths of an inch.

Nominal Size Range, Inches Over To	Class RC1 Clearance*	Class RC1 Hole H5	Class RC1 Shaft g4	Class RC2 Clearance*	Class RC2 Hole H6	Class RC2 Shaft g5	Class RC3 Clearance*	Class RC3 Hole H7	Class RC3 Shaft f6	Class RC4 Clearance*	Class RC4 Hole H8	Class RC4 Shaft f7
0–0.12	0.1 / 0.45	+0.2 / 0	−0.1 / −0.25	0.1 / 0.55	+0.25 / 0	−0.1 / −0.3	0.3 / 0.95	+0.4 / 0	−0.3 / −0.55	0.3 / 1.3	+0.6 / 0	−0.3 / −0.7
0.12–0.24	0.15 / 0.5	+0.2 / 0	−0.15 / −0.3	0.15 / 0.65	+0.3 / 0	−0.15 / −0.35	0.4 / 1.12	+0.5 / 0	−0.4 / −0.7	0.4 / 1.6	+0.7 / 0	−0.4 / −0.9
0.24–0.40	0.2 / 0.6	+0.25 / 0	−0.2 / −0.35	0.2 / 0.85	+0.4 / 0	−0.2 / −0.45	0.5 / 1.5	+0.6 / 0	−0.5 / −0.9	0.5 / 1.0	+0.9 / 0	−0.5 / −1.1
0.40–0.71	0.25 / 0.75	+0.3 / 0	−0.25 / −0.45	0.25 / 0.95	+0.4 / 0	−0.25 / −0.55	0.6 / 1.7	+0.7 / 0	−0.6 / −1.0	0.6 / 2.3	+1.0 / 0	−0.6 / −1.3
0.71–1.19	0.3 / 0.95	+0.4 / 0	−0.3 / −0.55	0.3 / 1.2	+0.5 / 0	−0.3 / −0.7	0.8 / 2.1	+0.8 / 0	−0.8 / −1.3	0.8 / 2.8	+1.2 / 0	−0.8 / −1.6
1.19–1.97	0.4 / 1.1	+0.4 / 0	−0.4 / −0.7	0.4 / 1.4	+0.6 / 0	−0.4 / −0.8	1.0 / 2.6	+1.0 / 0	−1.0 / −1.6	1.0 / 3.6	+1.6 / 0	−1.0 / −2.0
1.97–3.15	0.4 / 1.2	+0.5 / 0	−0.4 / −0.7	0.4 / 1.6	+0.7 / 0	−0.4 / −0.9	1.2 / 3.1	+1.2 / 0	−1.2 / −1.9	1.2 / 4.2	+1.8 / 0	−1.2 / −2.4
3.15–4.73	0.5 / 1.5	+0.6 / 0	−0.5 / −0.9	0.5 / 2.0	+0.9 / 0	−0.5 / −1.1	1.4 / 3.7	+1.4 / 0	−1.4 / −2.3	1.4 / 5.0	+2.2 / 0	−1.4 / −2.8
4.73–7.09	0.6 / 1.8	+0.7 / 0	−0.6 / −1.1	0.6 / 2.3	+1.0 / 0	−0.6 / −1.3	1.6 / 4.2	+1.6 / 0	−1.6 / −2.6	1.6 / 5.7	+2.5 / 0	−1.6 / −3.2
7.09–9.85	0.6 / 2.0	+0.8 / 0	−0.6 / −1.2	0.6 / 2.6	+1.2 / 0	−0.6 / −1.4	2.0 / 5.0	+1.8 / 0	−2.0 / −3.2	2.0 / 6.6	+2.8 / 0	−2.0 / −3.8
9.85–12.41	0.8 / 2.3	+0.9 / 0	−0.8 / −1.4	0.8 / 2.9	+1.2 / 0	−0.8 / −1.7	2.5 / 5.7	+2.0 / 0	−2.5 / −3.7	2.5 / 7.5	+3.0 / 0	−2.5 / −4.5
12.41–15.75	1.0 / 2.7	+1.0 / 0	−1.0 / −1.7	1.0 / 3.4	+1.4 / 0	−1.0 / −2.0	3.0 / 6.6	+2.2 / 0	−3.0 / −4.4	3.0 / 8.7	+3.5 / 0	−3.0 / −5.2
15.75–19.69	1.2 / 3.0	+1.0 / 0	−1.2 / −2.0	1.2 / 3.8	+1.6 / 0	−1.2 / −2.2	4.0 / 8.1	+2.5 / 0	−4.0 / −5.6	4.0 / 10.5	+4.0 / 0	−4.0 / −6.5

See footnotes at end of table.

Values shown below are in thousandths of an inch.

Nominal Size Range, Inches Over–To	Class RC5 Clearance*	Class RC5 Hole H8	Class RC5 Shaft e7	Class RC6 Clearance*	Class RC6 Hole H9	Class RC6 Shaft e8	Class RC7 Clearance*	Class RC7 Hole H9	Class RC7 Shaft d8	Class RC8 Clearance*	Class RC8 Hole H10	Class RC8 Shaft c9	Class RC9 Clearance*	Class RC9 Hole H11	Class RC9 Shaft
0–0.12	0.6 / 1.6	+0.6 / 0	−0.6 / −1.0	0.6 / 2.2	+1.0 / 0	−0.6 / −1.2	1.0 / 2.6	+1.0 / 0	−1.0 / −1.6	2.5 / 5.1	+1.6 / 0	−2.5 / −3.5	4.0 / 8.1	+2.5 / 0	−4.0 / −5.6
0.12–0.24	0.8 / 2.0	+0.7 / 0	−0.8 / −1.3	0.8 / 2.7	+1.2 / 0	−0.8 / −1.5	1.2 / 3.1	+1.2 / 0	−1.2 / −1.9	2.8 / 5.8	+1.8 / 0	−2.8 / −4.0	4.5 / 9.0	+3.0 / 0	−4.5 / −6.0
0.24–0.40	1.0 / 2.5	+0.9 / 0	−1.0 / −1.6	1.0 / 3.3	+1.4 / 0	−1.0 / −1.9	1.6 / 3.9	+1.4 / 0	−1.6 / −2.5	3.0 / 6.6	+2.2 / 0	−3.0 / −4.4	5.0 / 10.7	+3.5 / 0	−5.0 / −7.2
0.40–0.71	1.2 / 2.9	+1.0 / 0	−1.2 / −1.9	1.2 / 3.8	+1.6 / 0	−1.2 / −2.2	2.0 / 4.6	+1.6 / 0	−2.0 / −3.0	3.5 / 7.9	+2.8 / 0	−3.5 / −5.1	6.0 / 12.8	+4.0 / 0	−6.0 / −8.8
0.71–1.19	1.6 / 3.6	+1.2 / 0	−1.6 / −2.4	1.6 / 4.8	+2.0 / 0	−1.6 / −2.8	2.5 / 5.7	+2.0 / 0	−2.5 / −3.7	4.5 / 10.0	+3.5 / 0	−4.5 / −6.5	7.0 / 15.5	+5.0 / 0	−7.0 / −10.5
1.19–1.97	2.0 / 4.6	+1.6 / 0	−2.0 / −3.0	2.0 / 6.1	+2.5 / 0	−2.0 / −3.6	3.0 / 7.1	+2.5 / 0	−3.0 / −4.6	5.0 / 11.5	+4.0 / 0	−5.0 / −7.5	8.0 / 18.0	+6.0 / 0	−8.0 / −12.0
1.97–3.15	2.5 / 5.5	+1.8 / 0	−2.5 / −3.7	2.5 / 7.3	+3.0 / 0	−2.5 / −4.3	4.0 / 8.8	+3.0 / 0	−4.0 / −5.8	6.0 / 13.5	+4.5 / 0	−6.0 / −9.0	9.0 / 20.5	+7.0 / 0	−9.0 / −13.5
3.15–4.73	3.0 / 6.6	+2.2 / 0	−3.0 / −4.4	3.0 / 8.7	+3.5 / 0	−3.0 / −5.2	5.0 / 10.7	+3.5 / 0	−5.0 / −7.2	7.0 / 15.5	+5.0 / 0	−7.0 / −10.5	10.0 / 24.0	+9.0 / 0	−10.0 / −15.0
4.73–7.09	3.5 / 7.6	+2.5 / 0	−3.5 / −5.1	3.5 / 10.0	+4.0 / 0	−3.5 / −6.0	6.0 / 12.5	+4.0 / 0	−6.0 / −8.5	8.0 / 18.0	+6.0 / 0	−8.0 / −12.0	12.0 / 28.0	+10.0 / 0	−12.0 / −18.0
7.09–9.85	4.0 / 8.6	+2.8 / 0	−4.0 / −5.8	4.0 / 11.3	+4.5 / 0	−4.0 / −6.8	7.0 / 14.3	+4.5 / 0	−7.0 / −9.8	10.0 / 21.5	+7.0 / 0	−10.0 / −14.5	15.0 / 34.0	+12.0 / 0	−15.0 / −22.0
9.85–12.41	5.0 / 10.0	+3.0 / 0	−5.0 / −7.0	5.0 / 13.0	+5.0 / 0	−5.0 / −8.0	8.0 / 16.0	+5.0 / 0	−8.0 / −11.0	12.0 / 25.0	+8.0 / 0	−12.0 / −17.0	18.0 / 38.0	+12.0 / 0	−18.0 / −26.0
12.41–15.75	6.0 / 11.7	+3.5 / 0	−6.0 / −8.2	6.0 / 15.5	+6.0 / 0	−6.0 / −9.5	10.0 / 19.5	+6.0 / 0	−10.0 / −13.5	14.0 / 29.0	+9.0 / 0	−14.0 / −20.0	22.0 / 45.0	+14.0 / 0	−22.0 / −31.0
15.75–19.69	8.0 / 14.5	+4.0 / 0	−8.0 / −10.5	8.0 / 18.0	+6.0 / 0	−8.0 / −12.0	12.0 / 22.0	+6.0 / 0	−12.0 / −16.0	16.0 / 32.0	+10.0 / 0	−16.0 / −22.0	25.0 / 51.0	+16.0 / 0	−25.0 / −35.0

Data in boldface type are in accordance with American–British–Canadian (ABC) Agreements. Symbols H5, g4, etc. are hole and shaft designations in ABC system. Limits for sizes above 19.69 inches are also given in the ANSI Standard.

*Pairs of values shown represent minimum and maximum amounts of clearance resulting from application of standard tolerance limits.

Source: Reprinted courtesy of The American Society of Mechanical Engineers.

ANSI RUNNING AND SLIDING FITS (RC)—ENGLISH UNITS

ANSI CLEARANCE LOCATION FITS (LC)—ENGLISH UNITS

American National Standard Clearance Locational Fits (ANSI B4.1–1967, R1979)

Tolerance limits given in body of table are added or subtracted to basic size (as indicated by + or − sign) to obtain maximum and minimum sizes of mating parts.

Values shown below are in thousandths of an inch.

Nominal Size Range, Inches Over To	Class LC1 Clearance*	Class LC1 Hole H6	Class LC1 Shaft h5	Class LC2 Clearance*	Class LC2 Hole H7	Class LC2 Shaft h6	Class LC3 Clearance*	Class LC3 Hole H8	Class LC3 Shaft h7	Class LC4 Clearance*	Class LC4 Hole H10	Class LC4 Shaft h9	Class LC5 Clearance*	Class LC5 Hole H7	Class LC5 Shaft g6
0–0.12	0	+0.25	0	0	+0.4	0	0	+0.6	0	0	+1.6	0	0.1	+0.4	−0.1
	0.45	0	−0.2	0.65	0	−0.25	1	0	−0.4	2.6	0	−1.0	0.75	0	−0.35
0.12–0.24	0	+0.3	0	0	+0.5	0	0	+0.7	0	0	+1.8	0	0.15	+0.5	−0.15
	0.5	0	−0.2	0.8	0	−0.3	1.2	0	−0.5	3.0	0	−1.2	0.95	0	−0.45
0.24–0.40	0	+0.4	0	0	+0.6	0	0	+0.9	0	0	+2.2	0	0.2	+0.6	−0.2
	0.65	0	−0.25	1.0	0	−0.4	1.5	0	−0.6	3.6	0	−1.4	1.2	0	−0.6
0.40–0.71	0	+0.4	0	0	+0.7	0	0	+1.0	0	0	+2.8	0	0.25	+0.7	−0.25
	0.7	0	−0.3	1.1	0	−0.4	1.7	0	−0.7	4.4	0	−1.6	1.35	0	−0.65
0.71–1.19	0	+0.5	0	0	+0.8	0	0	+1.2	0	0	+3.5	0	0.3	+0.8	−0.3
	0.9	0	−0.4	1.3	0	−0.5	2	0	−0.8	5.5	0	−2.0	1.6	0	−0.8
1.19–1.97	0	+0.6	0	0	+1.0	0	0	+1.6	0	0	+4.0	0	0.4	+1.0	−0.4
	1.0	0	−0.4	1.6	0	−0.6	2.6	0	−1	6.5	0	−2.5	2.0	0	−1.0
1.97–3.15	0	+0.7	0	0	+1.2	0	0	+1.8	0	0	+4.5	0	0.4	+1.2	−0.4
	1.2	0	−0.5	1.9	0	−0.7	3	0	−1.2	7.5	0	−3	2.3	0	−1.1
3.15–4.73	0	+0.9	0	0	+1.4	0	0	+2.2	0	0	+5.0	0	0.5	+1.4	−0.5
	1.5	0	−0.6	2.3	0	−0.9	3.6	0	−1.4	8.5	0	−3.5	2.8	0	−1.4
4.73–7.09	0	+1.0	0	0	+1.6	0	0	+2.5	0	0	+6.0	0	0.6	+1.6	−0.6
	1.7	0	−0.7	2.6	0	−1.0	4.1	0	−1.6	10.0	0	−4	−3.2	0	−1.6
7.09–9.85	0	+1.2	0	0	+1.8	0	0	+2.8	0	0	+7.0	0	0.6	+1.8	−0.6
	2.0	0	−0.8	3.0	0	−1.2	4.6	0	−1.8	11.5	0	−4.5	3.6	0	−1.8
9.85–12.41	0	+1.2	0	0	+2.0	0	0	+3.0	0	0	+8.0	0	0.7	+2.0	−0.7
	2.1	0	−0.9	3.2	0	−1.2	5	0	−2.0	13.0	0	−5	3.9	0	−1.9
12.41–15.75	0	+1.4	0	0	+2.2	0	0	+3.5	0	0	+9.0	0	0.7	+2.2	−0.7
	2.4	0	−1.0	3.6	0	−1.4	5.7	0	−2.2	15.0	0	−6	4.3	0	−2.1
15.75–19.69	0	+1.6	0	0	+2.5	0	0	+4	0	0	+10.0	0	0.8	+2.5	−0.8
	2.6	0	−1.0	4.1	0	−1.6	6.5	0	−2.5	16.0	0	−6	4.9	0	−2.4

See footnotes at end of table.

ANSI CLEARANCE LOCATION FITS (LC)—ENGLISH UNITS

Values shown below are in thousandths of an inch.

Nominal Size Range, Inches, Over To	Class LC6			Class LC7			Class LC8			Class LC9			Class LC10			Class LC11		
	Clearance*	Hole H9	Shaft f8	Clearance*	Hole H10	Shaft e9	Clearance*	Hole H10	Shaft d9	Clearance*	Hole H11	Shaft c10	Clearance*	Hole H12	Shaft	Clearance*	Hole H13	Shaft
0–0.12	0.3, 1.9	+1.0, 0	−0.3, −0.9	0.6, 3.2	+1.6, 0	−0.6, −1.6	1.0, 2.0	+1.6, 0	−1.0, −2.0	2.5, 6.6	+2.5, 0	−2.5, −4.1	4, 12	+4, 0	−4, −8	5, 17	+6, 0	−5, −11
0.12–0.24	0.4, 2.3	+1.2, 0	−0.4, −1.1	0.8, 3.8	+1.8, 0	−0.8, −2.0	1.2, 4.2	+1.8, 0	−1.2, −2.4	2.8, 7.6	+3.0, 0	−2.8, −4.6	4.5, 14.5	+5, 0	−4.5, −9.5	6, 20	+7, 0	−6, −13
0.24–0.40	0.5, 2.8	+1.4, 0	−0.5, −1.4	1.0, 4.6	+2.3, 0	−1.0, −2.4	1.6, 5.2	+2.2, 0	−1.6, −3.0	3.0, 8.7	+3.5, 0	−3.0, −5.2	5, 17	+6, 0	−5, −11	7, 25	+9, 0	−7, −16
0.40–0.71	0.6, 3.2	+1.6, 0	−0.6, −1.6	1.3, 5.6	+2.8, 0	−1.2, −2.8	2.0, 6.4	+2.8, 0	−2.0, −3.6	3.5, 10.3	+4.0, 0	−3.5, −6.3	6, 20	+7, 0	−6, −13	8, 28	+10, 0	−8, −18
0.71–1.19	0.8, 4.0	+2.0, 0	−0.8, −2.0	1.6, 7.1	+3.5, 0	−1.6, −3.6	3.5, 8.0	+3.5, 0	−2.5, −4.5	4.5, 13.0	+5.0, 0	−4.5, −8.0	7, 23	+8, 0	−7, −15	10, 34	+12, 0	−10, −22
1.19–1.97	1.0, 5.1	+2.5, 0	−1.0, −2.6	2.0, 8.5	+4.0, 0	−2.0, −4.5	3.6, 9.5	+4.0, 0	−3.0, −5.5	5.0, 15.0	+6, 0	−5.0, −9.0	8, 28	+10, 0	−8, −18	12, 44	+16, 0	−12, −28
1.97–3.15	1.2, 6.0	+3.0, 0	−1.0, −3.0	2.5, 10.0	+4.5, 0	−2.5, −5.5	4.0, 11.5	+4.5, 0	−4.0, −7.0	6.0, 17.5	+7, 0	−6.0, −10.5	10, 34	+12, 0	−10, −22	14, 50	+18, 0	−14, −32
3.15–4.73	1.4, 7.1	+3.5, 0	−1.4, −3.6	3.0, 11.5	+5.0, 0	−3.0, −6.5	5.0, 13.5	+5.0, 0	−5.0, −8.5	7, 21	+9, 0	−7, −12	11, 39	+14, 0	−11, −25	16, 60	+22, 0	−16, −38
4.73–7.09	1.6, 8.1	+4.0, 0	−1.6, −4.1	3.5, 13.5	+6.0, 0	−3.5, −7.5	6, 16	+6, 0	−6, −10	8, 24	+10, 0	−8, −14	12, 44	+16, 0	−12, −28	18, 68	+25, 0	−18, −43
7.09–9.85	2.0, 9.3	+4.5, 0	−2.0, −4.8	4.0, 15.5	+7.0, 0	−4.0, −8.5	7, 18.5	+7, 0	−7, −11.5	10, 29	+12, 0	−10, −17	16, 52	+18, 0	−16, −34	22, 78	+28, 0	−22, −50
9.85–12.41	2.2, 10.2	+5.0, 0	−2.2, −5.2	4.5, 17.5	+8.0, 0	−4.5, −9.5	7, 20	+8, 0	−7, −12	12, 32	+12, 0	−12, −20	20, 60	+20, 0	−20, −40	28, 88	+30, 0	−28, −58
12.41–15.75	2.5, 12.0	+6.0, 0	−2.5, −6.0	5.0, 20.0	+9.0, 0	−5, −11	8, 23	+9, 0	−8, −14	14, 37	+14, 0	−12, −23	22, 66	+22, 0	−22, −44	30, 100	+35, 0	−30, −65
15.75–19.69	2.8, 12.8	+6.0, 0	−2.8, −6.8	5.0, 21.0	+10.0, 0	−5, −11	9, 25	+10, 0	−9, −15	16, 42	+16, 0	−16, −26	25, 75	+25, 0	−25, −50	35, 115	+40, 0	−35, −75

Data in boldface type are in accordance with American–British–Canadian (ABC) agreements. Symbols H6, H7, f6, etc. are hole and shaft designations in ABC system. Limits for sizes above 19.69 inches are not covered by ABC agreements but are given in the ANSI Standard.

*Pairs of values shown represent minimum and maximum amounts of interference resulting from application of standard tolerance limits.

Source: Reprinted courtesy of The American Society of Mechanical Engineers.

APPENDIX A **ANSI PREFERRED ENGLISH LIMITS AND FITS**

ANSI TRANSITION LOCATION FITS (LT)—ENGLISH UNITS
ANSI Standard Transition Location Fits (ANSI B4.1–1967, R1979)

Values shown below are in thousandths of an inch.

Nominal Size Range, Inches, Over To	Class LT1 Fit*	Class LT1 Hole H7	Class LT1 Shaft js6	Class LT2 Fit*	Class LT2 Hole H8	Class LT2 Shaft js7	Class LT3 Fit*	Class LT3 Hole H7	Class LT3 Shaft k6	Class LT4 Fit*	Class LT4 Hole H7	Class LT4 Shaft n6	Class LT5 Fit*	Class LT5 Hole H7	Class LT5 Shaft n6	Class LT6 Fit*	Class LT6 Hole H7	Class LT6 Shaft n7
0–0.12	−0.12 / +0.52	+0.4 / 0	+0.12 / −0.12	−0.2 / +0.8	+0.6 / 0	+0.2 / −0.2							−0.5 / +0.15	+0.4 / 0	+0.5 / +0.25	−0.65 / +0.15	+0.4 / 0	+0.65 / +0.25
0.12–0.24	−0.15 / +0.65	+0.5 / 0	+0.15 / −0.15	−0.25 / +0.95	+0.7 / 0	+0.25 / −0.25							−0.6 / +0.2	+0.5 / 0	+0.6 / +0.3	−0.8 / +0.2	+0.5 / 0	+0.8 / +0.3
0.24–0.40	−0.2 / +0.8	+0.6 / 0	+0.2 / −0.2	−0.3 / +1.2	+0.9 / 0	+0.3 / −0.3	−0.5 / +0.5	+0.6 / 0	+0.5 / +0.1	−0.7 / +0.8	+0.9 / 0	+0.7 / +0.1	−0.8 / +0.2	+0.6 / 0	+0.8 / +0.4	−1.0 / +0.2	+0.6 / 0	+1.0 / +0.4
0.40–0.71	−0.2 / +0.9	+0.7 / 0	+0.2 / −0.2	−0.35 / +1.35	+1.0 / 0	+0.35 / −0.35	−0.5 / +0.6	+0.7 / 0	+0.5 / +0.1	−0.8 / +0.9	+1.0 / 0	+0.8 / +0.1	−0.9 / +0.2	+0.7 / 0	+0.9 / +0.5	−1.2 / +0.2	+0.7 / 0	+1.2 / +0.5
0.71–1.19	−0.25 / +1.05	+0.8 / 0	+0.25 / −0.25	−0.4 / +1.6	+1.2 / 0	+0.4 / −0.4	−0.6 / +0.7	+0.8 / 0	+0.6 / +0.1	−0.9 / +1.1	+1.2 / 0	+0.9 / +0.1	−1.1 / +0.2	+0.8 / 0	+1.1 / +0.6	−1.4 / +0.2	+0.8 / 0	+1.4 / +0.6
1.19–1.97	−0.3 / +1.3	+1.0 / 0	+0.3 / −0.3	−0.5 / +2.1	+1.6 / 0	+0.5 / −0.5	−0.7 / +0.9	+1.0 / 0	+0.7 / +0.1	−1.1 / +1.5	+1.6 / 0	+1.1 / +0.1	−1.3 / +0.3	+1.0 / 0	+1.3 / +0.7	−1.7 / +0.3	+1.0 / 0	+1.7 / +0.7
1.97–3.15	−0.3 / +1.5	+1.2 / 0	+0.3 / −0.3	−0.6 / +2.4	+1.8 / 0	+0.6 / −0.6	−0.8 / +1.1	+1.2 / 0	+0.8 / +0.1	−1.3 / +1.7	+1.8 / 0	+1.3 / +0.1	−1.5 / +0.4	+1.2 / 0	+1.5 / +0.8	−2.0 / +0.4	+1.2 / 0	+2.0 / +0.8
3.15–4.73	−0.4 / +1.8	+1.4 / 0	+0.4 / −0.4	−0.7 / +2.9	+2.2 / 0	+0.7 / −0.7	−1.0 / +1.3	+1.4 / 0	+1.0 / +0.1	−1.5 / +2.1	+2.2 / 0	+1.5 / +0.1	−1.9 / +0.4	+1.4 / 0	+1.9 / +1.0	−2.4 / +0.4	+1.4 / 0	+2.4 / +1.0
4.73–7.09	−0.5 / +2.1	+1.6 / 0	+0.5 / −0.5	−0.8 / +3.3	+2.5 / 0	+0.8 / −0.8	−1.1 / +1.5	+1.6 / 0	+1.1 / +0.1	−1.7 / +2.4	+2.5 / 0	+1.7 / +0.1	−2.2 / +0.4	+1.6 / 0	+2.2 / +1.2	−2.8 / +0.4	+1.6 / 0	+2.8 / +1.2
7.09–9.85	−0.6 / +2.4	+1.8 / 0	+0.6 / −0.6	−0.9 / +3.7	+2.8 / 0	+0.9 / −0.9	−1.4 / +1.6	+1.8 / 0	+1.4 / +0.2	−2.0 / +2.6	+2.8 / 0	+2.0 / +0.2	−2.6 / +0.4	+1.8 / 0	+2.6 / +1.4	−3.2 / +0.4	+1.8 / 0	+3.2 / +1.4
9.85–12.41	−0.6 / +2.6	+2.0 / 0	+0.6 / −0.6	−1.0 / +4.0	+3.0 / 0	+1.0 / −1.0	−1.4 / +1.8	+2.0 / 0	+1.4 / +0.2	−2.2 / +2.8	+3.0 / 0	+2.2 / +0.2	−2.6 / +0.6	+2.0 / 0	+2.6 / +1.4	−3.4 / +0.6	+2.0 / 0	+3.4 / +1.4
12.41–15.75	−0.7 / +2.9	+2.2 / 0	+0.7 / −0.7	−1.0 / +4.5	+3.5 / 0	+1.0 / −1.0	−1.6 / +2.0	+2.2 / 0	+1.6 / +0.2	−2.4 / +3.3	+3.5 / 0	+2.4 / +0.2	−3.0 / +0.6	+2.2 / 0	+3.0 / +1.6	−3.8 / +0.6	+2.2 / 0	+3.8 / +1.6
15.75–19.69	−0.8 / +3.3	+2.5 / 0	+0.8 / −0.8	−1.2 / +5.2	+4.0 / 0	+1.2 / −1.2	−1.8 / +2.3	+2.5 / 0	+1.8 / +0.2	−2.7 / +3.8	+4.0 / 0	+2.7 / +0.2	−3.4 / +0.7	+2.5 / 0	+3.4 / +1.8	−4.3 / +0.7	+2.5 / 0	+4.3 / +1.8

Data in boldface type are in accordance with American–British–Canadian (ABC) Agreements. Symbols H7, js6, etc. are hole and shaft designations in ABC system.

*Pairs of values shown represent maximum amount of interference (−) and maximum amount of clearance (+) resulting from application of standard tolerance limits.

Source: Reprinted courtesy of The American Society of Mechanical Engineers.

Nominal Size Range, Inches Over To	Class LN1 Standard Limits			Class LN2 Standard Limits			Class LN3 Standard Limits		
	Limits of Inter-ference*	Hole H6	Shaft n5	Limits of Inter-ference*	Hole H7	Shaft p6	Limits of Inter-ference*	Hole H7	Shaft r6
	Values shown below are in thousandths of an inch.								
0–0.12	0 0.45	+0.25 0	+0.45 +0.25	0 0.65	+0.4 0	+0.65 +0.4	0.1 0.75	+0.4 0	+0.75 +0.5
0.12–0.24	0 0.5	+0.3 0	+0.5 +0.3	0 0.8	+0.5 0	+0.8 +0.5	0.1 0.9	+0.5 0	+0.9 +0.6
0.24–0.40	0 0.65	+0.4 0	+0.65 +0.4	0 1.0	+0.6 0	+1.0 +0.6	0.2 1.2	+0.6 0	+1.2 +0.8
0.40–0.71	0 0.8	+0.4 0	+0.8 +0.4	0 1.1	+0.7 0	+1.1 +0.7	0.3 1.4	+0.7 0	+1.4 +1.0
0.71–1.19	0 1.0	+0.5 0	+1.0 +0.5	0 1.3	+0.8 0	+1.3 +0.8	0.4 1.7	+0.8 0	+1.7 +1.2
1.19–1.97	0 1.1	+0.6 0	+1.1 +0.6	0 1.6	+1.0 0	+1.6 +1.0	0.4 2.0	+1.0 0	+2.0 +1.4
1.97–3.15	0.1 1.3	+0.7 0	+1.3 +0.8	0.2 2.1	+1.2 0	+2.1 +1.4	0.4 2.3	+1.2 0	+2.3 +1.6
3.15–4.73	0.1 1.6	+0.9 0	+1.6 +1.0	0.2 2.5	+1.4 0	+2.5 +1.6	0.6 2.9	+1.4 0	+2.9 +2.0
4.73–7.09	0.2 1.9	+1.0 0	+1.9 +1.2	0.2 2.8	+1.6 0	+2.8 +1.8	0.9 3.5	+1.6 0	+3.5 +2.5
7.09–9.85	0.1 2.2	+1.2 0	+2.2 +1.4	0.2 3.2	+1.8 0	+3.2 +2.0	1.2 4.2	+1.8 0	+4.2 +3.0
9.85–12.41	0.2 2.3	+1.2 0	+2.3 +1.4	0.2 3.4	+2.0 0	+3.4 +2.2	1.5 4.7	+2.0 0	+4.7 +3.5
12.41–15.75	0.2 2.6	+1.4 0	+2.6 +1.6	0.3 3.9	+2.2 0	+3.9 +2.5	2.3 5.9	+2.2 0	+5.9 +4.5
15.75–19.69	0.2 1.8	+1.6 0	+2.8 +1.8	0.3 4.4	+2.5 0	+4.4 +2.8	2.5 6.6	+2.5 0	+6.6 +5.0

All data in this table are in accordance with American–British–Canadian (ABC) agreements.

Limits for sizes above 19.69 inches are not covered by ABC agreements but are given in the ANSI Standard. Symbols H7, p6, etc. are hole and shaft designations in ABC system.

*Pairs of values shown represent minimum and maximum amounts of interference resulting from application of standard tolerance limits.

*Source: Reprinted courtesy of The American Society of Mechanical Engineers.

■ ANSI FORCE AND SHRINK FITS (FN)–ENGLISH UNITS
ANSI Standard Force and Shrink Fits (ANSI B4.1-1967, R1979)

Values shown below are in thousandths of an inch.

Nominal Size Range, Inches (Over–To)	Class FN1 Inter-ference*	Class FN1 Hole H6	Class FN1 Shaft	Class FN2 Inter-ference*	Class FN2 Hole H7	Class FN2 Shaft s6	Class FN3 Inter-ference*	Class FN3 Hole H7	Class FN3 Shaft 16	Class FN4 Inter-ference*	Class FN4 Hole H7	Class FN4 Shaft u6	Class FN5 Inter-ference*	Class FN5 Hole H8	Class FN5 Shaft x7
0–0.12	0.05 / 0.5	+0.25 / 0	+0.5 / +0.3	0.2 / 0.85	+0.4 / 0	+0.85 / +0.6				0.3 / 0.95	+0.4 / 0	+0.95 / +0.7	0.3 / 1.3	+0.6 / 0	+1.3 / +0.9
0.12–0.24	0.1 / 0.6	+0.3 / 0	+0.6 / +0.4	0.2 / 1.0	+0.5 / 0	+1.0 / +0.7				0.4 / 1.2	+0.5 / 0	+1.2 / +0.9	0.5 / 1.7	+0.7 / 0	+1.7 / +1.2
0.24–0.40	0.1 / 0.75	+0.4 / 0	+0.75 / +0.5	0.4 / 1.4	+0.6 / 0	+1.4 / +1.0				0.6 / 1.6	+0.6 / 0	+1.6 / +1.2	0.5 / 2.0	+0.9 / 0	+2.0 / +1.4
0.40–0.56	0.1 / 0.8	+0.4 / 0	+0.8 / +0.5	0.5 / 1.6	+0.7 / 0	+1.6 / +1.2				0.7 / 1.8	+0.7 / 0	+1.8 / +1.4	0.6 / 2.3	+1.0 / 0	+2.3 / +1.6
0.56–0.71	0.2 / 0.9	+0.4 / 0	+0.9 / +0.6	0.5 / 1.6	+0.7 / 0	+1.6 / +1.2				0.7 / 1.8	+0.7 / 0	+1.8 / +1.4	0.8 / 2.5	+1.0 / 0	+2.5 / +1.8
0.71–0.95	0.2 / 1.1	+0.5 / 0	+1.1 / +0.7	0.6 / 1.9	+0.8 / 0	+1.9 / +1.4				0.8 / 2.1	+0.8 / 0	+2.1 / +1.6	1.0 / 3.0	+1.2 / 0	+3.0 / +2.2
0.95–1.19	0.3 / 1.2	+0.5 / 0	+1.2 / +0.8	0.6 / 1.9	+0.8 / 0	+1.9 / +1.4	0.8 / 2.1	+0.8 / 0	+2.1 / +1.6	1.0 / 2.3	+0.8 / 0	+2.3 / +1.8	1.3 / 3.3	+1.2 / 0	+3.3 / +2.5
1.19–1.58	0.3 / 1.3	+0.6 / 0	+1.3 / +0.9	0.8 / 2.4	+1.0 / 0	+2.4 / +1.8	1.0 / 2.6	+1.0 / 0	+2.6 / +2.0	1.5 / 3.1	+1.0 / 0	+3.1 / +2.5	1.4 / 4.0	+1.6 / 0	+4.0 / +3.0
1.58–1.97	0.4 / 1.4	+0.6 / 0	+1.4 / +1.0	0.8 / 2.4	+1.0 / 0	+2.4 / +1.8	1.2 / 2.8	+1.0 / 0	+2.8 / +2.2	1.8 / 3.4	+1.0 / 0	+3.4 / +2.8	2.4 / 5.0	+1.6 / 0	+5.0 / +4.0
1.97–2.56	0.6 / 1.8	+0.7 / 0	+1.8 / +1.3	0.8 / 2.7	+1.2 / 0	+2.7 / +2.0	1.3 / 3.2	+1.2 / 0	+3.2 / +2.5	2.3 / 4.2	+1.2 / 0	+4.2 / +3.5	3.2 / 6.2	+1.8 / 0	+6.2 / +5.0
2.56–3.15	0.7 / 1.9	+0.7 / 0	+1.9 / +1.4	1.0 / 2.9	+1.2 / 0	+2.9 / +2.2	1.8 / 3.7	+1.2 / 0	+3.7 / +3.0	2.8 / 4.7	+1.2 / 0	+4.7 / +4.0	4.2 / 7.2	+1.8 / 0	+7.2 / +6.0
3.15–3.94	0.9 / 2.4	+0.9 / 0	+2.4 / +1.8	1.4 / 3.7	+1.4 / 0	+3.7 / +2.8	2.1 / 4.4	+1.4 / 0	+4.4 / +3.5	3.6 / 5.9	+1.4 / 0	+5.9 / +5.0	4.8 / 8.4	+2.2 / 0	+8.4 / +7.0
3.94–4.73	1.1 / 2.6	+0.9 / 0	+2.6 / +2.0	1.6 / 3.9	+1.4 / 0	+3.9 / +3.0	2.6 / 4.9	+1.4 / 0	+4.9 / +4.0	4.6 / 6.9	+1.4 / 0	+6.9 / +6.0	5.8 / 9.4	+2.2 / 0	+9.4 / +8.0

See footnotes at end of table.

Values shown below are in thousandths of an inch.

Nominal Size Range, Inches Over To	Class FN — Interference*	Class FN — Hole H6	Class FN — Shaft	Class FN — Interference*	Class FN — Hole H7	Class FN — Shaft s6	Class FN — Interference*	Class FN — Hole H7	Class FN — Shaft t6	Class FN — Interference*	Class FN — Hole H7	Class FN — Shaft u6	Class FN — Interference*	Class FN — Hole H8	Class FN — Shaft x7
4.73–5.52	1.2 / 2.9	+1.0 / 0	+2.9 / +2.2	1.9 / 4.5	+1.6 / 0	+4.5 / +3.5	3.4 / 6.0	+1.6 / 0	+6.0 / +5.0	5.4 / 8.0	+1.6 / 0	+8.0 / +7.0	7.5 / 11.6	+2.5 / 0	+11.6 / +10.0
5.52–6.30	1.5 / 3.2	+1.0 / 0	+3.2 / +2.5	2.4 / 5.0	+1.6 / 0	+5.0 / +4.0	3.4 / 6.0	+1.6 / 0	+6.0 / +5.0	5.4 / 8.0	+1.6 / 0	+8.0 / +7.0	9.5 / 13.6	+2.5 / 0	+13.6 / +12.0
6.30–7.09	1.8 / 3.5	+1.0 / 0	+3.5 / +2.8	2.9 / 5.5	+1.6 / 0	+5.5 / +4.5	4.4 / 7.0	+1.6 / 0	+7.0 / +6.0	6.4 / 9.0	+1.6 / 0	+9.0 / +8.0	9.5 / 13.6	+2.5 / 0	+13.6 / +12.0
7.09–7.88	1.8 / 3.8	+1.2 / 0	+3.8 / +3.0	3.2 / 6.2	+1.8 / 0	+6.2 / +5.0	5.2 / 8.2	+1.8 / 0	+8.2 / +7.0	7.2 / 10.2	+1.8 / 0	+10.2 / +9.0	11.2 / 15.8	+2.8 / 0	+15.8 / +14.0
7.88–8.86	2.3 / 4.3	+1.2 / 0	+4.3 / +3.5	3.2 / 6.2	+1.8 / 0	+6.2 / +5.0	5.2 / 8.2	+1.8 / 0	+8.2 / +7.0	8.2 / 11.2	+1.8 / 0	+11.2 / +10.0	13.2 / 17.8	+2.8 / 0	+17.8 / +16.0
8.86–9.85	2.3 / 4.3	+1.2 / 0	+4.3 / +3.5	4.2 / 7.2	+1.8 / 0	+7.2 / +6.0	6.2 / 9.2	+1.8 / 0	+9.2 / +8.0	10.2 / 13.2	+1.8 / 0	+13.2 / +12.0	13.2 / 17.8	+2.8 / 0	+17.8 / +16.0
9.85–11.03	2.8 / 4.9	+1.2 / 0	+4.9 / +4.0	4.0 / 7.2	+2.0 / 0	+7.2 / +6.0	7.0 / 10.2	+2.0 / 0	+10.2 / +9.0	10.0 / 13.2	+2.0 / 0	+13.2 / +12.0	15.0 / 20.0	+3.0 / 0	+20.0 / +18.0
11.03–12.41	2.8 / 4.9	+1.2 / 0	+4.9 / +4.0	5.0 / 8.2	+2.0 / 0	+8.2 / +7.2	7.0 / 10.2	+2.0 / 0	+10.2 / +9.0	12.0 / 15.2	+2.0 / 0	+15.2 / +14.0	17.0 / 22.0	+3.0 / 0	+22.0 / +20.0
12.41–13.98	3.1 / 5.5	+1.4 / 0	+5.5 / +4.5	5.8 / 9.4	+2.2 / 0	+9.4 / +8.0	7.8 / 11.4	+2.2 / 0	+11.4 / +10.0	13.8 / 17.4	+2.2 / 0	+17.4 / +16.0	18.5 / 24.2	+3.5 / 0	+24.2 / +22.0
13.98–15.75	3.6 / 6.1	+1.4 / 0	+6.1 / +5.0	5.8 / 9.4	+2.2 / 0	+9.4 / +8.0	9.8 / 13.4	+2.2 / 0	+13.4 / +12.0	15.8 / 19.4	+2.2 / 0	+19.4 / +18.0	21.5 / 27.2	+3.5 / 0	+27.2 / +25.0
15.75–17.72	4.4 / 7.0	+1.6 / 0	+7.0 / +6.0	6.5 / 10.6	+2.5 / 0	+10.6 / +9.0	9.5 / 13.6	+2.5 / 0	+13.6 / +12.0	17.5 / 21.6	+2.5 / 0	+21.6 / +20.0	24.0 / 30.5	+4.0 / 0	+30.5 / +28.0
17.72–19.69	4.4 / 7.0	+1.6 / 0	+7.0 / +6.0	7.5 / 11.6	+2.5 / 0	+11.6 / +10.0	11.5 / 15.6	+2.5 / 0	+15.6 / +14.0	19.5 / 23.6	+2.5 / 0	+23.6 / +22.0	26.0 / 32.5	+4.0 / 0	+32.5 / +30.0

Data in boldface type are in accordance with American–British–Canadian (ABC) agreements. Symbols H6, H7, s6, etc. are hole and shaft designations in ABC system. Limits for sizes above 19–69 inches are not covered by ABC agreements but are given in the ANSI standard.

*Pairs of values shown represent minimum and maximum amounts of interference resulting from application of standard tolerance limits.

Source: Reprinted courtesy of The American Society of Mechanical Engineers.

ANSI FORCE AND SHRINK FITS (FN)–ENGLISH UNITS

B ANSI PREFERRED METRIC LIMITS AND FITS

ANSI PREFERRED HOLE BASIS METRIC CLEARANCE FITS—METRIC UNITS

American National Standard Preferred Hole Basis Metric Clearance Fits (ANSI B4.2–1978, R1984)

Basic Size		Loose-Running			Free-Running			Close-Running			Sliding			Locational Clearance		
		Hole H11	Shaft c11	Fit†	Hole H9	Shaft d9	Fit†	Hole H8	Shaft f7	Fit†	Hole H7	Shaft g6	Fit†	Hole H7	Shaft h6	Fit†
1	Max	1.060	0.940	0.180	1.025	0.980	0.070	1.014	0.994	0.030	1.010	0.998	0.018	1.010	1.000	0.016
	Min	1.000	0.880	0.060	1.000	0.955	0.020	1.000	0.984	0.006	1.000	0.992	0.002	1.000	0.994	0.000
1.2	Max	1.260	1.140	0.180	1.225	1.180	0.070	1.214	1.194	0.030	1.210	1.198	0.018	1.210	1.200	0.016
	Min	1.200	1.080	0.060	1.200	1.155	0.020	1.200	1.184	0.006	1.200	1.192	0.002	1.200	1.194	0.000
1.6	Max	1.660	1.540	0.180	1.625	1.580	0.070	1.614	1.594	0.030	1.610	1.598	0.018	1.610	1.600	0.016
	Min	1.600	1.480	0.060	1.600	1.555	0.020	1.600	1.584	0.006	1.600	1.592	0.002	1.600	1.594	0.000
2	Max	2.060	1.940	0.180	2.025	1.980	0.070	2.014	1.994	0.030	2.010	1.998	0.018	2.010	2.000	0.016
	Min	2.000	1.880	0.060	2.000	1.955	0.020	2.000	1.984	0.006	2.000	1.992	0.002	2.000	1.994	0.000
2.5	Max	2.560	2.440	0.180	2.525	2.480	0.070	2.514	2.494	0.030	2.510	2.498	0.018	2.510	2.500	0.016
	Min	2.500	2.380	0.060	2.500	2.455	0.020	2.500	2.484	0.006	2.500	2.492	0.002	2.500	2.494	0.000
3	Max	3.060	2.940	0.180	3.025	2.980	0.070	3.014	2.994	0.030	3.010	2.998	0.018	3.010	3.000	0.016
	Min	3.000	2.880	0.060	3.000	2.955	0.020	3.000	2.984	0.006	3.000	2.992	0.002	3.000	2.994	0.000
4	Max	4.075	3.930	0.220	4.030	3.970	0.090	4.018	3.990	0.040	4.012	3.996	0.024	4.012	4.000	0.020
	Min	4.000	3.855	0.070	4.000	3.940	0.030	4.000	3.978	0.010	4.000	3.988	0.004	4.000	3.992	0.000
5	Max	5.075	4.930	0.220	5.030	4.970	0.090	5.018	4.990	0.040	5.012	4.996	0.024	5.012	5.000	0.020
	Min	5.000	4.855	0.070	5.000	4.940	0.030	5.000	4.978	0.010	5.000	4.988	0.004	5.000	4.992	0.000
6	Max	6.075	5.930	0.220	6.030	5.970	0.090	6.018	5.990	0.040	6.012	5.996	0.024	6.012	6.000	0.020
	Min	6.000	5.855	0.070	6.000	5.940	0.030	6.000	5.978	0.010	6.000	5.988	0.004	6.000	5.992	0.000
8	Max	8.090	7.920	0.260	8.036	7.960	0.112	8.022	7.987	0.050	8.015	7.995	0.029	8.015	8.000	0.024
	Min	8.000	7.830	0.080	8.000	7.924	0.040	8.000	7.972	0.013	8.000	7.986	0.005	8.000	7.991	0.000
10	Max	10.090	9.920	0.260	10.036	9.960	0.112	10.022	9.987	0.050	10.015	9.995	0.029	10.015	10.000	0.024
	Min	10.000	9.830	0.080	10.000	9.924	0.040	10.000	9.972	0.013	10.000	9.986	0.005	10.000	9.991	0.000
12	Max	12.110	11.905	0.315	12.043	11.956	0.136	12.027	11.984	0.061	12.018	11.994	0.035	12.018	12.000	0.029
	Min	12.000	11.795	0.095	12.000	11.907	0.050	12.000	11.966	0.016	12.000	11.983	0.006	12.000	11.989	0.000
16	Max	16.110	15.905	0.315	16.043	15.950	0.136	16.027	15.984	0.061	16.018	15.994	0.035	16.018	16.000	0.029
	Min	16.000	15.795	0.095	16.000	15.907	0.050	16.000	15.966	0.016	16.000	15.983	0.006	16.000	15.989	0.000
20	Max	20.130	19.890	0.370	20.052	19.935	0.169	20.033	19.980	0.074	20.021	19.993	0.042	20.021	20.000	0.034
	Min	20.000	19.760	0.110	20.000	19.883	0.065	20.000	19.959	0.020	20.000	19.980	0.007	20.000	19.987	0.000

Basic Size		Loose Running			Free Running			Close Running			Sliding			Locational Clearance		
		Hole H11	Shaft c11	Fit[†]	Hole H9	Shaft d9	Fit[†]	Hole H8	Shaft f7	Fit[†]	Hole H7	Shaft g6	Fit[†]	Hole H7	Shaft h6	Fit[†]
25	Max	25.130	24.890	0.370	25.052	24.935	0.169	25.033	24.980	0.074	25.021	24.993	0.041	25.021	25.000	0.034
	Min	25.000	24.760	0.110	25.000	24.883	0.065	25.000	24.959	0.010	25.000	24.980	0.007	25.000	24.987	0.000
30	Max	30.130	29.890	0.370	30.052	29.935	0.169	30.033	29.980	0.074	30.021	29.993	0.041	30.021	30.000	0.034
	Min	30.000	29.760	0.110	30.000	29.883	0.065	30.000	29.959	0.020	30.000	29.980	0.007	30.000	29.987	0.000
40	Max	40.160	39.880	0.440	40.062	39.920	0.204	40.039	39.975	0.089	40.025	39.991	0.050	40.025	40.000	0.041
	Min	40.000	39.720	0.120	40.000	39.858	0.080	40.000	39.950	0.025	40.000	39.975	0.009	40.000	39.984	0.000
50	Max	50.160	49.870	0.450	50.062	49.920	0.204	50.039	49.975	0.089	50.025	49.991	0.050	50.025	50.000	0.041
	Min	50.000	49.710	0.130	50.000	49.858	0.080	50.000	49.950	0.025	50.000	49.975	0.009	50.000	49.984	0.000
60	Max	60.190	59.860	0.520	60.074	59.900	0.248	60.046	59.970	0.106	60.030	59.990	0.059	60.030	60.000	0.049
	Min	60.000	59.670	0.140	60.000	59.826	0.100	60.000	59.940	0.030	60.000	59.971	0.010	60.000	59.981	0.000
80	Max	80.190	79.850	0.530	80.074	79.900	0.248	80.046	79.970	0.106	80.030	79.990	0.059	80.030	80.000	0.049
	Min	80.000	79.660	0.150	80.000	79.826	0.100	80.000	79.940	0.030	80.000	79.971	0.010	80.000	79.981	0.000
100	Max	100.220	99.830	0.610	100.087	99.880	0.294	100.054	99.964	0.125	100.035	99.988	0.069	100.035	100.000	0.057
	Min	100.000	99.610	0.170	100.000	99.793	0.120	100.000	99.929	0.036	100.000	99.966	0.012	100.000	99.978	0.000
120	Max	120.220	119.820	0.620	120.087	119.880	0.294	120.054	119.964	0.125	120.035	119.988	0.069	120.035	120.000	0.057
	Min	120.000	119.600	0.180	120.000	119.793	0.120	120.000	119.929	0.036	120.000	119.966	0.012	120.000	119.978	0.000
160	Max	160.250	159.790	0.710	160.100	159.855	0.345	160.063	159.957	0.146	160.040	159.986	0.079	160.040	160.000	0.065
	Min	160.000	159.540	0.210	160.000	159.755	0.145	160.000	159.917	0.043	160.000	159.961	0.014	160.000	159.975	0.000
200	Max	200.290	199.760	0.820	200.115	199.830	0.400	200.072	199.950	0.168	200.046	199.985	0.090	200.046	200.000	0.071
	Min	200.000	199.470	0.240	200.000	199.715	0.170	200.000	199.904	0.050	200.000	199.956	0.015	200.000	199.971	0.000
250	Max	250.290	249.720	0.860	250.115	249.830	0.400	250.072	249.950	0.168	250.046	249.985	0.090	250.046	250.000	0.075
	Min	250.000	249.430	0.230	250.000	249.715	0.170	250.000	249.904	0.050	250.000	249.956	0.015	250.000	249.971	0.000
300	Max	300.320	299.670	0.970	300.130	299.810	0.450	300.081	299.944	0.189	300.052	299.983	0.101	300.052	300.000	0.084
	Min	300.000	299.350	0.330	300.000	299.680	0.190	300.000	299.892	0.056	300.000	299.951	0.017	300.000	299.968	0.000
400	Max	400.360	399.600	1.120	400.140	399.790	0.490	400.089	399.938	0.208	400.057	399.982	0.111	400.057	400.000	0.093
	Min	400.000	399.240	0.400	400.000	399.650	0.210	400.000	399.881	0.063	400.000	399.946	0.018	400.000	399.964	0.000
500	Max	500.400	499.520	1.280	500.155	499.770	0.540	500.097	499.932	0.228	500.063	499.980	0.123	500.063	500.000	0.103
	Min	500.000	499.120	0.480	500.000	499.615	0.230	500.000	499.869	0.068	500.000	499.940	0.020	500.000	499.960	0.000

All dimensions are in millimeters.

Preferred fits for other sizes can be calculated from data given in ANSI B4.2–1978 (R1984).

[†] All fits shown in this table have clearance.

Source: Reprinted courtesy of The American Society of Mechanical Engineers.

■ ANSI PREFERRED HOLE BASIS TRANSITION AND INTERFERENCE FITS—METRIC UNITS

American National Standard Preferred Hole Basis Metric Transition and Interference Fits (ANSI B4.2–1978, R1984)

Basic Size		Locational Transition Hole H7	Shaft k6	Fit†	Locational Transition Hole H7	Shaft n6	Fit†	Locational Interference Hole H7	Shaft p6	Fit†	Medium Drive Hole H7	Shaft s6	Fit†	Force Hole H7	Shaft u6	Fit†
1	Max	1.010	1.006	+0.010	1.010	1.010	+0.006	1.010	1.012	+0.004	1.010	1.020	−0.004	1.010	1.024	−0.008
	Min	1.000	1.000	−0.006	1.000	1.004	−0.010	1.000	1.006	−0.012	1.000	1.014	−0.020	1.000	1.018	−0.024
1.2	Max	1.210	1.206	+0.010	1.210	1.210	+0.006	1.210	1.212	+0.004	1.210	1.220	−0.004	1.210	1.224	−0.008
	Min	1.200	1.200	−0.006	1.200	1.204	−0.010	1.200	1.206	−0.012	1.200	1.214	−0.020	1.200	1.218	−0.024
1.6	Max	1.610	1.606	+0.010	1.610	1.610	+0.006	1.610	1.612	+0.004	1.610	1.620	−0.004	1.610	1.624	−0.008
	Min	1.600	1.600	−0.006	1.600	1.604	−0.010	1.600	1.606	−0.012	1.600	1.614	−0.020	1.600	1.618	−0.024
2	Max	2.010	2.006	+0.010	2.010	2.010	+0.006	2.010	2.012	+0.004	2.010	2.020	−0.004	2.010	2.024	−0.008
	Min	2.000	2.000	−0.006	2.000	2.004	−0.010	2.000	2.006	−0.012	2.000	2.014	−0.020	2.000	2.018	−0.024
2.5	Max	2.510	2.506	+0.010	2.510	2.510	+0.006	2.510	2.512	+0.004	2.510	2.520	−0.004	2.510	2.524	−0.008
	Min	2.500	2.500	−0.006	2.500	2.504	−0.010	2.500	2.506	−0.012	2.500	2.514	−0.020	2.500	2.518	−0.024
3	Max	3.010	3.006	+0.010	3.010	3.010	+0.006	3.010	3.012	+0.004	3.010	3.020	−0.004	3.010	3.024	−0.008
	Min	3.000	3.000	−0.006	3.000	3.004	−0.010	3.000	3.006	−0.012	3.000	3.014	−0.020	3.000	3.018	−0.024
4	Max	4.012	4.009	+0.011	4.012	4.016	+0.004	4.012	4.020	0.000	4.012	4.027	−0.007	4.012	4.031	−0.011
	Min	4.000	4.001	−0.009	4.000	4.008	−0.016	4.000	4.012	−0.020	4.000	4.019	−0.027	4.000	4.023	−0.031
5	Max	5.012	5.009	+0.011	5.012	5.016	+0.004	5.012	5.020	0.000	5.012	5.027	−0.007	5.012	5.031	−0.011
	Min	5.000	5.001	−0.009	5.000	5.008	−0.016	5.000	5.012	−0.020	5.000	5.019	−0.027	5.000	5.023	−0.031
6	Max	6.012	6.009	+0.011	6.012	6.016	+0.004	6.012	6.020	0.000	6.012	6.027	−0.007	6.012	6.031	−0.011
	Min	6.000	6.001	−0.009	6.000	6.008	−0.016	6.000	6.012	−0.020	6.000	6.019	−0.027	6.000	6.023	−0.031
8	Max	8.015	8.010	+0.014	8.015	8.019	+0.005	8.015	8.024	0.000	8.015	8.032	−0.008	8.015	8.037	−0.013
	Min	8.000	8.001	−0.010	8.000	8.010	−0.019	8.000	8.015	−0.024	8.000	8.023	−0.032	8.000	8.028	−0.037
10	Max	10.015	10.010	+0.014	10.015	10.019	+0.005	10.015	10.024	0.000	10.015	10.032	−0.008	10.015	10.037	−0.013
	Min	10.000	10.001	−0.010	10.000	10.010	−0.019	10.000	10.015	−0.024	10.000	10.023	−0.032	10.000	10.028	−0.037
12	Max	12.018	12.012	+0.017	12.018	12.023	+0.006	12.018	12.029	0.000	12.018	12.039	−0.010	12.018	12.044	−0.015
	Min	12.000	12.001	−0.012	12.000	12.012	−0.023	12.000	12.018	−0.029	12.000	12.028	−0.039	12.000	12.033	−0.044
16	Max	16.018	16.012	+0.017	16.018	16.023	+0.006	16.018	16.029	0.000	16.018	16.039	−0.010	16.018	16.044	−0.015
	Min	16.000	16.001	−0.012	16.000	16.012	−0.023	16.000	16.018	−0.029	16.000	16.028	−0.039	16.000	16.033	−0.044
20	Max	20.021	20.015	+0.019	20.021	20.028	+0.005	20.021	20.035	−0.001	20.021	20.048	−0.014	20.021	20.054	−0.020
	Min	20.000	20.002	−0.015	20.000	20.015	−0.028	20.000	20.022	−0.035	20.000	20.035	−0.048	20.000	20.041	−0.054

Basic Size		Locational Transition Hole H7	Shaft k6	Fit[†]	Locational Transition Hole H7	Shaft n6	Fit[†]	Locational Interference Hole H7	Shaft p6	Fit[†]	Medium Drive Hole H7	Shaft s6	Fit[†]	Force Hole H7	Shaft u6	Fit[†]
25	Max	25.021	25.015	+0.019	25.021	25.028	+0.006	25.021	25.035	−0.001	25.021	25.048	−0.014	25.021	25.061	−0.027
	Min	25.000	25.002	−0.015	25.000	25.015	−0.028	25.000	25.022	−0.035	25.000	25.035	−0.048	25.000	25.048	−0.061
30	Max	30.021	30.015	+0.019	30.021	30.028	+0.006	30.021	30.035	−0.001	30.021	30.048	−0.014	30.021	30.061	−0.027
	Min	30.000	30.002	−0.015	30.000	30.015	−0.028	30.000	30.022	−0.035	30.000	30.035	−0.048	30.000	30.048	−0.061
40	Max	40.025	40.018	+0.023	40.025	40.033	+0.008	40.025	40.042	−0.001	40.025	40.059	−0.018	40.025	40.076	−0.035
	Min	40.000	40.002	−0.018	40.000	40.017	−0.033	40.000	40.026	−0.042	40.000	40.043	−0.059	40.000	40.060	−0.076
50	Max	50.025	50.018	+0.023	50.025	50.033	+0.008	50.025	50.042	−0.002	50.025	50.059	−0.018	50.025	50.086	−0.045
	Min	50.000	50.002	−0.018	50.000	50.017	−0.033	50.000	50.026	−0.042	50.000	50.043	−0.059	50.000	50.070	−0.086
60	Max	60.030	60.021	+0.028	60.030	60.039	+0.010	60.030	60.051	−0.002	60.030	60.072	−0.023	60.030	60.106	−0.057
	Min	60.000	60.002	−0.021	60.000	60.020	−0.039	60.000	60.032	−0.052	60.000	60.053	−0.072	60.000	60.087	−0.106
80	Max	80.030	80.021	+0.028	80.030	80.039	+0.010	80.030	80.051	−0.002	80.030	80.078	−0.029	80.030	80.121	−0.072
	Min	80.000	80.002	−0.021	80.000	80.020	−0.039	80.000	80.032	−0.051	80.000	80.059	−0.078	80.000	80.102	−0.121
100	Max	100.035	100.025	+0.032	100.035	100.045	+0.012	100.035	100.059	−0.002	100.035	100.093	−0.036	100.035	100.146	−0.089
	Min	100.000	100.003	−0.025	100.000	100.023	−0.045	100.000	100.037	−0.059	100.000	100.071	−0.093	100.000	100.124	−0.146
120	Max	120.035	120.025	+0.032	120.035	120.045	+0.012	120.035	120.059	−0.002	120.035	120.101	−0.044	120.035	120.166	−0.109
	Min	120.000	120.003	−0.025	120.000	120.023	−0.045	120.000	120.037	−0.059	120.000	120.079	−0.101	120.000	120.144	−0.166
160	Max	160.040	160.028	+0.037	160.040	160.052	+0.013	160.040	160.068	−0.003	160.040	160.125	−0.060	160.040	160.215	−0.150
	Min	160.000	160.003	−0.028	160.000	160.027	−0.052	160.000	160.043	−0.068	160.000	160.100	−0.125	160.000	160.190	−0.215
200	Max	200.046	200.033	+0.042	200.046	200.060	+0.015	200.046	200.079	−0.004	200.046	200.151	−0.076	200.046	200.265	−0.190
	Min	200.000	200.004	−0.033	200.000	200.031	−0.060	200.000	200.050	−0.079	200.000	200.122	−0.151	200.000	200.236	−0.265
250	Max	250.046	250.033	+0.042	250.046	250.060	+0.015	250.046	250.079	−0.004	250.046	250.169	−0.094	250.046	250.313	−0.238
	Min	250.000	250.004	−0.033	250.000	250.031	−0.060	250.000	250.050	−0.079	250.000	250.140	−0.169	250.000	250.284	−0.313
300	Max	300.052	300.036	+0.048	300.052	300.066	+0.018	300.052	300.088	−0.004	300.052	300.202	−0.118	300.052	300.382	−0.298
	Min	300.000	300.004	−0.036	300.000	300.034	−0.066	300.000	300.056	−0.088	300.000	300.170	−0.102	300.000	300.350	−0.382
400	Max	400.057	400.040	0.053	400.057	400.073	+0.020	400.057	400.098	−0.005	400.057	400.244	−0.151	400.057	400.471	−0.378
	Min	400.000	400.004	−0.040	400.000	400.037	−0.073	400.000	400.062	−0.098	400.000	400.208	−0.244	400.000	400.435	−0.471
500	Max	500.063	500.045	+0.058	500.063	500.080	+0.023	500.063	500.108	−0.005	500.063	500.292	−0.189	500.063	500.580	−0.477
	Min	500.000	500.005	−0.045	500.000	500.040	−0.080	500.000	500.068	−0.108	500.000	500.252	−0.292	500.000	500.540	−0.580

All dimensions are in millimeters.

Preferred fits for other sizes can be calculated from data given in ANSI B4.2–1978 (R1984).

[†]A plus sign indicates clearance; a minus sign indicates interference.

Source: Reprinted courtesy of The American Society of Mechanical Engineers.

■ ANSI PREFERRED SHAFT BASIS METRIC CLEARANCE FITS—METRIC UNITS

American National Standard Preferred Shaft Basis Metric Clearance Fits (ANSI B4.2–1978, R1984)

Basic Size		Loose-Running			Free-Running			Close-Running			Sliding			Locational Clearance		
		Hole C11	Shaft h11	Fit†	Hole D9	Shaft h9	Fit†	Hole F5	Shaft h7	Fit†	Hole G7	Shaft h6	Fit†	Hole H7	Shaft h6	Fit†
1	Max	1.120	1.000	0.180	1.045	1.000	0.070	1.020	1.000	0.030	1.012	1.000	0.018	1.010	1.000	0.016
	Min	1.060	0.940	0.060	1.020	0.975	0.020	1.006	0.990	0.006	1.002	0.994	0.002	1.000	0.994	0.000
1.2	Max	1.320	1.200	0.180	1.245	1.200	0.070	1.220	1.200	0.030	1.212	1.200	0.018	1.210	1.200	0.016
	Min	1.260	1.140	0.060	1.220	1.175	0.020	1.206	1.190	0.006	1.202	1.194	0.002	1.200	1.194	0.000
1.6	Max	1.720	1.600	0.180	1.645	1.600	0.070	1.620	1.600	0.030	1.612	1.600	0.018	1.610	1.600	0.016
	Min	1.660	1.540	0.060	1.620	1.575	0.020	1.606	1.590	0.006	1.602	1.594	0.002	1.600	1.594	0.000
2	Max	2.120	2.000	0.180	2.045	2.000	0.070	2.020	2.000	0.030	2.012	2.000	0.018	2.010	2.000	0.016
	Min	2.060	1.940	0.060	2.020	1.975	0.020	2.006	1.990	0.006	2.007	1.994	0.002	2.000	1.994	0.000
2.5	Max	2.620	2.500	0.180	2.545	2.500	0.070	2.520	2.500	0.030	2.512	2.500	0.018	2.510	2.500	0.016
	Min	2.560	2.440	0.060	2.520	2.475	0.020	2.506	2.490	0.006	2.502	2.494	0.002	2.500	2.494	0.000
3	Max	3.120	3.000	0.180	3.045	3.000	0.070	3.020	3.000	0.030	3.012	3.000	0.018	3.010	3.000	0.016
	Min	3.060	2.940	0.060	3.020	2.975	0.020	3.006	2.990	0.006	3.002	2.994	0.002	3.000	2.994	0.000
4	Max	4.145	4.000	0.220	4.060	4.000	0.090	4.028	4.000	0.040	4.016	4.000	0.024	4.012	4.000	0.020
	Min	4.070	3.925	0.070	4.030	3.970	0.030	4.010	3.988	0.010	4.004	3.992	0.004	4.000	3.992	0.000
5	Max	5.145	5.000	0.220	5.060	5.000	0.090	5.028	5.000	0.040	5.016	5.000	0.024	5.012	5.000	0.020
	Min	5.070	4.925	0.070	5.030	4.970	0.030	5.010	4.988	0.010	5.004	4.992	0.004	5.000	4.992	0.000
6	Max	6.145	6.000	0.220	6.060	6.000	0.090	6.028	6.000	0.040	6.016	6.000	0.024	6.012	6.000	0.020
	Min	6.070	5.925	0.070	6.030	5.970	0.030	6.010	5.988	0.010	6.004	5.992	0.004	6.000	5.992	0.000
8	Max	8.170	8.000	0.260	8.076	8.000	0.112	8.035	8.000	0.050	8.020	8.000	0.029	8.015	8.000	0.024
	Min	8.080	7.910	0.080	8.040	7.964	0.040	8.013	7.985	0.013	8.005	7.991	0.005	8.000	7.991	0.000
10	Max	10.170	10.000	0.260	10.076	10.000	0.112	10.035	10.000	0.050	10.020	10.000	0.029	10.015	10.000	0.024
	Min	10.080	9.910	0.080	10.040	9.964	0.040	10.013	9.985	0.013	10.005	9.991	0.005	10.000	9.991	0.000
12	Max	12.205	12.000	0.315	12.093	12.000	0.136	12.043	12.000	0.061	12.024	12.000	0.035	12.018	12.000	0.029
	Min	12.095	11.890	0.095	12.050	11.957	0.050	12.016	11.982	0.026	12.006	11.989	0.006	12.000	11.989	0.000
16	Max	16.205	16.000	0.315	16.093	16.000	0.136	16.043	16.000	0.061	16.024	16.000	0.035	16.018	16.000	0.029
	Min	16.095	15.890	0.095	16.050	15.957	0.050	16.016	15.982	0.016	16.006	15.989	0.006	16.000	15.989	0.000
20	Max	20.240	20.000	0.370	20.117	20.000	0.169	20.053	20.000	0.074	20.028	20.000	0.041	20.021	20.000	0.034
	Min	20.110	19.870	0.110	20.065	19.948	0.065	20.020	19.979	0.020	20.007	19.987	0.007	20.000	19.987	0.000

ANSI PREFERRED SHAFT BASIS METRIC CLEARANCE FITS—METRIC UNITS

Basic Size		Loose-Running			Free-Running			Close-Running			Sliding			Locational Clearance		
		Hole C11	Shaft h11	Fit†	Hole D9	Shaft h9	Fit†	Hole F8	Shaft h7	Fit†	Hole G7	Shaft h6	Fit†	Hole H7	Shaft h6	Fit†
25	Max	25.240	25.000	0.370	25.117	25.000	0.169	25.053	25.000	0.074	25.028	25.000	0.041	25.021	25.000	0.034
	Min	25.110	24.870	0.110	25.065	24.948	0.065	25.020	24.979	0.020	25.007	24.987	0.007	25.000	24.987	0.000
30	Max	30.240	30.000	0.370	30.117	30.000	0.169	30.053	30.000	0.074	30.028	30.000	0.041	30.021	30.000	0.034
	Min	30.110	29.870	0.110	30.065	29.948	0.065	30.020	29.979	0.020	30.007	29.987	0.007	30.000	29.987	0.000
40	Max	40.280	40.000	0.440	40.142	40.000	0.204	40.064	40.000	0.089	40.034	40.000	0.050	40.025	40.000	0.041
	Min	40.120	39.840	0.120	40.080	39.938	0.080	40.025	39.975	0.025	40.009	39.984	0.009	40.000	39.984	0.000
50	Max	50.290	50.000	0.450	50.142	50.000	0.204	50.064	50.000	0.089	50.034	50.000	0.050	50.025	50.000	0.041
	Min	50.130	49.840	0.130	50.080	49.938	0.080	50.025	49.975	0.025	50.009	49.984	0.009	50.000	49.984	0.000
60	Max	60.330	60.000	0.520	60.174	60.000	0.248	60.076	60.000	0.106	60.040	60.000	0.059	60.030	60.000	0.049
	Min	60.140	59.810	0.140	60.100	59.926	0.100	60.030	59.970	0.030	60.010	59.981	0.010	60.000	59.981	0.000
80	Max	80.340	80.000	0.530	80.174	80.000	0.248	80.076	80.000	0.106	80.040	80.000	0.059	80.030	80.000	0.049
	Min	80.150	79.810	0.150	80.100	79.926	0.100	80.030	79.970	0.030	80.010	79.981	0.010	80.000	79.981	0.000
100	Max	100.390	100.000	0.610	100.207	100.000	0.294	100.090	100.000	0.125	100.047	100.000	0.069	100.035	100.000	0.057
	Min	100.270	99.780	0.170	100.120	99.913	0.120	100.036	99.965	0.036	100.012	99.978	0.012	100.000	99.978	0.000
120	Max	120.400	120.000	0.620	120.207	120.000	0.294	120.090	120.000	0.125	120.047	120.000	0.069	120.035	120.000	0.057
	Min	120.180	119.780	0.180	120.120	119.913	0.120	120.036	119.965	0.036	120.012	119.978	0.012	120.000	119.978	0.000
160	Max	160.460	160.000	0.710	160.245	160.000	0.345	160.106	160.000	0.146	160.054	160.000	0.079	160.040	160.000	0.063
	Min	160.210	159.750	0.210	160.145	159.900	0.145	160.043	159.960	0.043	160.014	159.975	0.014	160.000	159.975	0.000
200	Max	200.530	200.000	0.820	200.285	200.000	0.400	200.122	200.000	0.168	200.061	200.000	0.090	200.046	200.000	0.075
	Min	200.240	199.710	0.240	200.170	199.885	0.170	200.050	199.954	0.050	200.015	199.971	0.015	200.000	199.911	0.000
250	Max	250.570	250.000	0.860	250.285	250.000	0.400	250.122	250.000	0.168	250.061	250.000	0.090	250.046	250.000	0.075
	Min	250.280	249.710	0.280	250.170	249.885	0.170	250.050	249.954	0.050	250.015	249.971	0.015	250.000	249.971	0.000
300	Max	300.650	300.000	0.970	300.320	300.000	0.450	300.137	300.000	0.189	300.069	300.000	0.101	300.052	300.000	0.084
	Min	300.330	299.680	0.330	300.190	299.870	0.190	300.056	299.948	0.056	300.017	299.968	0.017	300.000	299.968	0.000
400	Max	400.760	400.000	1.120	400.350	400.000	0.490	400.151	400.000	0.208	400.075	400.000	0.111	400.057	400.000	0.093
	Min	400.400	399.640	0.400	400.210	399.860	0.210	400.062	399.943	0.062	400.018	399.964	0.018	400.000	399.964	0.000
500	Max	500.880	500.000	1.280	500.385	500.000	0.540	500.165	500.000	0.228	500.083	500.000	0.123	500.063	500.000	0.103
	Min	500.480	499.600	0.480	500.230	499.845	0.230	500.068	499.937	0.068	500.020	499.960	0.020	500.000	499.960	0.000

All dimensions are in millimeters.

Preferred fits for other sizes can be calculated from data given in ANSI B4.2–1978 (R1984)

†All fits shown in this table have clearance.

Source: Reprinted courtesy of The American Society of Mechanical Engineers.

■ ANSI PREFERRED SHAFT BASIS METRIC TRANSITION AND INTERFERENCE FITS—METRIC UNITS

American National Standard Preferred Shaft Basis Metric Transition and Interference Fits (ANSI B4.2–1978, R1984)

Basic Size		Locational Transition			Locational Transition			Locational Interference			Medium Drive			Force		
		Hole K7	Shaft h6	Fit†	Hole N7	Shaft h6	Fit†	Hole P7	Shaft h6	Fit†	Hole S7	Shaft h6	Fit†	Hole U7	Shaft h6	Fit†
1	Max	1.000	1.000	+0.006	0.996	1.000	+0.002	0.994	1.000	0.000	0.986	1.000	−0.008	0.982	1.000	−0.012
	Min	0.990	0.994	−0.010	0.986	0.994	−0.014	0.984	0.994	−0.016	0.976	0.994	−0.024	0.972	0.994	−0.028
1.2	Max	1.200	1.200	+0.006	1.196	1.200	+0.002	1.194	1.200	0.000	1.186	1.200	−0.008	1.182	1.200	−0.012
	Min	1.190	1.194	−0.010	1.186	1.194	−0.014	1.184	1.194	−0.016	1.176	1.194	−0.024	1.172	1.194	−0.028
1.6	Max	1.600	1.600	+0.006	1.596	1.600	+0.002	1.594	1.600	0.000	1.586	1.600	0.008	1.582	1.600	−0.012
	Min	1.590	1.594	−0.010	1.586	1.594	−0.014	1.584	1.594	−0.016	1.576	1.594	−0.024	1.572	1.594	−0.028
2	Max	2.000	2.000	+0.006	1.996	2.000	+0.002	1.994	2.000	0.000	1.986	2.000	0.008	1.982	2.000	−0.012
	Min	1.990	1.994	−0.010	1.986	1.994	−0.014	1.984	1.994	−0.016	1.976	1.994	−0.024	1.972	1.994	−0.028
2.5	Max	2.500	2.500	+0.006	2.496	2.500	+0.002	2.494	2.500	0.000	2.486	2.500	−0.008	2.482	2.500	−0.012
	Min	2.490	2.494	−0.010	2.486	2.494	−0.014	2.484	2.494	−0.016	2.476	2.494	−0.024	2.472	2.494	−0.028
3	Max	3.000	3.000	+0.006	2.996	3.000	+0.002	2.994	3.000	0.000	2.986	3.000	−0.008	2.982	3.000	−0.012
	Min	2.990	2.994	−0.010	2.986	2.994	−0.014	2.984	2.994	−0.016	2.976	2.994	−0.024	2.972	2.994	−0.028
4	Max	4.003	4.000	+0.011	3.996	4.000	+0.004	3.992	4.000	0.000	3.985	4.000	−0.007	3.981	4.000	−0.011
	Min	3.991	3.992	−0.009	3.984	3.992	−0.016	3.980	3.992	−0.020	3.973	3.992	−0.027	3.969	3.992	−0.031
5	Max	5.003	5.000	+0.011	4.996	5.000	+0.004	4.992	5.000	0.000	4.985	5.000	−0.007	4.981	5.000	−0.011
	Min	4.991	4.992	−0.009	4.984	4.992	−0.016	4.980	4.992	−0.020	4.973	4.992	−0.027	4.969	4.992	−0.031
6	Max	6.003	6.000	+0.011	5.996	6.000	+0.004	5.992	6.000	0.000	5.985	6.000	−0.007	5.981	6.000	−0.011
	Min	5.991	5.992	−0.009	5.984	5.992	−0.016	5.980	5.992	−0.020	5.973	5.992	−0.027	5.969	5.992	−0.031
8	Max	8.005	8.000	+0.014	7.996	8.000	+0.005	7.991	8.000	0.000	7.983	8.000	−0.008	7.978	8.000	−0.013
	Min	7.990	7.991	−0.010	7.981	7.991	−0.019	7.976	7.991	−0.024	7.968	7.991	−0.032	7.963	7.991	−0.037
10	Max	10.005	10.000	+0.014	9.996	10.000	+0.005	9.991	10.000	0.000	9.983	10.000	−0.008	9.978	10.000	−0.013
	Min	9.990	9.991	−0.010	9.981	9.991	−0.019	9.976	9.991	−0.024	9.968	9.991	−0.032	9.963	9.991	−0.037
12	Max	12.006	12.000	+0.017	11.995	12.000	+0.006	11.989	12.000	0.000	11.979	12.000	−0.010	11.974	12.000	−0.015
	Min	11.988	11.989	−0.012	11.977	11.989	−0.023	11.971	11.989	−0.029	11.961	11.989	−0.039	11.956	11.989	−0.044
16	Max	16.006	16.000	+0.017	15.995	16.000	+0.006	15.989	16.000	0.000	15.979	16.000	−0.010	15.974	16.000	−0.015
	Min	15.988	15.989	−0.012	15.977	15.989	−0.023	15.971	15.989	−0.029	15.961	15.989	−0.039	15.956	15.989	−0.044
20	Max	20.006	20.000	+0.019	19.993	20.000	+0.006	19.986	20.000	0.001	19.973	20.000	−0.014	19.967	20.000	−0.020
	Min	19.985	19.987	−0.015	19.972	19.987	−0.028	19.965	19.987	−0.035	19.952	19.987	−0.045	19.946	19.987	−0.054

Basic Size		Locational Transition			Locational Transition			Locational Interference			Medium Drive			Force		
		Hole K7	Shaft h6	Fit†	Hole N7	Shaft h5	Fit†	Hole P7	Shaft h6	Fit†	Hole S7	Shaft h6	Fit†	Hole U7	Shaft h6	Fit†
25	Max	25.006	25.000	+0.019	24.993	25.000	+0.006	24.986	25.000	−0.001	24.973	25.000	−0.014	24.960	25.000	−0.027
	Min	24.985	24.987	−0.015	24.972	24.987	−0.028	24.965	24.987	−0.035	24.952	24.987	−0.048	24.939	24.987	−0.061
30	Max	30.006	30.000	+0.019	29.993	30.000	+0.006	29.986	30.000	−0.001	29.973	30.000	−0.014	29.960	30.000	−0.027
	Min	29.985	29.987	−0.015	29.972	29.987	−0.028	29.965	29.987	−0.035	29.952	29.987	−0.048	29.939	29.987	−0.061
40	Max	40.007	40.000	+0.023	39.992	40.000	+0.008	39.983	40.000	−0.001	39.966	40.000	−0.018	39.949	40.000	−0.035
	Min	39.982	39.984	−0.018	39.967	39.984	−0.033	39.958	39.984	−0.042	39.941	39.984	−0.059	39.914	39.984	−0.076
50	Max	50.007	50.000	+0.023	49.992	50.000	+0.008	49.983	50.000	−0.001	49.966	50.000	−0.018	49.939	50.000	−0.055
	Min	49.982	49.984	−0.018	49.967	49.984	−0.033	49.958	49.984	−0.042	49.941	49.984	−0.059	49.914	49.984	−0.086
60	Max	60.009	60.000	+0.028	59.991	60.000	+0.010	59.979	60.000	−0.002	59.958	60.000	−0.023	59.924	60.000	−0.087
	Min	59.979	59.981	−0.021	59.961	59.981	−0.039	59.949	59.981	−0.051	59.928	59.981	−0.072	59.894	59.981	−0.106
80	Max	80.009	80.000	+0.028	79.991	80.000	+0.010	79.979	80.000	−0.002	79.952	80.000	−0.029	79.909	80.000	−0.072
	Min	79.979	79.981	−0.021	79.961	79.981	−0.039	79.949	79.981	−0.051	79.922	79.981	−0.078	79.879	79.981	−0.121
100	Max	100.010	100.000	+0.032	99.990	100.000	+0.012	99.976	100.000	−0.002	99.942	100.000	−0.036	99.889	100.000	−0.089
	Min	99.975	99.978	−0.025	99.955	99.978	−0.045	99.941	99.978	−0.059	99.907	99.978	−0.093	99.854	99.978	−0.146
120	Max	120.010	120.000	+0.032	119.990	120.000	+0.012	119.976	120.000	−0.002	119.934	120.000	−0.044	119.869	120.000	−0.109
	Min	119.975	119.978	−0.025	119.955	119.978	−0.045	119.941	119.978	−0.059	119.899	119.978	−0.101	119.834	119.978	−0.166
160	Max	160.012	160.000	+0.037	159.988	160.000	+0.013	159.972	160.000	−0.003	159.915	160.000	−0.060	159.825	160.000	−0.150
	Min	159.972	159.975	−0.028	159.948	159.975	−0.053	159.932	159.975	−0.068	159.875	159.975	−0.125	159.785	159.975	−0.213
200	Max	200.013	200.000	+0.042	199.986	200.000	+0.015	199.967	200.000	−0.004	199.895	200.000	−0.076	199.781	200.000	−0.190
	Min	199.967	199.971	−0.033	199.940	199.971	−0.060	199.921	199.971	−0.079	199.849	199.971	−0.151	199.735	199.971	−0.265
250	Max	250.013	250.000	+0.042	249.986	250.000	+0.015	249.967	250.000	−0.004	249.877	250.000	−0.094	249.733	250.000	−0.238
	Min	249.967	249.971	−0.033	249.940	249.971	−0.060	249.921	249.971	−0.079	249.831	249.971	−0.169	249.687	249.971	−0.313
300	Max	300.016	300.000	+0.048	299.986	300.000	+0.018	299.964	300.000	−0.004	299.850	300.000	−0.118	299.670	300.000	−0.298
	Min	299.964	299.968	−0.036	299.934	299.968	−0.066	299.912	299.968	−0.088	299.798	299.968	−0.202	299.618	299.968	−0.382
400	Max	400.017	400.000	+0.053	399.984	400.000	+0.020	399.959	400.000	−0.005	399.813	400.000	−0.151	399.586	400.000	−0.378
	Min	399.960	399.964	−0.040	399.927	399.964	−0.073	399.902	399.964	−0.098	399.756	399.964	−0.244	399.529	399.954	−0.471
500	Max	500.018	500.000	+0.058	499.983	500.000	+0.023	499.955	500.000	−0.005	499.771	500.000	−0.189	499.483	500.000	−0.477
	Min	499.955	499.960	−0.045	499.920	499.950	−0.080	499.892	499.960	−0.108	499.708	499.960	−0.292	499.420	499.960	−0.580

All dimensions are in millimeters.

Preferred fits for other sizes can be calculated from data given in ANSI B4.2–1978 (R1984).

†A plus sign indicates clearance; a minus sign indicates interference.

Source: Reprinted courtesy of The American Society of Mechanical Engineers.

☐ INDEX

Isometric grid paper

Oblique right grid paper

Oblique left grid paper

Rectangular grid paper

VP ○

C _____ D

A _____ B

One-point perspective

VPR

C _____ D

A _____ B

VPL

Two-point perspective

EMBRY-RIDDLE
Aeronautical University

COLLEGE OF ENGINEERING
Engineering Fundamentals

NAME:

TITLE:

SECTION/INST:

UNITS:

DATE:

SCALE:

SHEET: OF

NAME :

TITLE:

SECTION / INST :

UNITS: DATE:

SCALE: SHEET: OF

EMBRY-RIDDLE
Aeronautical University.

COLLEGE OF ENGINEERING
Engineering Fundamentals

NAME:

TITLE:

SECTION/INST:

UNITS: DATE:

SCALE: SHEET: OF

NAME:

TITLE:

SECTION/INST:

UNITS: DATE:

SCALE: SHEET: OF

NAME :

TITLE :

SECTION / INST :

UNITS :   DATE :

SCALE :   SHEET :   OF

EMBRY-RIDDLE
Aeronautical University.

COLLEGE OF ENGINEERING
Engineering Fundamentals

NAME:

TITLE:

SECTION/INST:

UNITS: DATE:

SCALE: SHEET: OF

EMBRY-RIDDLE Aeronautical University.

COLLEGE OF ENGINEERING
Engineering Fundamentals

NAME:

TITLE:

SECTION/INST:

UNITS: DATE:

SCALE: SHEET: OF

EMBRY-RIDDLE
Aeronautical University.

COLLEGE OF ENGINEERING
Engineering Fundamentals

NAME:

TITLE:

SECTION/INST:

UNITS: DATE:

SCALE: SHEET: OF

EMBRY-RIDDLE
Aeronautical University

COLLEGE OF ENGINEERING
Engineering Fundamentals

NAME:

TITLE:

SECTION/INST:

UNITS: DATE:

SCALE: SHEET: OF

NAME :

TITLE :

SECTION / INST :

UNITS :   DATE :

SCALE :   SHEET :   OF

EMBRY-RIDDLE
Aeronautical University.

COLLEGE OF ENGINEERING
Engineering Fundamentals

NAME:

TITLE:

SECTION/INST:

UNITS: DATE:

SCALE: SHEET: OF

NAME:

TITLE:

SECTION/INST:

UNITS: DATE:

SCALE: SHEET: OF

EMBRY-RIDDLE
Aeronautical University.

COLLEGE OF ENGINEERING
Engineering Fundamentals

NAME:

TITLE:

SECTION/INST:

UNITS: DATE:

SCALE: SHEET: OF

EMBRY-RIDDLE
Aeronautical University™

COLLEGE OF ENGINEERING
Engineering Fundamentals

NAME:

TITLE:

SECTION/INST:

UNITS: DATE:

SCALE: SHEET: OF

NAME :

TITLE :

SECTION / INST :

UNITS :   DATE :

SCALE :   SHEET :   OF

EMBRY-RIDDLE

Aeronautical University.

COLLEGE OF ENGINEERING
Engineering Fundamentals

NAME:

TITLE:

SECTION/INST:

UNITS: | DATE:

SCALE: | SHEET: OF

NAME:

TITLE:

SECTION/INST:

UNITS: | DATE:

SCALE: | SHEET: OF

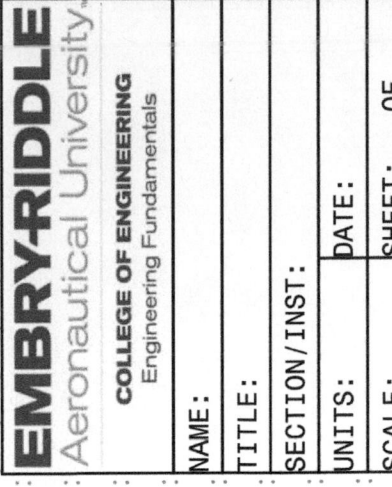

EMBRY-RIDDLE Aeronautical University™

COLLEGE OF ENGINEERING
Engineering Fundamentals

NAME:

TITLE:

SECTION / INST:

UNITS: DATE:

SCALE: SHEET: OF

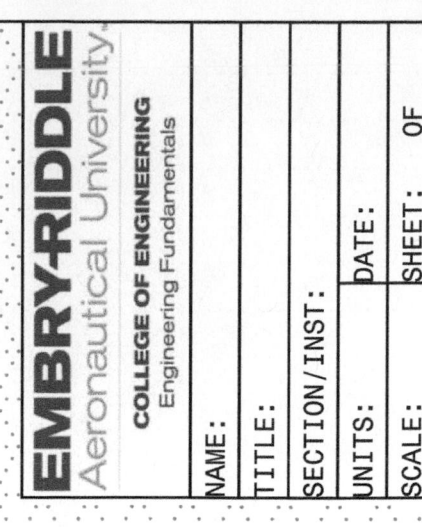

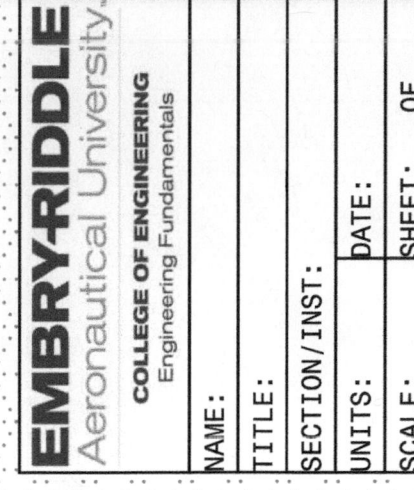

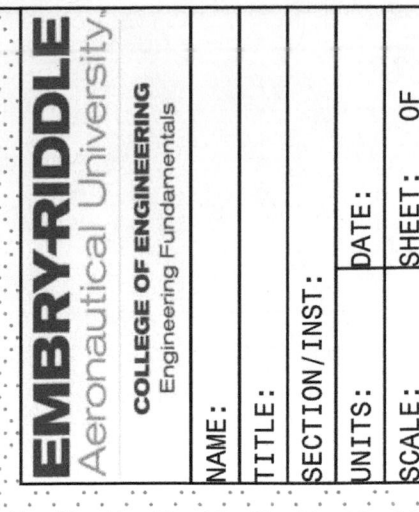

EMBRY-RIDDLE Aeronautical University.

COLLEGE OF ENGINEERING
Engineering Fundamentals

NAME:

TITLE:

SECTION/INST:

UNITS: DATE:

SCALE: SHEET: OF

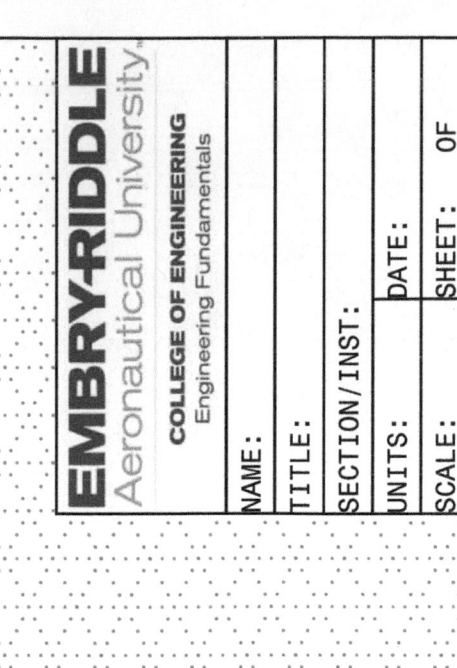